Methods in Molecular Biology™

Series Editor
John M. Walker
School of Life Sciences
University of Hertfordshire
Hatfield, Hertfordshire, AL10 9AB, UK

For further volumes:
http://www.springer.com/series/7651

Plant Chemical Genomics

Methods and Protocols

Edited by

Glenn R. Hicks

Department of Botany and Plant Sciences,
University of California, Riverside, CA, USA

Stéphanie Robert

Department of Forest Genetics and Plant Physiology, Umeå Plant Science Centre,
Swedish University of Agricultural Sciences (SLU), Umeå, Sweden

Editors
Glenn R. Hicks
Department of Botany and Plant Sciences
University of California, Riverside
Riverside, CA, USA

Stéphanie Robert
Department of Forest Genetics and Plant Physiology
Umeå Plant Science Centre, Swedish University of Agricultural Sciences (SLU)
Umeå, Sweden

ISSN 1064-3745 ISSN 1940-6029 (electronic)
ISBN 978-1-62703-591-0 ISBN 978-1-62703-592-7 (eBook)
DOI 10.1007/978-1-62703-592-7
Springer New York Heidelberg Dordrecht London

Library of Congress Control Number: 2013955601

Printed on acid-free paper

Humana Press is a brand of Springer
Springer is part of Springer Science+Business Media (www.springer.com)

Foreword

Chemical biology, the use of small molecules to dissect biological processes, is a relatively new and exciting field in the plant sciences with some great examples of its usefulness in recent years. Interestingly, the basic concept of discovering novel small molecules or natural products that produce a phenotype in plants has been in existence for many years. The agrichemical industry has for decades utilized small molecules based on known natural molecules such as auxins, cytokinins, abscissic acid, and other growth regulators. The primary focus of this work was and still is to produce viable modes of action aimed at pesticide discovery and improvement. Of course, in virtually all cases, the primary desired phenotype was lethality of distinct plant types such as broad-leafed weeds by distinct and desirable modes of action. Such work has even been highly automated in terms of plant application of different compounds for the development of herbicides and other agriculturally useful compounds. Lead compounds are then developed and systematically improved by iterative cycles of chemistry, screening, and phenotyping, a workflow not so fundamentally different than pharmaceutical discovery. In all cases, the key elements are an understanding of the biology and biochemistry of the organism combined with the application of analytical and synthetic chemistry. These days, the advent of genomics approaches has enhanced the ability to discover new modes of chemical activity by identifying genetic redundancy (common in plants) and enhancing the efficiency of discovering novel gene targets of commercial value. In contrast with commercial applications, basic research has utilized small molecules to generate interesting phenotypes in a manner analogous to forward mutational genetics. The desire is then to identify the cognate gene targets of such bioactive chemicals in order to discover novel genes, gene functions, and interactions between pathways. From the viewpoint of basic biological research which integrates chemistry, we would define this as chemical biology.

What is new in the realm of modern plant chemical biology? The approach has been widened to include larger numbers of compounds that are structurally diverse to capture as much target space as possible as part of the discovery process. Although not without exception, the general concept has been to uncover as many structures as possible leading to desirable phenotypes. The presumption is that as many proteins and pathways as possible can be perturbed in this way leading to new genes and functions. From this point on, structures may be refined using the iterative process. The breadth of phenotypes that can be examined have also been greatly expanded as small molecules have become a tool for basic research. These phenotypes can include plant morphology and development and even intracellular phenotypes which are made possible by the increased miniaturization, increased scaling, and enhanced automation of chemical screens in the hands of plant biologists. Some examples are the use of automated microscopes, high-throughput fluid robotics, automated or semi-automated methods to screen for desirable phenotypes, and more recently the introduction of image analysis to characterize plant traits in a quantifiable manner.

These tools cannot replace the importance of designing robust and precise chemical screens that produce the desired outcomes. As with forward genetic screens, one gets what is asked for in terms of chemical screen design. If the screen is not designed thoughtfully with secondary confirmatory screens and is not well controlled, it is likely that compounds affecting undesirable or multiple targets will be identified. For example, growth inhibition is a broad and commonly used phenotype to look for resistant mutants and gene targets. Such a broad phenotype has a chemical target space of potentially thousands of genes. This requires the use of a secondary screen that is much more specific to focus the results on a more precise set of targets. This two-step strategy for chemical screening is found in a number of chapters in this volume. Beyond screening for bioactive compounds, chemical biology also employs important tools including chemical synthesis, bioinformatics, and genetic screening for targets which employ genomics approaches.

In this volume, our intent was to produce a collection of techniques for the identification of bioactive compounds from a large selection of fields in plant biology including plant pathogenesis, immune responses, small RNA processing, endomembrane trafficking, lipids, plant hormone signaling, and cell wall. The presentation of these examples should provide readers with an overview of the practical aspects and range of chemical screens that are possible.

Beyond screening methods, we have incorporated chapters highlighting other critical aspects for the successful application of small molecules for research. These include methods for automation and imaging of plant samples for use in chemical screens. A section on methods for the use of informatics for the analysis of chemical structure that is approachable for biologists is included utilizing the ChemMine database as an example as well as approaches for designing compounds for follow-up screening. The use of clustering to visualize and compare complex phenotypic datasets is included as an example method that enables the analysis and visualization of hundreds of compounds producing multiple phenotypes. As databases of chemical phenotypes grow, this type of more sophisticated analysis will be essential to classify and focus on desirable compounds systematically.

To address the overwhelming need to more rapidly identify cognate gene targets of bioactive compounds, we have included several chapters on prioritizing compounds for target selection as well as reserve genetics to speed the identification of target genes. There are more extensive approaches toward target identification utilizing next generation sequencing that are being developed and deployed. However, these tools are more generalized for identifying mutations and beyond the scope of this volume.

Finally, an often underrepresented area of chemical biology is the understanding of plant responses to chemicals at the metabolic level. This is particularly applicable to the endogenous plant hormones such as auxins. Auxins are transported and are fundamental for plant growth and development. In plant chemical biology auxin perception and response pathways are often affected by bioactive chemicals. Thus we have included methods for examining auxin transport and metabolism. The intersection of metabolites and bioactive compounds can be quite specialized in terms of analytical chemistry. For the typical biologist this may require collaborative research in order to access the range of possible knowledge gained. However, we have included several methods that provide methodology in mass spectrometry and NMR that can be applied to chemical biology experiments.

We deeply thank the contributors to this volume. Our hope is that this will be an excellent reference for biologists who are new to chemical biology as well as experienced researchers who would like to reference specific protocols of interest.

Riverside, CA, USA *Glenn R. Hicks*
Umeå, Sweden *Stephanie Robert*

Contents

Foreword v
Contributors ix

PART I AUTOMATION AND IMAGING

1 Fully Automated Compound Screening in *Arabidopsis thaliana* Seedlings 3
Dominique Audenaert, Long Nguyen, Bert De Rybel, and Tom Beeckman

2 Time-Profiling Fluorescent Reporters in the *Arabidopsis* Root 11
Antoine P. Larrieu, Andrew P. French, Tony P. Pridmore, Malcolm J. Bennett, and Darren M. Wells

3 Screening for Bioactive Small Molecules by In Vivo Monitoring of Luciferase-Based Reporter Gene Expression in *Arabidopsis thaliana* 19
Christian Meesters and Erich Kombrink

PART II CHEMICAL SCREENING

4 Application of Yeast-Two Hybrid Assay to Chemical Genomic Screens: A High-Throughput System to Identify Novel Molecules Modulating Plant Hormone Receptor Complexes 35
Andrea Chini

5 High-Throughput Screening of Small-Molecule Libraries for Inducers of Plant Defense Responses 45
Colleen Knoth and Thomas Eulgem

6 Using a Reverse Genetics Approach to Investigate Small-Molecule Activity . . . 51
Siamsa M. Doyle and Stéphanie Robert

7 Investigating the Phytohormone Ethylene Response Pathway by Chemical Genetics 63
Lee-Chung Lin, Chiao-Mei Chueh, and Long-Chi Wang

8 Screening for Inhibitors of Chloroplast Galactolipid Synthesis Acting in Membrano and in Planta 79
Laurence Boudière and Eric Maréchal

9 Forward Chemical Screening of Small RNA Pathways 95
Yifan Lii and Hailing Jin

10 Identification and Use of Fluorescent Dyes for Plant Cell Wall Imaging Using High-Throughput Screening 103
Charles T. Anderson and Andrew Carroll

11 High-Throughput Identification of Chemical Endomembrane Cycling Disruptors Utilizing Tobacco Pollen 111
Michelle Q. Brown, Nolan Ung, Natasha V. Raikhel, and Glenn R. Hicks

12 Plant Chemical Genomics: Gravity Sensing and Response 115
Marci Surpin

13 Screening Chemical Libraries for Compounds That Affect Protein Sorting to the Yeast Vacuole 125
Jan Zouhar

PART III CHEMINFORMATICS

14 The Use of Multidrug Approach to Uncover New Players of the Endomembrane System Trafficking Machinery 131
Daniela Urbina, Patricio Pérez-Henríquez, and Lorena Norambuena

15 Cheminformatic Analysis of High-Throughput Compound Screens 145
Tyler W.H. Backman and Thomas Girke

16 Endomembrane Dissection Using Chemically Induced Bioactive Clusters. . . . 159
Natasha Worden, Thomas Girke, and Georgia Drakakaki

17 Statistical Molecular Design: A Tool to Follow Up Hits from Small-Molecule Screening 169
Anders E.G. Lindgren, Andreas Larsson, Anna Linusson, and Mikael Elofsson

PART IV TARGET IDENTIFICATION

18 Early Stage Hit Triage for Plant Chemical Genetic Screens and Target Site Identification 191
Terence A. Walsh

19 Screening for Gene Function Using the FOX (*F*ull-Length cDNA *O*vere*X*pressor Gene) Hunting System. 201
Mieko Higuchi-Takeuchi and Minami Matsui

PART V HORMONE TRANSPORT AND METABOLITE PROFILING

20 Quantification of Stable Isotope Label in Metabolites via Mass Spectrometry 213
Jan Huege, Jan Goetze, Frederik Dethloff, Bjoern Junker, and Joachim Kopka

21 ^{1}H NMR-Based Metabolomics Methods for Chemical Genomics Experiments. 225
Daniel J. Orr, Gregory A. Barding Jr., Christiana E. Tolley, Glenn R. Hicks, Natasha V. Raikhel, and Cynthia K. Larive

22 Determination of Auxin Transport Parameters on the Cellular Level. 241
Jan Petrášek, Martina Laňková, and Eva Zažímalová

23 Analyzing the In Vivo Status of Exogenously Applied Auxins: A HPLC-Based Method to Characterize the Intracellularly Localized Auxin Transporters. 255
Sibu Simon, Petr Skůpa, Petre I. Dobrev, Jan Petrášek, Eva Zažímalová, and Jiří Friml

Index 265

Contributors

CHARLES T. ANDERSON • *Department of Biology, The Pennsylvania State University, University Park, PA, USA*

DOMINIQUE AUDENAERT • *Department of Plant Systems Biology, VIB, Gent, Belgium; Department of Plant Biotechnology and BioinformaticsVIB, Gent, Belgium*

TYLER W.H. BACKMAN • *Department of Bioengineering, University of California Riverside, Riverside, CA, USA*

GREGORY A. BARDING, JR. • *Department of Chemistry, University of California, Riverside, CA, USA*

TOM BEECKMAN • *Department of Plant Systems Biology, VIB, Gent, Belgium; Department of Plant Biotechnology and BioinformaticsVIB, Gent, Belgium*

MALCOLM J. BENNETT • *Centre for Plant Integrative Biology, School of Biosciences, University of Nottingham, Sutton Bonington, UK*

LAURENCE BOUDIÈRE • *Laboratoire de Physiologie Cellulaire et Végétale, UMR 5168 CNRS-CEA-INRA-Université J. Fourier Grenoble 1, Institut de Recherches en Technologies et Sciences pour le Vivant, CEA-Grenoble, Grenoble, France*

MICHELLE Q. BROWN • *Department of Botany and Plant Sciences, Center for Plant Cell Biology, University of California, Riverside, CA, USA*

ANDREW CARROLL • *Joint Bioenergy Institute, Emeryville, CA, USA*

ANDREA CHINI • *Departamento de Genética Molecular de Plantas, Centro Nacional de Biotecnología—CSIC, Campus UAM, Madrid, Spain*

CHIAO-MEI CHUEH • *Institute of Plant and Microbial Biology, Academia Sinica, Taipei, Taiwan*

BERT DE RYBEL • *Laboratory of Biochemistry, Wageningen University, Wageningen, The Netherlands*

FREDERIK DETHLOFF • *Department of Molecular Physiology, Max-Planck-Institute of Molecular Plant Physiology (MPIMP), Potsdam-Golm, Germany*

PETRE I. DOBREV • *Institute of Experimental Botany, Academy of Sciences of the Czech Republic, Prague, Czech Republic*

SIAMSA M. DOYLE • *Department of Forest Genetics and Plant Physiology, Umeå Plant Science Centre, Swedish University of Agricultural Sciences (SLU), Umeå, Sweden*

GEORGIA DRAKAKAKI • *Department of Plant Sciences, University of California, Davis, CA, USA*

MIKAEL ELOFSSON • *Laboratories for Chemical Biology, Department of Chemistry, Umeå Centre for Microbial Research and Laboratories for Infection Biology, Umeå University, Umeå, Sweden*

THOMAS EULGEM • *ChemGen Integrative Education and Research Traineeship Program, Department of Botany and Plant Sciences, Center for Plant Cell Biology, Institute for Integrative Genome Biology, University of California at Riverside, Riverside, CA, USA*

ANDREW P. FRENCH • *Centre for Plant Integrative Biology, School of Biosciences, University of Nottingham, Sutton Bonington, UK*

JIŘÍ FRIML • *Institute of Science and Technology Austria (IST Austria), Klosterneuburg, Austria*
THOMAS GIRKE • *Department of Botany and Plant Sciences, University of California Riverside, Riverside, CA, USA*
JAN GOETZE • *Department of Theory, Max-Planck-Institute of Coal Research, Mülheim an der Ruhr, Germany*
GLENN R. HICKS • *Department of Botany and Plant Sciences, Center for Plant Cell Biology, University of California, Riverside, CA, USA*
MIEKO HIGUCHI-TAKEUCHI • *RIKEN Center for Sustainable Resource Science, Yokohama, Kanagawa, Japan*
JAN HUEGE • *Systems Biology Research Group, Department of Physiology and Cell Biology, Leibniz Institute of Plant Genetics and Crop Plant Research (IPK), Gatersleben, Germany*
HAILING JIN • *Department of Plant Pathology and Microbiology, Institute for Integrative Genome Biology, University of California, Riverside, CA, USA*
BJOERN JUNKER • *Systems Biology Research Group, Department of Physiology and Cell Biology, Leibniz Institute of Plant Genetics and Crop Plant Research (IPK), Gatersleben, Germany*
COLLEEN KNOTH • *Focus Diagnostics, Cypress, CA, USA*
ERICH KOMBRINK • *Department of Plant-Microbe Interactions, Max Planck Institute for Plant Breeding Research, Köln, Germany*
JOACHIM KOPKA • *Department of Molecular Physiology, Max-Planck-Institute of Molecular Plant Physiology (MPIMP), Potsdam-Golm, Germany*
MARTINA LAŇKOVÁ • *Institute of Experimental Botany, ASC, Prague, Czech Republic*
CYNTHIA K. LARIVE • *Department of Chemistry, Center for Plant Cell Biolog, University of California, Riverside, CA, USA*
ANTOINE P. LARRIEU • *Centre for Plant Integrative Biology, School of Biosciences, University of Nottingham, Sutton Bonington, UK*
ANDREAS LARSSON • *Department of Chemistry, Umeå University, Umeå, Sweden*
YIFAN LII • *Department of Plant Pathology and Microbiology, Institute for Integrative Genome Biology, University of California, Riverside, CA, USA*
LEE-CHUNG LIN • *Institute of Plant and Microbial Biology, Academia Sinica, Taipei, Taiwan*
ANDERS E.G. LINDGREN • *Department of Chemistry, Umeå University, Umeå, Sweden*
ANNA LINUSSON • *Laboratories for Chemical Biology, Department of Chemistry, Umeå University, Umeå, Sweden*
ERIC MARÉCHAL • *Laboratoire de Physiologie Cellulaire et Végétale, UMR 5168 CNRS-CEA-INRA-Université J. Fourier Grenoble 1, Institut de Recherches en Technologies et Sciences pour le Vivant, CEA-Grenoble, Grenoble, France*
MINAMI MATSUI • *RIKEN Center for Sustainable Resource Science, Yokohama, Kanagawa, Japan*
CHRISTIAN MEESTERS • *Department of Plant-Microbe Interactions, Max Planck I nstitute for Plant Breeding Research, Köln, Germany*
CHRISTIANA E. TOLLEY • *Department of Chemistry, University of California, Riverside, Riverside, CA, USA*
LONG NGUYEN • *VIB Compound Screening Facility (VIB-CSF), Gent, Belgium*

LORENA NORAMBUENA • *Plant Molecular Biology Laboratory, Department of Biology, Faculty of Sciences, University of Chile, Santiago, Chile*
DANIEL J. ORR • *Department of Chemistry, University of California, Riverside, Riverside, CA, USA*
PATRICIO PÉREZ-HENRÍQUEZ • *Plant Molecular Biology Laboratory, Department of Biology, Faculty of Sciences, University of Chile, Santiago, Chile*
JAN PETRÁŠEK • *Institute of Experimental Botany, ASC, Prague, Czech Republic*
TONY P. PRIDMORE • *Centre for Plant Integrative Biology, School of Biosciences, University of Nottingham, Sutton Bonington, UK*
NATASHA V. RAIKHEL • *Department of Botany and Plant Sciences, Center for Plant Cell Biology, University of California, Riverside, CA, USA*
STÉPHANIE ROBERT • *Department of Forest Genetics and Plant Physiology, Umeå Plant Science Centre, Swedish University of Agricultural Sciences (SLU), Umeå, Sweden*
SIBU SIMON • *Institute of Science and Technology Austria (IST Austria), Klosterneuburg, Austria*
PETR SKŮPA • *Institute of Experimental Botany, Academy of Sciences of the Czech Republic, Prague, Czech Republic*
MARCI SURPIN • *Valent BioSciences Corporation, Long Grove, IL, USA*
NOLAN UNG • *Department of Botany and Plant Sciences, Center for Plant Cell Biology, University of California, Riverside, CA, USA*
DANIELA URBINA • *Plant Molecular Biology Laboratory, Department of Biology, Faculty of Sciences, University of Chile, Santiago, Chile*
TERENCE A. WALSH • *Discovery Research, Dow AgroSciences LLC, Indianapolis, IN, USA*
LONG-CHI WANG • *Institute of Plant and Microbial Biology, Taipei, Taiwan*
DARREN M. WELLS • *Centre for Plant Integrative Biology, School of Biosciences, University of Nottingham, Sutton Bonington, UK*
NATASHA WORDEN • *Department of Plant Sciences, University of California, Davis, CA, USA*
EVA ZAŽÍMALOVÁ • *Institute of Experimental Botany, ASC, Prague, Czech Republic*
JAN ZOUHAR • *Centro de Biotecnología y Genómica de Plantas, Universidad Politécnica de Madrid, Pozuelo de Alarcón, España*

Part I

Automation and Imaging

Chapter 1

Fully Automated Compound Screening in *Arabidopsis thaliana* Seedlings

Dominique Audenaert, Long Nguyen, Bert De Rybel, and Tom Beeckman

Abstract

High-throughput small molecule screenings in model plants are of great value to identify compounds that interfere with plant developmental processes. In academic research, the plant *Arabidopsis thaliana* is the most commonly used model organism for this purpose. However, compared to plant cellular systems, *A. thaliana* plants are less amenable to develop high-throughput screening assays. In this chapter, we describe a screening procedure that is compatible with liquid handling systems and increases the throughput of compound screenings in *A. thaliana* seedlings.

Key words *Arabidopsis thaliana*, High-throughput screening, Automation

1 Introduction

In chemical biology, low molecular mass molecules are applied as conditional tools to reveal the underlying mechanisms that control growth and development. Chemical biology approaches have been well established to study human biology and disease mechanisms [1, 2]. Also in plant sciences, several studies have shown that chemical biology represents a powerful tool to study plant developmental processes [3, 4].

The identification of small molecules that interfere with a biological process of interest requires screening of a large number of compounds in an assay that is compatible with high-throughput screening. During the past two decades, affordable collections of small molecules have become commercially available. This has allowed academic researchers to perform chemical screenings on large sets of compounds. But although pre-plated compound collections are readily available, academic researchers still face the major challenge of developing a proper and robust screening assay. In plants, cellular systems such as *Arabidopsis thaliana* protoplasts or tobacco Bright Yellow-2 cells are relatively easy to adapt to a screening format due to its compatibility with high-throughput

Glenn R. Hicks and Stéphanie Robert (eds.), *Plant Chemical Genomics: Methods and Protocols*, Methods in Molecular Biology, vol. 1056, DOI 10.1007/978-1-62703-592-7_1, © Springer Science+Business Media New York 2014

screening robotics [5]. However, because cellular systems lack a proper physiological context, they are less applicable to study developmental processes. The identification of compounds that affect specific developmental and physiological processes requires the availability of a screening assay in a plant model system. *A. thaliana* is a model organism that is frequently used to study plant growth and development because of its available genetics and genomics tools. In addition, *Arabidopsis* can germinate and grow until the young seedling stage in 96-well plates and consequently *Arabidopsis* assays are amenable to high-throughput screening purposes. Indeed, this has been demonstrated by several studies in which a large number of compounds were screened in *Arabidopsis* in 96-well plate format [6, 7].

Although adding compounds to the assay plates can be automated relatively easily using a liquid handling platform, a major limitation of *Arabidopsis* assays is the high degree of manual handling for distributing the seeds and performing the read-out. Consequently, the throughput of screenings in *Arabidopsis* is significantly inferior compared to cell-based assays. In this chapter, we describe a method to fully automate compound screenings in *A. thaliana* seedlings. We have developed a procedure to add seeds to 96-well plates via liquid handling robotics and to perform a plate reader-based read-out. This method greatly improves the throughput of screenings in *A. thaliana* seedlings.

2 Materials

2.1 Plant Material

1. Transgenic *A. thaliana* seeds.

2.2 Seed Sterilization

1. Sterile distilled water.
2. 70 % (v/v) ethanol solution.
3. 5 % (v/v) NaOCl/0.05 % (v/v) Tween 20 solution.

2.3 Seed Distribution and Germination

1. 0.1 % (w/v) sterile agar solution.
2. White 96-well filter plates (MSBVN1B50, Millipore) with plastic lids.
3. Half-strength Murashige and Skoog (MS) liquid growth medium consisting of 0.215 g/L MS salts supplemented with 10 g/L sucrose, 0.1 g/L myo-inositol and 0.5 g/L 2-(N-morpholino) ethanesulfonic acid (MES) monohydrate in distilled water. Adjust the final pH to 5.7 by adding 1 M KOH. Autoclave the medium at 1 bar overpressure for 20 min. Growth medium can be stored at 4 °C.
4. Liquid handling robotic system (*see* **Note 1**).

5. Porous tape for air-permeable sealing (1530-0, 3M Micropore™).
6. *A. thaliana* growth chamber with controlled light conditions and temperature.
7. Orbital shaker (IKA KS 260 basic).

2.4 Compound Distribution and Incubation

1. Collection of small molecules, dissolved in 100 % DMSO at a concentration of 5 mM, pre-plated in 96-well plates (Corning Life Sciences) and stored at −20 °C (*see* **Note 2**).
2. Liquid handling robotic system (*see* **Note 3**).
3. Vacuum manifold (MSVMHTS00, Millipore) and vacuum/pressure pump (WP6122050, Millipore) (*see* **Note 4**).
4. Half-strength MS liquid growth medium.
5. Porous tape for air-permeable sealing (1530-0, 3M Micropore™).
6. *A. thaliana* growth chamber with controlled light conditions and temperature.
7. Orbital shaker (IKA KS 260 basic).

2.5 Read-out

1. ONE-Glo Luciferase Assay System (E6130, Promega) (*see* **Note 5**).
2. Liquid handling robotic system (*see* **Note 6**).
3. Plate reader (*see* **Note 7**).

3 Methods

3.1 General Considerations

For automated plate reader-based read-outs, the method requires the use of a transgenic *A. thaliana* line that contains a construct consisting of the promoter of a marker gene (gene-of-interest) fused to the gene encoding firefly luciferase or green fluorescent protein (GFP) to visualize expression of the marker gene. Using GFP does not require the addition of a substrate solution, which makes the screening faster and less expensive. Furthermore, unlike luminescence-based read-outs which are endpoint measurements, GFP can be used for multiple time-point measurements. However, a major disadvantage of using GFP as a reporter is that autofluorescent compounds might interfere with the assay read-out.

Typically, the marker gene needs to meet several criteria to be applicable in this method.

1. Up- or downregulation of the marker gene is associated with the biological process under study. This requires thorough characterization of a marker gene that is as specific as possible for the process. Identification of a suitable marker gene can be achieved by previously performed microarray experiments.

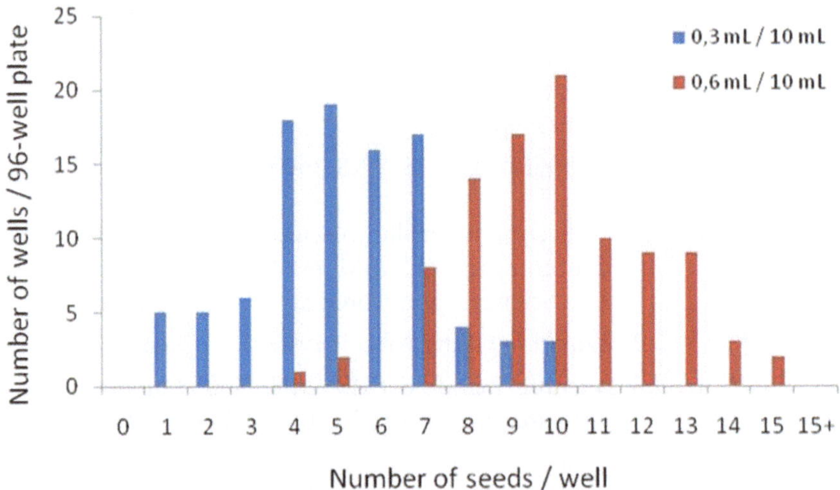

Fig. 1 Number of wells that contain the indicated number of seeds after dispensing 15 μL of the seed distribution solution. With a seed concentration of 0.3 mL/10 mL, the final seed number is between 3 and 7 seeds per well for 75 % of the wells. With a seed concentration of 0.6 mL/10 mL, the final seed number is between 7 and 11 seeds per well for 75 % of the wells

2. Changes in the expression level of the marker gene need to be as large as possible to allow changes to be detectable via a plate reader.
3. Preferably, expression of the marker gene is increased upon changes of the biological process. Using marker genes that show a reduced expression level may lead to false positive hits due to toxicity. For any setup, hit compounds should be further characterized using complementary assays.

3.2 Seed Sterilization

1. Add dry transgenic *A. thaliana* seeds in a sterile 15 mL Falcon tube (*see* **Note 8**).
2. Add 10 mL of a 70 % ethanol solution for 2 min.
3. Remove the 70 % ethanol solution and add 10 mL of a 5 % NaOCl/0.05 % Tween 20 solution for 15 min.
4. Remove the 5 % NaOCl/0.05 % Tween 20 solution and wash five times with sterile distilled water. Leave water in the tubes after the last wash, put at 4 °C and let the seeds settle for 24 h.

3.3 Seed Distribution and Germination

1. Add half-strength MS growth medium to a container that is compatible with the liquid handling platform (*see* **Note 1**).
2. Add 135 μL of the growth medium to the white 96-well filter plates.
3. Prepare the seed distribution solution by adding 0.3 or 0.6 mL of the sterilized seeds to 10 mL of a 0.1 % agar solution (Fig. 1 and **Note 9**).
4. Add the seed distribution solution to a container that is compatible with the liquid handling platform (*see* **Note 1**).

5. Add 15 μL of the seed distribution solution to the white 96-well filter plates.
6. Seal the plates with plastic lids and air-permeable tape.
7. Put the plates in the continuous light growth chamber at 21 °C under continuous shaking at 150 rpm (*see* **Note 10**).
8. Incubate for 5 days (*see* **Note 11**).

3.4 Compound Distribution and Incubation

1. Remove the growth medium by vacuum filtration (*see* **Note 12**).
2. Remove the air-permeable tape and plastic lids from the plates.
3. Add 148.5 μL of growth medium to the white 96-well filter plates (*see* **Notes 1** and **13**).
4. Add 1.5 μL of a negative control to column 1 and 1.5 μL of a positive control to column 12 (*see* **Note 14**).
5. Add 1.5 μL of the compound screening collection (5 mM, 100 % DMSO) to columns 2–11 (*see* **Notes 3** and **15**).
6. Seal the plates with plastic lids and air-permeable tape.
7. Put the plates in the continuous light growth chamber at 21 °C under continuous shaking at 150 rpm (*see* **Note 10**).
8. Incubate for 24 h (*see* **Note 16**).

3.5 Analysis

1. For luminescence-based read-outs, prepare the ONE-Glo luciferase substrate solution according to the manufacturer's protocol (*see* **Note 17**).
2. Add 100 μL of the prepared luciferase substrate solution and incubate for about 5 min.
3. Load the plates into the plate reader and perform the read-out (*see* **Note 18**).
4. Identify wells with compounds that interfere with the expression level of the marker gene under study.

4 Notes

1. For distribution of seeds and growth medium, our laboratory has access to a Beckman Coulter Biomek 2000 Laboratory Automation Workstation. To allow subsequent seed germination under sterile conditions, the platform is placed in a custom-made laminar flow. The platform has the capacity to process eight plates in one run. The addition of the growth medium and seeds to these plates takes about 25 min in total. Per screening batch, we perform three runs, which adds up to a total of 24 plates.
2. Several companies supply pre-plated diverse sets of DMSO-dissolved small molecules that can be used for screening purposes.

These companies include ChemBridge Corporation, Enamine, Life Chemicals, Maybridge, and Asinex among others.

3. For compound distribution, our laboratory has access to a Tecan Freedom EVO200 platform with 96-Multi Channel Arm option, a Robotic Manipulator Arm, and an integrated carousel with barcode scanner. This setup allows full walk-away automation during compound distribution, which leads to a significant increase in throughput.
4. The vacuum manifold and vacuum/pressure pump are integrated in the Tecan Freedom EVO200 platform.
5. Only required for luminescence-based read-outs.
6. The Tecan Freedom EVO200 platform is used to add the ONE-Glo reagent to the assay plates.
7. At the laboratory, we have a FLUOstar OPTIMA (BMG Labtech) available for fluorescence intensity measurements and a LUMIstar OPTIMA (BMG Labtech) for luminescence measurements.
8. Automated distribution of *Arabidopsis* seeds requires about 1 mL of dry seeds for 24 96-well plates.
9. Sterilized seeds should be aspirated slowly from the bottom of the Falcon tube with a 2 mL sterile plastic pipette. The amount of seeds to aspirate and dispense in the 0.1 % agar solution is dependent on the desired final number of seeds per well as described in Fig. 1. A total of 10 mL of the seed distribution solution suffices to perform one run (eight plates) on the Biomek 2000 platform. The 0.1 % agar solution prevents the seeds from settling down in the container during dispensing in the 96-well plates.
10. We have experienced optimal germination under these conditions. However, light, temperature, and shaking conditions can be adjusted dependent on the assay.
11. Germination and initial growth occur in the absence of chemicals to prevent excessive toxicity during the early developmental stages. However, the time of growth in the absence of compounds can be decreased or increased depending on the developmental process under study. For example, early root growth studies may require shorter times whereas leaf development would require longer times.
12. Integration of the vacuum manifold and vacuum/pressure pump in the Freedom EVO200 platform allows removal of the growth medium of the entire screening batch (i.e. 24 plates) in a fully automated fashion.
13. At this stage, the growth medium might contain additional components (e.g., plant hormones) to induce or inhibit a developmental process.

14. As a negative control, we add 1.5 μL of a 100 % DMSO solution. The positive control is dependent on the biological process under study. For example, auxin is used as a positive control for the process of lateral root development. Control compounds are applied with the Biomek 2000 platform.
15. This will yield a final compound concentration of 50 μM in the plates. The screening collection is applied with the Freedom EVO200 platform.
16. The usual compound incubation time that we apply is 24 h. However, dependent on the biological process under study, incubation times can be adjusted.
17. This step can be omitted for fluorescence-based read-outs.
18. Integrating the plate reader with a plate stacker would allow to fully automate the read-out. Importantly, with luminescence-based read-outs, the batch size to process should not be too large to avoid the luminescent signal to decrease over time. Therefore, after the substrate is added with the liquid handling robotics, the read-out should occur within 1 h. Taken that measuring an entire 96-well plate takes about 3 min, the total batch for luminescence measurements should not exceed 20 plates.

References

1. Collins I, Workman P (2006) New approaches to molecular cancer therapeutics. Nat Chem Biol 2:689–700
2. Gangadhar NM, Stockwell BR (2007) Chemical genetic approaches to probing cell death. Curr Opin Chem Biol 11:83–87
3. Robert S, Chary SN, Drakakaki G, Li S, Yang Z, Raikhel NV, Hicks GR (2008) Endosidin1 defines a compartment involved in endocytosis of the brassinosteroid receptor BRI1 and the auxin transporters PIN2 and AUX1. Proc Natl Acad Sci U S A 105:8464–8469
4. Kim TH, Hauser F, Ha T, Xue S, Böhmer M, Nishimura N, Munemasa S, Hubbard K, Peine N, Lee BH, Lee S, Robert N, Parker JE, Schroeder JI (2011) Chemical genetics reveals negative regulation of abscisic acid signaling by a plant immune response pathway. Curr Biol 21:990–997
5. De Sutter V, Vanderhaeghen R, Tilleman S, Lammertyn F, Vanhoutte I, Karimi M, Inzé D, Goossens A, Hilson P (2005) Exploration of jasmonate signalling via automated and standardized transient expression assays in tobacco cells. Plant J 44:1065–1076
6. Armstrong JI, Yuan S, Dale JM, Tanner VN, Theologis A (2004) Identification of inhibitors of auxin transcriptional activation by means of chemical genetics in Arabidopsis. Proc Natl Acad Sci U S A 101:14978–14983
7. De Rybel B, Audenaert D, Vert G, Rozhon W, Mayerhofer J, Peelman F, Coutuer S, Denayer T, Jansen L, Nguyen L, Vanhoutte I, Beemster GT, Vleminckx K, Jonak C, Chory J, Inzé D, Russinova E, Beeckman T (2009) Chemical inhibition of a subset of Arabidopsis thaliana GSK3-like kinases activates brassinosteroid signaling. Chem Biol 16:594–604

Chapter 2

Time-Profiling Fluorescent Reporters in the *Arabidopsis* Root

Antoine P. Larrieu, Andrew P. French, Tony P. Pridmore, Malcolm J. Bennett, and Darren M. Wells

Abstract

Confocal laser scanning microscopy is a useful nondestructive approach for the visualization of fluorescent reporters *in planta*. Samples are usually placed between a slide and a cover slip which, although suited to single time-point imaging, does not allow long-term observation. Here, we describe a technique to monitor changes in fluorescence in the *Arabidopsis* root over a long period of time. Treatment can easily be performed, and this approach is suitable for use in low-throughput chemical screens. We also present a rapid method to analyze fluorescence intensity profiles generated using this protocol.

Key words Confocal microscopy, Fluorescent reporters, Arabidopsis, Chemical screens, Time-profiling

1 Introduction

Confocal laser scanning microscopy allows measurement of a fluorophore (a fluorescent protein or dye) in living tissues [1]. Numerous techniques have been developed to study the subcellular localization of proteins, protein–protein interactions (FRET [2] and FLIM [3]), or protein movements in a cell (FRAP [4]). These techniques rely on the same principle: a fluorophore is excited using a laser, and the emitted light is detected at specific wavelengths.

Confocal microscopy can be used on living plants, making it ideal for following the fate of a fluorescent reporter in vivo after one or multiple stimuli. There are several reasons why quantitative analysis of confocal images is problematic [5, 6]. In this protocol, normalized intensity is measured in a root over time while holding as many other microscope parameters constant as practically possible. This approach can detect clear differences between mock and treatment plants (e.g., Fig. 1b). Convincingly, the mock response remains stable over the time course of the experiment.

Glenn R. Hicks and Stéphanie Robert (eds.), *Plant Chemical Genomics: Methods and Protocols*, Methods in Molecular Biology, vol. 1056, DOI 10.1007/978-1-62703-592-7_2, © Springer Science+Business Media New York 2014

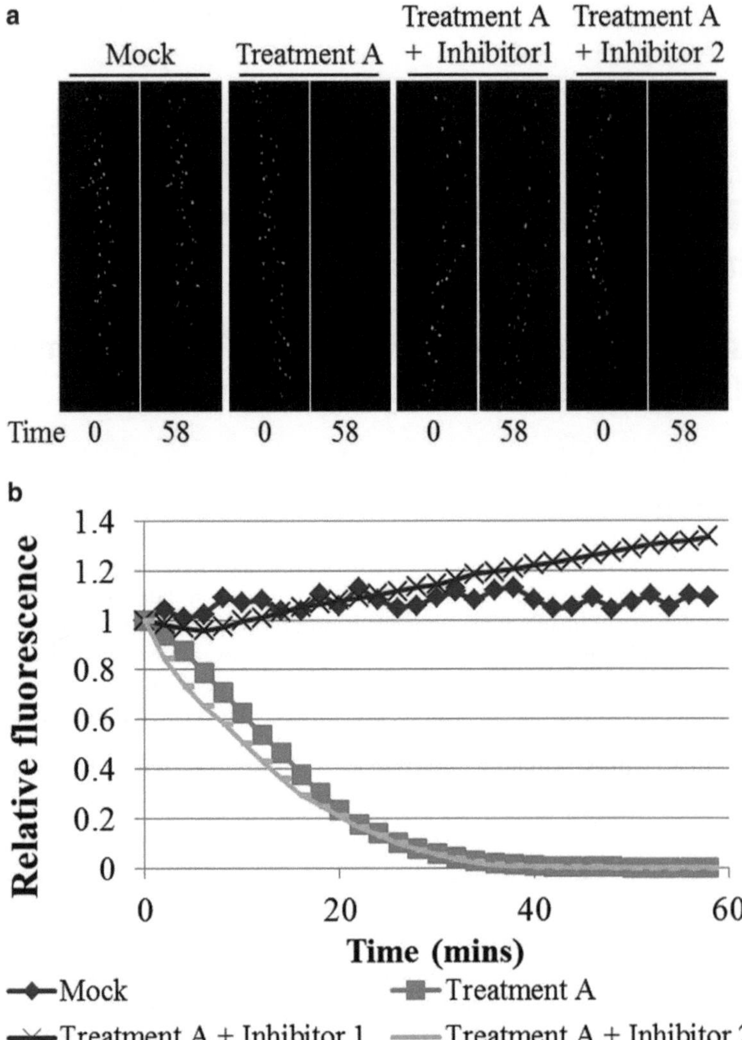

Fig. 1 Quantification of a series of confocal images to monitor fluorophore abundance. (**a**) Images showing the fluorescence (localized in the nucleus) immediately after the start of treatment (time point 0 min) and at the end of the time course (time point 58 min). (**b**) Quantification of the 31 images of the time course (one scan every 2 min) reveals that inhibitor 1 prevents the degradation of the reporter whereas inhibitor 2 has no effect on the reporter. Dynamics of degradation shows that inhibitor 2 does not delay or reduce the degradation of the reporter

Quantification of fluorophores is carried out on the data generated by the confocal microscope and provides an objective interpretation of changes in fluorescence over time. In this chapter, we describe a simple protocol for time course experiments on *Arabidopsis* seedlings and the subsequent quantification of fluorescence intensity using image analysis techniques and open-source software. The experimental setup allows monitoring of the fate of a fluorescent reporter on a confocal microscope within seconds of commencing treatment. This method allows simple and rapid

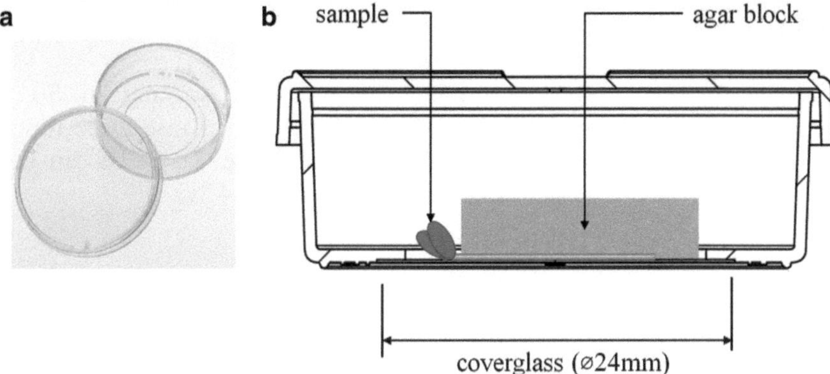

Fig. 2 Schematic diagram of glass-bottomed Petri dish used for time course experiments. (**a**) Glass-bottomed Petri dish (base diameter 35 mm). (**b**) Diagram showing arrangement of sample for imaging. Note that for long time course experiments, the lid may be sealed with gas-permeable tape [Micropore™ (3 M Co., USA)] to prevent the agar block from drying out. Original images copyright 2011: Greiner Bio-One GmbH, Frickenhausen, Germany

generation of fluorescence relative-intensity profiles from a time series of 2D confocal images. Importantly, the technique is not limited to *Arabidopsis* and may be applied to a wide variety of systems.

2 Materials

Follow all waste disposal regulations when disposing of transgenic *Arabidopsis* seeds and seedlings.

2.1 Arabidopsis thaliana Germination and Growth Medium Composition

1. Half-strength Murashige and Skoog (MS) medium: Weigh 2.15 g of MS salts, add approximately 800 mL of double-distilled water, and adjust the pH to 5.8 using 1 M KOH (*see* **Note 1**). Top up solution to 1 L and dispense into two 500 mL bottles (*see* **Note 2**). Add 5 g of bacto-agar to each bottle. Autoclave for 11 min at 121 °C, 20 psi. Autoclaved MS medium can be stored for at least 1 month at room temperature.
2. *Arabidopsis thaliana* seed surface sterilization solution: 50 % sodium hypochlorite (v/v) which should be prepared fresh each time.
3. Square or round Petri dishes.

2.2 Image Acquisition and Analysis

1. Inverted confocal laser scanning microscope. We used a Leica TCS SP5 for the measurements presented here.
2. Cell culture dish with glass bottom (part number 627961, Greiner Bio-One Inc.) (Fig. 2a).
3. Computer capable of running Fiji and spreadsheet program.
4. Fiji software, "Fiji Is Just ImageJ" [7] (ImageJ version 1.45 or later with LOCI Bio-Formats Plugin) [http://fiji.sc/wiki/index.php/Fiji].
5. Spreadsheet program (Microsoft Excel or OpenOffice Calc, for example).

3 Methods

3.1 Seed Germination

1. Surface sterilize the seeds by soaking in 50 % sodium hypochlorite solution for 4 min. Rinse three times in sterile double-distilled water (*see* **Note 3**). Seeds can be stored in sterile water for up to a week at 4 °C.
2. Prepare half-strength MS plates.
3. Sow the sterilized seeds in lines on solidified half-strength MS medium plates. Transfer the plates to 4 °C for 2 days (if not already stored in water). Transfer the plates to a tissue culture room (we use the following conditions: 24-h daylight, 22 °C, 100 μE light intensity) for 5 days.

3.2 Experimental Setup and Image Acquisition

1. On the day of the experiment, melt half-strength solid MS medium from the same batch as that used for seed germination using a steamer, autoclave, or microwave oven, and prepare plates as in step 2, Subheading 3.1. Before pouring the plates, add the compound(s) of interest at the appropriate concentration. Always remember to pour control plates without compound but with the solvent used to dissolve the compound (*see* **Note 4**).
2. Follow these steps to optimize the settings on the confocal microscope:
   (a) Choose the appropriate laser and detector combination for the fluorescent reporter used.
   (b) Transfer a seedling to the base of a glass-bottomed Petri dish (see Fig. 2b).
   (c) Immediately after transfer, gently lay down a piece (15 × 10 × 5 mm) of solid MS medium (from the control plate prepared in **step 2**, Subheading 3.1) on top of the seedling (*see* **Note 5**, Fig. 2b).
   (d) Place the Petri dish on the stage of the microscope, and select a region of interest.
   (e) Set the offset, laser power, detector sensitivity, pinhole, and frame and line averages to obtain the best signal-to-noise ratio, taking great care not to saturate the signal (*see* **Note 6**).
   (f) Choose the smallest time interval between each scan to give a suitable temporal resolution without bleaching the fluorophore (*see* **Note 7**).
   (g) Perform the analysis described in Subheading 3.3. If the results are satisfactory (less than 10 % variation in the control over the time course) repeat **steps 2b–e** above but this time using medium containing the chemicals. (If control profile *does* vary too much, *see* **Note 8**).

3.3 Data Analysis

1. Open the data file with an up-to-date version of Fiji (**File > Open**) (*see* **Note 9**).
2. On the import menu, uncheck all the options except "Split channels" and select "Hyperstack" in the "View stack with" menu, select XYCZT in the "Stack order" menu, and select "Default" in the Color mode menu.
3. When imported, choose the channel corresponding to the fluorophore. If not working on the whole picture, select the region of interest using the selection tools.
4. Set the measurements (**Analyse > Set Measurements**) to return only the integrated density. Make sure that the option "Limit to threshold" is selected. All the other options should be unchecked to perform the analysis described here. Nevertheless, depending on your applications, check the relevant options.
5. Plot the Z-axis profile (**Image > Stacks > Plot Z axis profile**). Copy and paste the **total fluorescence data** into a spreadsheet editor (*see* **Note 10**). To do this, **select Edit->Select All** in the table window and then **Edit->Copy**. Then paste into the spreadsheet.
6. Repeat the procedure (**step 5**), but set the threshold (**Image > Adjust > Threshold**) so that the background (everything except the fluorophore) is selected (*see* **Note 11**).
7. Plot the Z-axis profile, and copy and paste the **background fluorescence data** into a spreadsheet editor.
8. Subtract the **background raw integrated density** from the **total fluorescence raw integrated density** for each time point to obtain the **fluorophore-specific fluorescence**.
9. Normalize the values to the first time point: Divide each subsequent intensity measurement by the level at the first time point.
10. We recommend plotting the normalized mean of at least five replicates for each treatment to allow buffering for biological variation as well as variations in fluorescence caused by small movements of the sample during the experiment.

4 Notes

1. For 1 L of MS medium, three drops of 1 M KOH are usually sufficient (it is acceptable for the pH to be between 5.7 and 5.9). To accurately set the pH at 5.8, 0.1 M KOH should be used to prevent sudden pH changes.
2. We use 500 mL bottles as it allows rapid cooling of the media and facilitates subsequent remelting.
3. If using media containing sucrose, increase sterilization time to a maximum of 10 min.

4. Always use media from the same batch for plates for both seed germination and treatment in order to reduce nonspecific effects. Also, make sure that both media are at the same temperature.
5. This step is critical: Extra care must be taken to prevent the seedlings from drying out. To reduce drying out we place a drop of water at the bottom of the Petri dish, and we prepare the piece of MS medium before transferring the seedling.
6. To determine if non-signal (background) pixels are set to zero and to ensure that pixels corresponding to the fluorophore are not saturating (the signal is not being clipped at the upper limit of intensity), use a suitable lookup table (a mapping of colors onto the intensity range available). Usually, the microscope software has a setting which highlights both zero and clipped pixels.
7. A 2-min interval between scans represents a suitable compromise between temporal resolution and reporter bleaching. Shorter intervals can be used to monitor very rapid responses, dependent on the fluorescence bleaching characteristics of the specific reporter employed.
8. If there is too much variation in the control, then there could be too much variation in some of the inherent confocal hardware (*see* ref. 6). To try and eliminate this, repeat the control while trying to ensure that the confocal laser is warmed up, a lower laser power is used to prevent photo bleaching, etc. (for further reading *see* ref. 8). Testing the control protocol with a standard fixed slide allows the elimination of the sample and fluorophore themselves as a factor.
9. The Fiji package contains ImageJ and several plug-ins. To perform the analysis, the only necessary plug-in is LOCI Bio-Formats [9] (included with Fiji). This package allows files generated by confocal microscopy from many different microscope manufacturers to be imported. Make sure that you always perform the analysis of all your pictures using the same version of Fiji and the plug-ins.
10. Note that Fiji refers to Z-stack, but the stack is actually a time series. For each time point, Fiji returns two values: integrated density and raw integrated density (see the ImageJ User Guide [http://rsbweb.nih.gov/ij/docs/user-guide.pdf] for details). For quantification, always use the raw integrated density.
11. Select "red" in the right-hand box. Then, when setting the threshold, the region included in the measurement appears in red. The aim is to select background pixels for quantification, so we can exclude the background signal from the measurements. Do not click **Apply**, as this will convert the images to binary—simply leave the Threshold box open.

Acknowledgement

CPIB is a center for integrated systems biology funded by the Biotechnology and Biological Sciences Research Council (BBSRC) and Engineering and Physical Sciences Research Council (EPSRC). MJB and AL were also funded by the Belgian Scientific Policy (BELSPO contract BARN).

References

1. Wang Y-L, Taylor DL (eds) (1989) Methods in cell biology, vol. 30: fluorescence microscopy of living cells in culture. Academic, San Diego
2. Bacskai BJ, Skoch J, Hickey GA et al (2003) Fluorescence resonance energy transfer determinations using multiphoton fluorescence lifetime imaging microscopy to characterize amyloid-beta plaques. J Biomed Opt 8:368
3. Lakowicz JR, Szmacinski H, Nowaczyk K et al (1992) Fluorescence lifetime imaging. Anal Biochem 202:316–330
4. Axelrod D, Koppel D, Schlessinger J et al (1976) Mobility measurement by analysis of fluorescence photobleaching recovery kinetics. Biophys J 16:1055–1069
5. French AP, Mills S, Swarup R et al (2008) Colocalization of fluorescent markers in confocal microscope images of plant cells. Nat Protoc 3:619–628
6. Pawley J (2000) The 39 steps: a cautionary tale of quantitative 3-D fluorescence microscopy. Biotechniques 28:884–889
7. Schindelin J, Arganda-Carreras I, Frise E et al (2012) Fiji: an open-source platform for biological-image analysis. Nat Methods 9:676–682
8. North AJ (2006) Seeing is believing? A beginners' guide to practical pitfalls in image acquisition. J Cell Biol 172:9–18
9. Linkert M, Rueden CT, Allan C et al (2010) Metadata matters: access to image data in the real world. J Cell Biol 189:777–782

Chapter 3

Screening for Bioactive Small Molecules by In Vivo Monitoring of Luciferase-Based Reporter Gene Expression in *Arabidopsis thaliana*

Christian Meesters and Erich Kombrink

Abstract

Chemical genetics is a scientific strategy that utilizes bioactive small molecules as experimental tools to dissect biological processes. Bioactive compounds occurring in nature represent an enormous diversity of structures that potentially can be used as activators or inhibitors of biochemical pathways, transport processes, regulatory networks, or developmental programs. Screening methods to identify bioactive small molecules can vary greatly, ranging from visual evaluation of phenotypic alterations to quantifying biometric traits such as enzyme activities. Here, we describe a general methodology that permits identification of compounds modulating the expression of reporter genes in *Arabidopsis thaliana* seedlings. The selection of luciferase-based reporter systems has the advantage that it allows in vivo imaging of reporter gene activity in a semiquantitative manner without affecting plant viability. We chose an *Arabidopsis* line harboring the luciferase reporter under the control of the jasmonate-inducible *LOX2* promoter to screen for either activators or inhibitors of gene expression. The outlined assay conditions can readily be applied to *Arabidopsis* lines containing other reporter genes. Thereby screening for small molecules affecting different signaling pathways and/or phenotypic responses is possible.

Key words Chemical genetics, Chemical screen, Jasmonate, Luciferase activity, Luminescence-based assay, Reporter gene expression

1 Introduction

Traditional forward genetic approaches have been widely used to identify genes or sets of genes that are responsible for a particular phenotype. In model organisms this strategy often entails random or directed mutagenesis and gene mapping by breeding. However, there are limitations in the genes discoverable by forward genetics (*discussed in* Ref. 1). Small cell-permeable molecules with activating or inhibitory activity have the potential to complement mutational approaches for dissection of biological processes because chemical interference can be performed in a conditional, dose-dependent, and reversible manner [2].

Glenn R. Hicks and Stéphanie Robert (eds.), *Plant Chemical Genomics: Methods and Protocols*, Methods in Molecular Biology, vol. 1056, DOI 10.1007/978-1-62703-592-7_3, © Springer Science+Business Media New York 2014

Chemical genetic techniques have long been applied in animal systems in areas such as cancer research, cell death, and drug development [3–6] but more recently have also found application in plant biology [7–13]. In particular, the model plant *Arabidopsis thaliana* provides excellent prerequisites for scanning small-molecule libraries for compounds acting on cellular targets that are inaccessible or recalcitrant to conventional mutational analysis [14, 15]. It is suitable for cultivation in microplates and thus allows high-throughput screening using miniaturized bioassays [7, 8, 12, 16]. Studies that have systematically explored the potential of small molecules to interfere with plant-specific processes include the gravitropic response; auxin-, abscisic acid-, or brassinosteroid-mediated signaling; plant cell morphogenesis; and innate immunity [8, 10, 16–26]. The methods applied to screen for bioactivity of small molecules differ greatly, ranging from visual evaluation of phenotypic alterations such as seed germination or hypocotyl length [23, 26–30] to quantifying biometric traits such as enzyme activities or reporter gene expression [8, 10, 17, 31], but all use miniaturized assay formats in microplates.

Here we outline a chemical screening procedure that relies on inducible firefly luciferase (LUC) reporter constructs in *Arabidopsis*, specifically, using a transgenic line harboring LUC under the control of the jasmonate-inducible promoter of the *LIPOXYGENASE 2* (*LOX2*) gene [32]. The advantage of reporter-based screens is that they afford quantitative or at least semiquantitative data that allow discrimination between compounds with high and low bioactivity. The LUC reporter system, in contrast to the β-glucuronidase reporter, further allows in vivo monitoring of activity without compromising plant viability, thus facilitating subsequent genetic screens. Furthermore, the choice of an inducible reporter gene system permits bidirectional screening for either activators of gene expression or inhibitors that impair induced gene expression, which is demonstrated in Figs. 1 and 2. Sifting through a chemical library of about 1,700 small molecules of natural and semisynthetic origin (AnalytiCon Discovery, Potsdam, Germany), we identified a single compound, 1-propyl-2-carboxy-3,8-dihydroxy-9,10-anthracenedione **766**, that seemed to activate expression of the reporter gene *LOX2p::LUC* (Fig. 1a). Conversely, screening of the same library for inhibitors uncovered three small molecules that strongly impaired methyl jasmonate (MeJA)-induced expression of the *LOX2p::LUC* reporter gene (Fig. 2). The identified compounds, cycloheximide **21**, and the two trichothecene mycotoxins diacetoxyscirpenol **86** and neosolaniol **92**, have in common that they inhibit protein synthesis [26]. Any of these compounds may serve as suitable positive control in reporter-based screens for inhibitors. Note that the MeJA-stimulated *LOX2p::LUC* expression varies considerably across all samples of this primary screen. This is largely attributed to variable

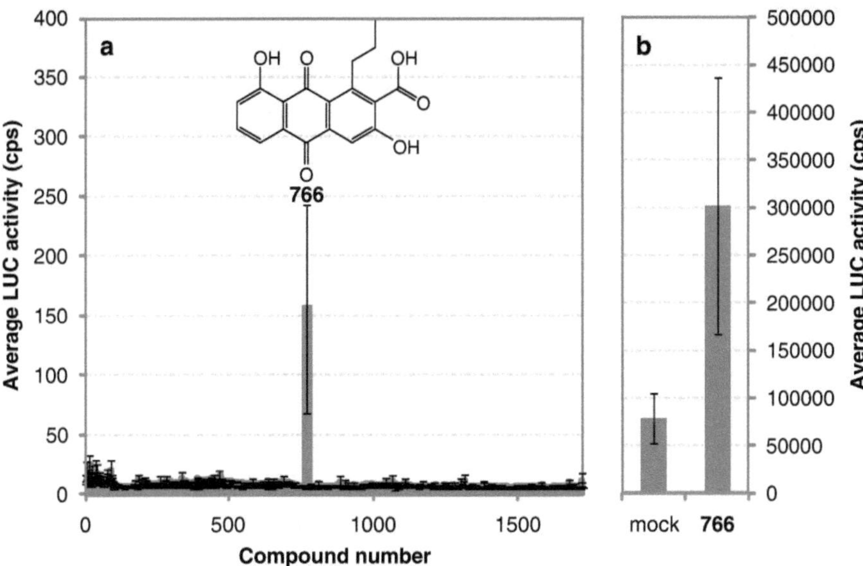

Fig. 1 Chemical screen for activators of reporter gene expression. (**a**) 12-day-old *Arabidopsis* seedlings harboring the *LOX2p::LUC* reporter gene were treated for 24 h with 1,728 different compounds (each at 25 μM). Only one of these compounds (**766**) caused an increase in LUC activity, which is presented as average (± standard deviation, $n=2$) from duplicate samples. (**b**) Counter screen to resolve impact of compound **766** on different reporter gene. *Arabidopsis* seedlings harboring the constitutively expressed reporter gene *CaMV35Sp::LUC* were treated for 24 h with **766** (25 μM) or DMSO only (mock). The increased activity (mean ± standard deviation, $n=8$) also in this line indicates that **766** directly affected LUC activity/stability rather than expression of the *LOX2p::LUC* reporter gene

seedling size and orientation within individual microplate wells, thus leading to variable luminescence detection, which effectively renders the assay only semiquantitative. However, stringent selection criteria for primary hits and their confirmation with increased sample numbers will reduce the false discovery rate.

It is important to note that identification of candidate compounds in a chemical screening campaign is only the first step of a successful chemical genetic strategy. Of course, bioactive compounds originating from the initial screen need to be critically validated. This includes verification of their activity in secondary screens (e.g., using a biological readout that is related but not identical to the primary screen), determination of IC_{50} values, and evaluation of their selectivity by comparing their effects on a variety of biological responses. The latter assays should also comprise dedicated counter screens, which are designed such that to identify false-positive hits. For example, firefly LUC has a relatively short half-life of 3 h, and its degradation is prone to inhibition by various chemicals; thus, an apparent activation may in fact reflect increased enzyme stability [33]. When applying such counter screens to the identified apparent activator **766** we observed that this compound

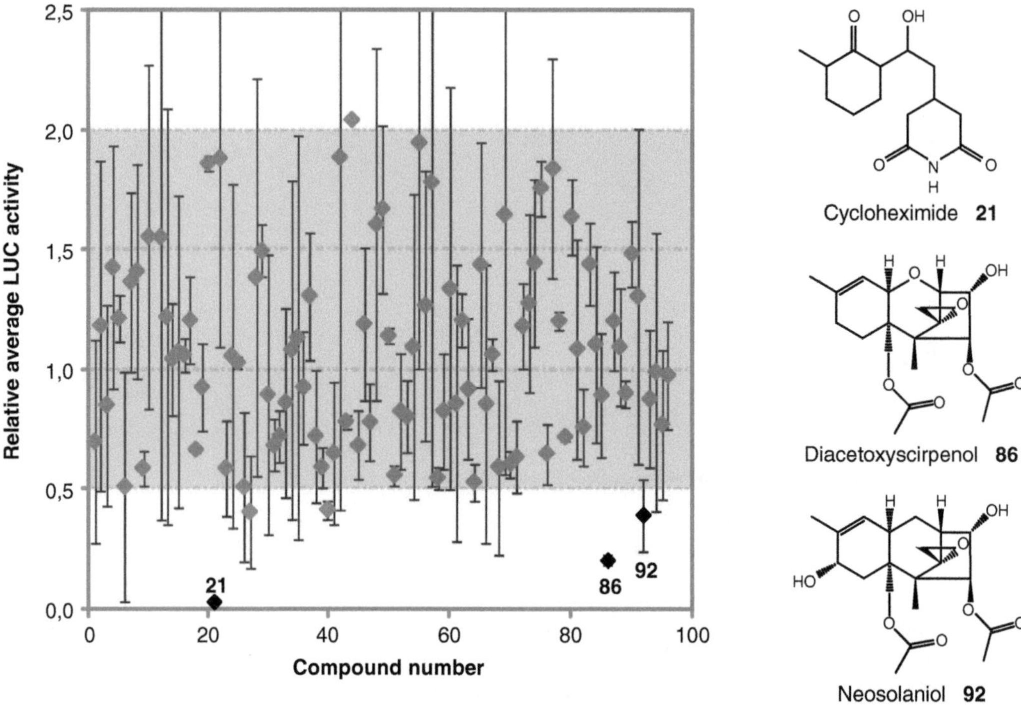

Fig. 2 Chemical screen for inhibitors of reporter gene expression. 12-day-old *LOX2p::LUC* seedlings were preincubated for 1 h with the indicated compounds (each at 25 μM) and expression of the reporter gene induced by 100 μM methyl jasmonate. LUC activity was determined after 24 h in duplicate samples; values (± standard deviation) are normalized to the average activity of the whole plate (96 samples). The shaded area shows the twofold upper and lower threshold of the average. Three candidate compounds (**21**, **86**, **92**; *black diamonds*) were confirmed in a secondary screen (data not shown) and identified as translational inhibitors, cycloheximide (**21**), diacetoxyscirpenol (**86**), and neosolaniol (**92**). Other candidates outside the indicated thresholds have not yet been confirmed (*gray diamonds*)

also strongly stimulated LUC activity in an *Arabidopsis* line containing the constitutively expressed reporter gene *CaMV35Sp::LUC* (Fig. 1b), whereas expression of the reporter gene *LOX2p::GUS* was not affected (not shown). This provides strong evidence that **766** acts on LUC rather than the *LOX2* promoter.

Ultimately, the identification of the protein target of a bioactive small molecule is of fundamental importance for understanding its mode of action. To this end, various experimental strategies can be applied, which to describe is beyond the scope of this chapter. However, the established screening conditions outlined here may be the first step into a chemical genetics project because they can readily be applied to *Arabidopsis* lines containing other LUC-based reporter genes, thus allowing the search for small molecules affecting different signaling pathways and/or phenotypic responses.

2 Materials

1. Reporter line: Select or generate an *A. thaliana* line expressing the LUC reporter under the control of a suitable promoter of your interest (*see* **Note 1**).
2. Growth medium: 0.5× Murashige and Skoog (MS) medium [34], 0.5 % (w/v) sucrose, pH 5.8. Dissolve 2.15 g Murashige and Skoog Basal Salt Mixture (M9274, Sigma, Taufkirchen, Germany) in 800 mL water. Add 5 g sucrose and mix. Adjust with KOH to pH 5.8. Add water to a final volume of 1 L. Autoclave solution and store at 4 °C.
3. Luciferin buffer: 1 M $KHPO_4$, pH 7.8. Dissolve 17.42 g K_2HPO_4 in 50 mL water and adjust to 100 mL to obtain a 1 M solution. Dissolve 13.61 g KH_2PO_4 in final volume of 100 mL water to obtain a 1 M solution. Mix 5 volumes of 1 M K_2HPO_4 with 1 volume of 1 M KH_2PO_4 and adjust with KOH to pH 7.8. Autoclave buffer.
4. Luciferin solution: 5 mM D-luciferin, 0.01 % Triton X-100. Dilute 500 μL luciferin buffer to 100 mL, add 140.16 mg D-luciferin (free acid), and dissolve (*see* **Note 2**). Add 10 μL Triton X-100. Store in aliquots at −80 °C.
5. Chemical library: Select a chemical library of your interest (*see* **Note 3**).
6. Microplates: 96-well microplates are the most frequently used format (*see* **Note 4**).
7. Clear sealing foil for microplates.
8. Luminescence detector: Use a device, which can detect luminescence at low sensitivity in the 96-well microplate format (*see* **Note 5**).
9. Plant growth cabinet with light and temperature control.

3 Methods

3.1 Plant Growth Conditions

Arabidopsis seedlings are grown in hydroponic culture under sterile conditions directly in the 96-well microplates.

1. Transfer seeds of your selected *Arabidopsis* reporter line into a microfuge spin column. 20 mg, corresponding to approximately 1,000 seeds, will be appropriate for five microplates.
2. Wash seeds twice with 600 μL each of 70 % ethanol and spin down for a few seconds using a tabletop centrifuge.
3. Repeat washing step with 600 μL 100 % ethanol, and centrifuge for 1 min at full speed.

4. Open spin column, and dry seeds under sterile conditions (*see* **Note 6**). If not immediately used, seeds should be stored in dry and cold place.
5. Distribute 200 μL growth medium into each well of a 96-well microplate.
6. Add two seeds to each well (*see* **Note 7**), cover the microplate with the clear lid, and seal with parafilm to reduce evaporation of liquid.
7. Keep microplates for 2 days at 4 °C in the dark. This stratification step will synchronize seed germination.
8. Transfer microplates to a growth cabinet with appropriate conditions. We use a photoperiod of 12-h light and 12-h dark at 21 °C.

3.2 Screening Chemical Libraries

For large-scale screening campaigns we recommend using five 96-well microplates at a time. This allows sifting through one microplate of chemicals using duplicate samples and scanning for both activators and inhibitors, because we used an inducible reporter line (*LOX2p::LUC*). One plate is available for various control treatments.

1. After 12–14 days of cultivation, take microplates from the growth cabinet.
2. Check seedlings for even growth, and mark small seedlings or empty wells for later data analysis (*see* **Note 8**).
3. Replace medium with 200 μL fresh growth medium (*see* **Note 9**).
4. Take one library plate with the working concentration of the chemicals (1 mM) (*see* **Note 10**). Using a multichannel pipette, transfer 2 μL each from row 1 of the chemicals to the first row of a microplate with seedlings (yielding a final concentration of 10 μM). Add the same chemicals to the second row of seedlings as replicate. Following this scheme will eventually yield two seedling-containing microplates per chemical plate. Repetition of the whole procedure will yield another set of two seedling-containing microplates for alternative treatment (*see* **Note 11**).
5. Prepare one microplate with appropriate controls (*see* **Note 12**).
6. Leave microplates at room temperature for 1 h to allow uptake and distribution of chemicals in the seedlings. When screening for activators of reporter gene expression, transfer microplates directly back to growth cabinet.
7. In a screen for inhibitors of reporter gene activity, induce expression by adding activator to a set of two microplates (*see* **Note 13**). Transfer microplates back to growth cabinet.
8. Keep microplates at growth conditions for an appropriate time period to afford sufficient reporter gene expression (LUC activity) (*see* **Note 14**).

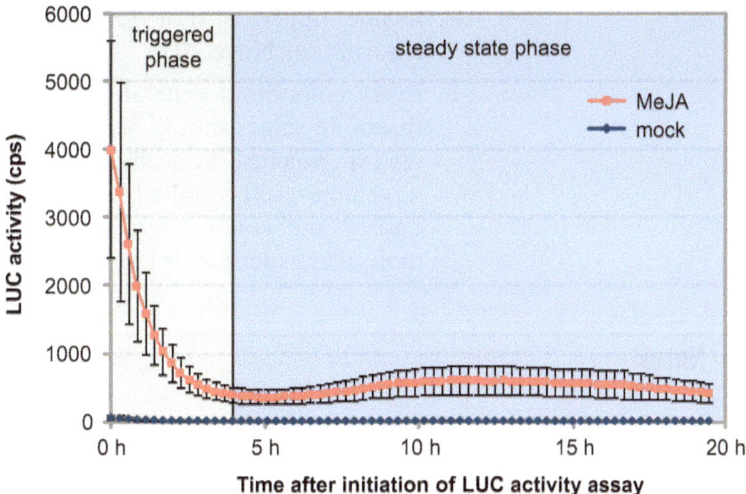

Fig. 3 Time course of LUC activity. 12-day-old *Arabidopsis LOX2p::LUC* seedlings were treated with 100 μM methyl jasmonate (MeJA) to induce reporter gene expression. After 24 h in vivo LUC activity measurement was initiated by addition of luciferin ($t = 0$ h) and recorded for the indicated time. Values represent average activity (± standard deviation, $n = 33$)

3.3 Luciferase Assay

1. Thaw luciferin solution at room temperature, and collect microplates from growth cabinet.
2. Add 10 μL luciferin solution to each well of the seedling-containing microplates, yielding a final concentration of 250 μM luciferin (*see* **Note 15**).
3. Place microplates into luminescence reader (*see* **Note 16**).
4. Record luminescence from each well for 2 s, and repeat in cycles for appropriate time period, usually several hours (*see* **Note 17**).

3.4 Data Analysis

1. Transfer collected activity data to a suitable spreadsheet (e.g., Microsoft Excel, Apple Numbers, OpenOffice Calc). Design of a master sheet for a whole microplate will facilitate data analysis, particularly for large-scale experiments.
2. Plot time-dependent activity of each single well for data comparison and evaluation (*see* Fig. 3).
3. Calculate average value for defined window of triggered LUC activity for all samples (*see* **Note 18**).
4. Calculate mean of average activity and standard deviation for the two replicates.
5. Proceed with calculating corresponding values for control treatments (*see* **Note 19**).
6. Plot activity values (absolute or normalized) in diagrams with standard deviation (*see* Figs. 1 and 2).

7. Define upper and/or lower threshold values relative to controls (*see* **Note 20**).
8. Every compound generating an activity beyond the defined threshold values should be considered a candidate for follow-up experiments. Typically, this includes confirmation of the screening result using a larger sample size, application of secondary and counter screens, and further physiological and molecular experiments (*see* **Note 21**).

4 Notes

1. The *Arabidopsis* line used to generate the data presented here harbors the jasmonate-inducible reporter gene *LOX2p::LUC* and, in addition, *LOX2p::GUS* as second reporter gene. On the one hand, this is useful because the same reporter line can be applied in a secondary screen, which is independent of LUC. On the other hand, we experienced significant silencing particularly of GUS expression in this line and hence for secondary screens eventually used other jasmonate-responsive reporter constructs (e.g., *VSPp::LUC*, *VSPp::GUS*, *OPCL1p::GUS*), many of which are available from stock collections (e.g., Nottingham Arabidopsis Stock Centre, UK).
2. We find that D-luciferin dissolves better in buffer of high molarity. Thus, start by dissolving D-luciferin in small volume (20–30 % of final), add a few drops of 1 M luciferin buffer in case of problems, and then dilute to final volume.
3. Chemical libraries are commercially available from different companies. They can be provided in different formats (e.g., microplates, individual tubes), may contain different types of small molecules (e.g., natural, synthetic, or semisynthetic compounds), and may be assembled to cover a large diversity of structures or to targeted selected functions (e.g., protein kinases) [35, 36]. Usually the chemicals of such libraries are dissolved in dimethyl sulfoxide (DMSO), often at 5–20 mM, and supplied in 96-well microplates. Depending on the supplier a library plate may contain 96 compounds or only 80 compounds with two columns of the microplate being available for controls (e.g., solvent and/or reference compound). We recommend using the master plates to prepare diluted copies at 1–2 mM in DMSO. This facilitates pipetting when using final working concentrations of 10–20 μM and reduces the danger of contaminating the master plate when multiple screens are performed. Store library as recommended by the supplier (at 4, −20, or −80 °C).

4. We use black 96-well microplates with transparent lid. Consult the manual of your luminescence reader for recommendation. Depending on the detector and/or the intended bioassay, microplates of other format and quality may be more appropriate.
5. For screening at a larger scale (>1,000 compounds) we recommend a luminescence detector, which can read multiple microplates automatically. We used the TopCount NXT™ Microplate Scintillation and Luminescence Counter (Perkin-Elmer, Waltham, USA). For small-scale luciferase assays simple luminescence readers handling single microplates may be sufficient. We used the luminometer Centro LB 960 (Berthold Technologies, Bad Wildbad, Germany) for such experiments.
6. In the opened column, seeds should be dry within 20 min. To check, flip closed column with your finger. Sticky seeds are still wet.
7. Using two rather than single seed per well reduces the possibility of having empty wells if a seed fails to germinate. By contrast, adding more than two seeds per well increases the variation of the recorded signal due to quenching of luciferase luminescence. To achieve reliable and reproducible seed distribution, we find it easiest to spread seeds on a piece of parafilm. A sterile wooden toothpick, briefly dipped into growth medium, is used to adsorb two of the spread seeds. Seeds are released from the wet tip by quickly dipping it into a selected well. If the movement is too slow or the toothpick too wet, the release of seeds is impaired. This is tedious work, but in contrast to liquid handling an automatic seed distribution device is not yet available.
8. Seedling size will depend on the specific growth conditions. Other development stages may be more appropriate for reporter genes other than *LOX2p::LUC*. In any case, seedlings should still be completely submerged in medium to facilitate even uptake of added compounds by all parts of the plants and to minimize well-to-well variation. It may be useful at this stage to replace small seedlings and fill empty wells with seedlings from the control plate. If doing so make sure that your chosen reporter does not respond to this stress.
9. We observed no differences in samples from changed or unchanged medium. However, loss of liquid due to condensation at the lid may be significant under certain circumstances, particularly in wells located at the edges.
10. If library is stored frozen, let it thaw at room temperature. Mix library plate gently by shaking it horizontally, and avoid spill and cross-contamination. Briefly centrifuge library plate to collect droplets from lid and well surface.

11. We find that final concentrations of 10–25 μM for the chemicals are useful and appropriate. The final DMSO concentration should be kept constant and as low as possible. We observed that increasing DMSO concentrations yielded higher luminescence units in the luciferase assay. Setting up the screen with duplicate samples (replicate measurements) has two major advantages: (1) The built-in verification of primary hits helps to minimize the number of false positive and false negatives, and (2) it allows to apply minimum statistical data analysis to improve the sensitivity and selectivity of the screening process, which cannot be met by technological and organizational improvements alone [37]. Thus, replicates increase the precision of activity measurements, allow the estimation of variability associated with the measurements, and make it easier to detect minimally or moderately active compounds.
12. An obvious control is that several seedlings are treated with solvent only. Addition of control chemicals with known effect may be useful (e.g., cycloheximide to prevent synthesis of expressed reporter gene product or MeJA to achieve maximum induction in case of *LOX2p::LUC*). Often chemical library plates contain one empty and/or one solvent-filled row. In this case these can be used for controls instead of a separate microplate. Alternatively, controls can be completely omitted if it is assumed that the majority of added chemicals had no effect on the readout. Under these circumstances it is established practice to use the average of all samples of a test plate for normalization [37].
13. For the *LOX2p::LUC* line add 2 μL 10 mM MeJA solution to obtain a final concentration of 100 μM.
14. For the *LOX2p::LUC* line we find an incubation period of 24 h suitable and convenient.
15. Addition of luciferin will immediately trigger the luciferase reaction. Since this initial activity declines relatively rapidly (*see* **Note 17**, Fig. 3), take the time period for pipetting several microplates into consideration to obtain reliable and comparable activity values for each sample.
16. For large-scale experiments measuring multiple microplates in series with the TopCount instrument, seal plates with foil and pierce holes into it to avoid condensation, which may impair luminescence detection. For small-scale assays using single microplates in a simple luminometer sealing is not required.
17. Initial luciferase activity is proportional to the amount of proteins expressed from the activated reporter gene (and the turnover of the protein). However, since luciferase is

inhibited by its product oxyluciferin, maximum enzyme activity will decline over time and reach a steady state after a certain time. For *LOX2p::LUC* this triggered phase is finished after 4–5 h (*see* Fig. 3). Thereafter the recorded activities remain rather constant reflecting a steady synthesis and turnover of luciferase.

18. For *LOX2p::LUC* seedlings we used the first 4 h after initiation of LUC measurement for averaging activity. This average value approximates the integrated activity over time and provides reproducible and reliable data sets despite the relatively rapid decline of maximum activity. Instead of the triggered phase, activity data could in principle also be collected from the steady-state phase (*see* Fig. 3). However, we find this less convenient and reliable because it requires extended time periods for recording luminescence and the values are strongly dependent on size (biomass) and orientation of seedlings in the wells.

19. For large-scale experiments, when extensive data sets from numerous microplates have to be compared, it is important to include positive and negative controls, which serve to normalize the data [37]. Alternatively, the average of all measurements (of one or several plates) may be used as reference value for normalization, thus completely omitting separate controls. However, this is only advised when it is expected that most added chemicals do not have any effect [37].

20. The threshold value directly dictates the selection of candidate compounds from the analyzed population. Thus, its appropriate choice is an important consideration; setting of a stringent (higher) threshold will lead to few, a relaxed threshold to many candidate compounds for subsequent analysis. However, compounds with lower potency may also be interesting if they act selectively or provide new lead structures. In practice the threshold value could be based on the control's standard deviation or could be defined as twofold difference from the normalized mean value (*see* Fig. 2).

21. The LUC reporter system is well established as readout in numerous bioassays because it combines easy handling with sensitive monitoring of in vivo activity. However, this activity, as recorded here, cannot easily be related to biomass, which may eventually result in relatively large variation of single measurements (*see* Fig. 2). This is not necessarily true for all assay conditions (*see* Fig. 1) and reporter gene constructs. In any case, it is important to critically evaluate each single measurement and scan for deviations from typical activity curves. Thereby the risk of following false-positive or -negative hits can be minimized.

Acknowledgements

This work was supported by the Max Planck Society and a PhD fellowship from the International Max Planck Research School program (C.M.). We thank Dr. Ferenc Nagy (Institute of Plant Biology, Szeged, Hungary) for providing the *Arabidopsis* line harboring the *CaMV35Sp::LUC* gene and Dr. John Mundy (University of Copenhagen, Denmark) for providing the *Arabidopsis LOX2p::LUC LOX2p::GUS* line.

References

1. Hicks GR, Raikhel NV (2012) Small molecules present large opportunities in plant biology. Annu Rev Plant Biol 63:261–282
2. Smukste I, Stockwell BR (2005) Advances in chemical genetics. Annu Rev Genom Human Genet 6:261–286
3. Schreiber SL (1998) Chemical genetics resulting from a passion for synthetic organic chemistry. Bioorg Med Chem 6:1127–1152
4. Stockwell BR (2000) Chemical genetics: ligand-based discovery of gene function. Nat Rev Genet 1:116–125
5. Mayer TU (2003) Chemical genetics: tailoring tools for cell biology. Trends Cell Biol 13: 270–277
6. Gangadhar NM, Stockwell BR (2007) Chemical genetic approaches to probing cell death. Curr Opin Chem Biol 11:83–87
7. Blackwell HE, Zhao Y (2003) Chemical genetic approaches to plant biology. Plant Physiol 133:448–455
8. Armstrong JI, Yuan S, Dale JM, Tanner VN, Theologis A (2004) Identification of inhibitors of auxin transcriptional activation by means of chemical genetics in *Arabidopsis*. Proc Natl Acad Sci U S A 101:14978–14983
9. Walsh TA (2007) The emerging field of chemical genetics: potential applications for pesticide discovery. Pest Manag Sci 63:1165–1171
10. Serrano M, Robatzek S, Torres M, Kombrink E, Somssich IE, Robinson M, Schulze-Lefert P (2007) Chemical interference of pathogen-associated molecular pattern-triggered immune responses in *Arabidopsis* reveals a potential role for fatty-acid synthase type II complex-derived lipid signals. J Biol Chem 282:6803–6811
11. Hicks GR, Raikhel NV (2009) Opportunities and challenges in plant chemical biology. Nat Chem Biol 5:268–272
12. Tóth R, van der Hoorn RAL (2010) Emerging principles in plant chemical genetics. Trends Plant Sci 15:81–88
13. McCourt P, Desveaux D (2010) Plant chemical genetics. New Phytol 185:15–26
14. Tornero P, Chao RA, Luthin WN, Goff SA, Dangl JL (2002) Large-scale structure-function analysis of the Arabidopsis RPM1 disease resistance protein. Plant Cell 14:435–450
15. Park S-Y, Fung P, Nishimura N, Jensen DR, Fujii H, Zhao Y, Lumba S, Santiago J, Chow TF, Alfred SE, Bonetta D, Finkelstein R, Provart NJ, Desveaux D, Rodriguez PL, McCourt P, Zhu J-K, Schroeder JI, Volkman BF, Cutler SR (2009) Abscisic acid inhibits type 2C protein phosphatases via the PYR/PYL family of START proteins. Science 324:1068–1071
16. Lin L-C, Hsu J-H, Wang L-C (2010) Identification of novel inhibitors of 1-aminocyclopropane-1-carboxylic acid synthase by chemical screening in *Arabidopsis thaliana*. J Biol Chem 285:33445–33456
17. Zhao Y, Dai X, Blackwell HE, Schreiber SL, Chory J (2003) SIR1, an upstream component in auxin signaling identified by chemical genetics. Science 301:1107–1110
18. Zouhar J, Hicks GR, Raikhel NV (2004) Sorting inhibitors (Sortins): chemical compounds to study vacuolar sorting in *Arabidopsis*. Proc Natl Acad Sci U S A 101:9497–9501
19. Surpin M, Rojas-Pierce M, Carter C, Hicks GR, Vasquez J, Raikhel NV (2005) The power of chemical genomics to study the link between endomembrane system components and the gravitropic response. Proc Natl Acad Sci U S A 102:4902–4907
20. Yoneda A, Higaki T, Kutsuna N, Kondo Y, Osada H, Hasezawa S, Matsui M (2007) Chemical genetic screening identifies a novel inhibitor of parallel alignment of cortical microtubules and cellulose microfibrils. Plant Cell Physiol 48:1393–1403
21. DeBolt S, Gutierrez R, Ehrhardt DW, Melo CV, Ross L, Cutler SR, Somerville C, Bonetta D (2007) Morlin, an inhibitor of cortical microtubule dynamics and cellulose synthase movement. Proc Natl Acad Sci U S A 104: 5854–5859

22. Gendron JM, Haque A, Gendron N, Chang T, Asami T, Wang Z-Y (2008) Chemical genetic dissection of brassinosteroid-ethylene interaction. Mol Plant 1:368–379
23. Robert S, Chary SN, Drakakaki G, Li S, Yang Z, Raikhel NV, Hicks GR (2008) Endosidin1 defines a compartment involved in endocytosis of the brassinosteroid receptor BRI1 and the auxin transporters PIN2 and AUX1. Proc Natl Acad Sci U S A 105:8464–8469
24. Schreiber K, Wenzislava C, Peek J, Desveaux D (2008) A high-throughput chemical screen for resistance to *Pseudomonas syringae* in Arabidopsis. Plant J 54:522–531
25. Knoth C, Salus MS, Girke T, Eulgem T (2009) The synthetic elicitor 3,5-dichloroanthranilic acid induces *NPR1*-dependent and *NPR1-independent* mechanisms of disease resistance in Arabidopsis. Plant Physiol 150:333–347
26. Serrano M, Hubert DA, Dangl JL, Schulze-Lefert P, Kombrink E (2010) A chemical screen for suppressors of the *avrRpm1-RPM1-dependent* hypersensitive cell death response in *Arabidopsis thaliana*. Planta 231:1013–1023
27. Zhao Y, Chow TF, Puckrin RS, Alfred SE, Korir AK, Larive CK, Cutler SR (2007) Chemical genetic interrogation of natural variation uncovers a molecule that is glycoactivated. Nat Chem Biol 3:716–721
28. Savaldi-Goldstein S, Baiga TJ, Pojer F, Dabi T, Butterfield C, Parry G, Santner A, Dharmasiri N, Tao Y, Estelle M, Noel JP, Chory J (2008) New auxin analogs with growth-promoting effects in intact plants reveal a chemical strategy to improve hormone delivery. Proc Natl Acad Sci U S A 105:15190–15195
29. Bassel GW, Fung P, Chow TF, Foong JA, Provart NJ, Cutler SR (2008) Elucidating the germination transcriptional program using small molecules. Plant Physiol 147:143–155
30. De Rybel B, Audenaert D, Vert G, Rozhon W, Mayerhofer J, Peelman F, Coutuer S, Denayer T, Jansen L, Nguyen L, Vanhoutte I, Beemster GTS, Vleminckx K, Jonak C, Chory J, Inzé D, Russinova E, Beeckman T (2009) Chemical inhibition of a subset of *Arabidopsis thaliana* GSK3-like kinases activates brassinosteroid signaling. Chem Biol 16:594–604
31. Kim T-H, Hauser F, Ha T, Xue S, Böhmer M, Nishimura N, Munemasa S, Hubbard K, Peine N, Lee B-H, Lee S, Robert N, Parker JE, Schroeder JI (2011) Chemical genetics reveals negative regulation of abscisic acid signaling by a plant immune response pathway. Curr Biol 21:990–997
32. Jensen AB, Raventos D, Mundy J (2002) Fusion genetic analysis of jasmonate-signalling mutants in *Arabidopsis*. Plant J 29:595–606
33. Thorne N, Auld DS, Inglese J (2010) Apparent activity in high-throughput screening: origins of compound-dependent assay interference. Curr Opin Chem Biol 14:315–324
34. Murashige T, Skoog F (1962) A revised medium for rapid growth and bioassays with tobacco tissue cultures. Physiol Plant 15:473–497
35. Schreiber SL (2000) Target-oriented and diversity-oriented organic synthesis in drug discovery. Science 287:1964–1969
36. Kaiser M, Wetzel S, Kumar K, Waldmann H (2008) Biology-inspired synthesis of compound libraries. Cell Mol Life Sci 65: 1186–1201
37. Malo N, Hanley JA, Cerquozzi S, Pelletier J, Nadon R (2006) Statistical practice in high-throughput screening data analysis. Nat Biotechnol 24:167–175

Part II

Chemical Screening

Chapter 4

Application of Yeast-Two Hybrid Assay to Chemical Genomic Screens: A High-Throughput System to Identify Novel Molecules Modulating Plant Hormone Receptor Complexes

Andrea Chini

Abstract

Phytohormones are endogenous signalling molecules that regulate plant development, adaptation to the environment, and survival. Upon internal or external stimuli, hormones are quickly accumulated and perceived, which in turn activates specific signalling cascades regulating the appropriate physiological responses. In the last decade, great advances in understanding plant hormone perception mechanisms have been achieved. Among different methodological approaches, yeast-two hybrid (Y2H) assays played a pivotal role in the identification and analysis of plant hormone perception complexes. The Y2H assay is a rapid and straightforward technique that can be easily employed to identify small molecules directly modulating plant hormone perception complexes in a high-throughput manner. However, an Y2H chemical screen tends to isolate false positive molecules, and therefore a secondary *in planta* screen is required to confirm the genuine bioactivity of putative positive hits. This two-step screening approach can substantially save time and manual labor. This chapter focuses on the prospects of Y2H-based chemical genomic high-throughput screens applied to plant hormone perception complexes. Specifically, the method employed to carry out a chemical genomic screen to identify agonist and antagonist molecules of the phytohormone jasmonic acid in its conjugated form jasmonic acid–isoleucine (JA-Ile) is described. An easy *in planta* confirmation assay is also illustrated. However, this methodology can be easily extended to detect novel chemical compounds perturbing additional plant hormone receptor complexes. Finally, the high-throughput approach described here can also be implemented for the identification of molecules interfering with protein–protein interaction of plant complexes other than hormone receptors.

Key words Chemical genomic screen, Y2H screen, Plant hormone perception, COI1–JAZ receptor complex, JA-Ile agonist and antagonist molecules, JAZ degradation

1 Introduction

Plants are sessile organisms that evolved a complex signalling network, orchestrated by phytohormones, which are essential for plant development, responses to different stresses, and adaptation to the environment. Due to their importance, plant hormone

Glenn R. Hicks and Stéphanie Robert (eds.), *Plant Chemical Genomics: Methods and Protocols*, Methods in Molecular Biology, vol. 1056, DOI 10.1007/978-1-62703-592-7_4,

biosynthesis has been extensively studied [1, 2]; however, only recently have hormone perception mechanisms been revealed by the identification of the molecular components of several plant hormone receptor complexes (reviewed in refs. 3, 4).

These recent advances showed several similarities in the mechanisms of perception of different plant hormones. For example, the perception mechanisms of JA-Ile and auxin are remarkably similar. In agreement with the "molecular glue" model, JA-Ile induces the interaction between the F-box COI1 and the co-receptor JAZs forming a tetrameric perception complex [5–7]. In this context, Y2H experiments confirmed the direct action of JA-Ile on promoting COI1–JAZ interaction in yeast, a heterologous system not responding to JA [8]. In addition, the Y2H results showed that JA-Ile, but neither additional JA-conjugated compounds nor JA-Ile precursors, specifically promotes COI1–JAZ interaction [6, 8].

Likewise, the F-boxes TIR1/AFBs and the co-receptors Aux/IAAs form the auxin receptor complexes; the interaction between these two co-receptors is induced by auxins acting as "molecular glue" [9–11]. Y2H experiments were employed to confirm the conserved function of the TIR1/AFBs–Aux/IAAs perception complex in phylogenetically distant plant species such as *Physcomitrella patens* [12].

Recently, the binding of abscisic acid (ABA) to the receptor family PYR/PYL/ACAR has been reported to induce a conformational change responsible for the interaction with PPC2, and therefore promoting tetramer complex. Y2H experiments were instrumental to identify the specific members of the PYR/PYL/ACAR family interacting in an ABA-dependent manner with the PPC2 HAB1 [13].

Finally, the observation that the gibberellic acid (GA) receptor GID1 and the DELLA repressor proteins physically interact in a GA-dependent manner was first achieved in a Y2H assay [14] and then confirmed by crystallography of GID1 forming a complex with GA and the DELLA protein [15, 16].

These recent reports show the value of Y2H in plant hormone perception analyses. The reconstruction of the plant COI1–JAZ9–JA-Ile perception complex in yeast is a robust and simple system suitable for a high-throughput chemical screen [17]. This chapter describes the use of chemical genomics to identify novel compounds with JA-Ile agonist or antagonist activities in an Y2H COI1–JAZ9 interaction frame. Although Y2H looks like a fine system for high-throughput chemical screens, in our experience, Y2H is prone to identify several false positive molecules. Therefore, an *in planta* confirmation assay, or secondary screen, is also required. Alternatively, pull-down experiments using plants extract can be employed to confirm the direct action of putative compounds on COI1–JAZ interaction [8, 18]. In the case of JA-Ile perception, the hormone-mediated COI1–JAZ interaction leads to the ubiquitination and subsequent degradation of the JAZ proteins [5, 6]. Therefore, a JAZ-GUS degradation assay in Arabidopsis seedlings was successfully employed as a secondary screen. This two-step screening approach could save significant manual labor and time. The method reported here can be easily adjusted to identify novel chemical compounds modulating other plant hormone receptors such as auxin, gibberellin, or abscisic acid (ABA) perception complexes. In addition, molecules interfering with additional protein–protein interaction plant complexes can be identified using the approach described here [19].

2 Materials

2.1 Yeast Media, Strain and Vectors

The required yeast media are synthetic dropout (YSD), adenine-supplemented YPD (YPDA), and amino acid dropout supplement (–His –Ade –Leu –Trp) (Clontech). Filter-sterilized histidine (1,000× stock consists of 20 g/l dissolved in 0.5 M HCl monohydrate) and adenine (10 g/l equals 500×) solutions are also necessary. Autoclaved stock solutions for yeast transformation are 1 M lithium acetate (10×), Tris–HCl EDTA (TE) buffer pH 7.5 (100 mM equal to 10×), and 50 % polyethylene glycol (PEG). Commercial dimethyl sulfoxide (DMSO) and carrier DNA such as salmon sperm are also required.

The *Saccharomyces cerevisiae* yeast AH109 strain was employed (Clontech). The AH109-compatible pGADT7 and pGBKT7 vectors were modified by inserting a Gateway (Invitrogen) cassette in the multiple cloning site as described [17].

2.2 Plant Material

Plant media is 1× Murashige and Skoog (MS) medium supplemented with 1 % sucrose and 1.5 % agar. The X-Gluc solution consists of 50 mM phosphate buffer pH = 7, 0.2 % Triton, 1 mM K-Ferricyanide, 1 mM K-Ferrocyanide, and 1 mg/ml 5-bromo-4-chloro-3-indolyl glucuronide (X-Gluc dissolved in dimethylformamide). JAZ1-GUS transgenic lines have been previously generated [6].

2.3 Chemical Treatments

Several chemical libraries (such as the ChemBridge DIVERSet and LifeChem) are commercially available in standard 384-cell format plates at a 10 mM concentration dissolved in DMSO. The choice of appropriate libraries is crucial for the success of the screen, and this issue has already been discussed elsewhere [20]. Access to a liquid-handling robot with pin tools to transfer small volume (such as 0.2 μl) is necessary. Coronatine (1 mM stock in ethanol) and jasmonic acid (100 mM stock in dimethylformamide) are commercially available (Sigma).

3 Methods

3.1 Preparation of Y2H Screen 96-Well Plates

A liquid-handling robot is required for the preparation of the chemical screen plates.

1. Add 10 μL of liquid YSD –Ade –His –Trp –Leu in each well of four 96-well plates.
2. Transfer 0.2 μL of compound from a 384-well chemical stock plate into four 96-well plates (*see* **Notes 1** and **2**).
3. Add 90 μL of agar YSD –Ade –His –Trp –Leu medium into each well and mix well to obtain screen plates for the direct agonist of JA-Ile assay. To prepare screen plate for the JA-Ile antagonist assay, add coronatine, a bacterial analogue of JA-Ile, to agar YSD –4 medium to a final concentration of 1 μM coronatine.
4. The 96-well plates containing 100 μL of solid YSD supplemented with approximately 50 μM of each chemical compound are now ready for Y2H chemical screen (*see* **Note 3**).

3.2 Y2H Transformation

The method described here is essentially the classic LiAc-mediated yeast transformation protocol with minor adaptations:

1. Inoculate 10 mL of YPDA with a yeast AH109 colony (2–5 mm in diameter) from a fresh plate (*see* **Note 4**), vortex and grow at 30 °C and 250 rpm shaking overnight.
2. Dilute overnight culture in 100 mL of YPDA obtaining an $OD_{(600)}$ of 0.2 and grow at 30 °C and 250 rpm until the culture $OD_{(600)}$ reaches between 0.4 and 0.6 (*see* **Note 5**).
3. Transfer yeast culture in 50 ml tubes and spin at room temperature at 500 × *g* for 5 min.
4. Discard the supernatants and gently resuspend the cell pellets in 50 mL of sterile H_2O. Centrifuge at room temperature at 500 × *g* for 5 min.
5. Repeat the previous step resuspending the cell pellets in 1 mL of H_2O and transfer the yeast culture in a 1.5 ml tube. Spin at room temperature at 500 × *g* for 5 min.
6. Discard the supernatants and gently resuspend the cell pellets in 400 μl of freshly prepared, sterile LiAc/TE solution (100 mM lithium acetate and 1× TE) per 50 ml initial yeast culture volume. Competent yeast cells can be stored at room temperature for up to 1 h.
7. In a separate 1.5 mL tube, prepare 100 ng of each plasmid DNA (e.g., pGAD-JAZ9 and pGBK-COI1) and 100 μg carrier DNA (*see* **Note 6**).
8. Add 100 μL of yeast competent cells to each tube.

9. Add 600 μL of freshly prepared sterile PEG/lithium acetate/TE (40 % PEG 3350, 100 mM lithium acetate and 1× TE) solution to each tube and mix well by vortexing.
10. Incubate at 30 °C for 30 min with shaking at 250 rpm.
11. Add 70 μl of DMSO and gently mix by inverting the tubes four times (avoid vortexing).
12. Heat shock at 42 °C for 15 min and immediately chill cells on ice for 2 min.
13. Centrifuge cells for 5 s at 15,000 ×*g* at room temperature. Remove the supernatant and resuspend the cells in 200 μl of sterile H_2O by vortexing.
14. Plate yeast cells on the selection plates YSD –Leu –Trp and incubate them at 30 °C for 2 days to select yeast colonies co-transformed with both pGAD (Leu auxotrophy) and pGBK (Trp auxotrophy).
15. After 2–3 days, pick 5–10 colonies (with a diameter of 2–4 mm) and dissolve them in 10 ml liquid YSD –Lue –Trp. Incubate at 30 °C for 5–6 h with shaking at 250 rpm.
16. Measure $OD_{(600)}$, centrifuge 5 min at 2,500 ×*g*, discard supernatant and resuspend cells in H_2O to a final $OD_{(600)}$) of 1 (approx. 3 × 10e7 cells/ml).
17. Transfer a 3 μl drop of yeast into each well of the 96-well plate (*see* **Note 7**) and incubate them at 30 °C (*see* **Note 8**).
18. The COI1–JAZ9 interaction in Y2H is best detected as yeast growth after 3–5 days incubation. Putative agonist JA-Ile molecules induce COI1–JAZ9 interaction measured as yeast growth on YSD media without coronatine (Fig. 1). Lack of growth on YSD media supplemented with coronatine identifies putative JA-Ile antagonistic compounds (Fig. 2).

3.3 Secondary Screen: Confirming the Activity of the Putative Chemical Hits In Planta

1. Sterilize 35S-JAZ1-GUS seeds and vernalize for 3 days.
2. Place sterile seeds on a MS plate (supplemented with 1.5 % agar) and grow them vertically at standard conditions (22 °C 16-h light and 8-h dark) for 6 or 7 days (*see* **Note 9**).
3. To confirm putative agonist molecules, add 2 ml of liquid MS to each cell of a 12-well plate and add the appropriate volume of the candidate compounds (for example, 20 μl of 10 mM stock solution in 2 ml), previously identified in the Y2H screen, to different wells to obtain a solution of 100 μM of the candidate compounds. Use an equal amount of DMSO as control.
4. To confirm the compounds antagonist to JA-Ile, prepare a 1 μM JA solution in liquid MS and add 2 ml to each cell of a 12-well plate. Alternatively, a 50 nM coronatine solution in liquid MS

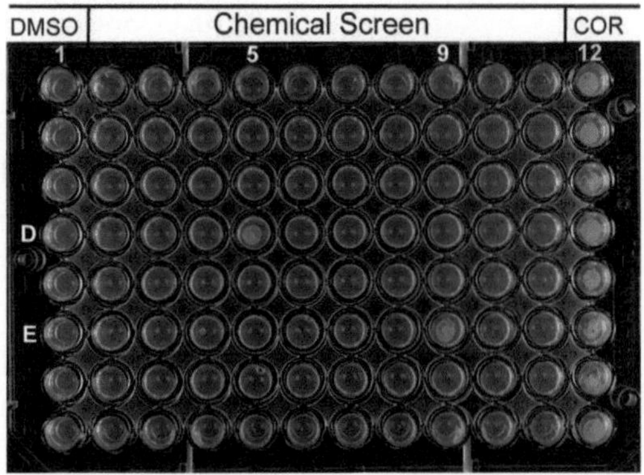

Fig. 1 Y2H screen plate for the identification of JA-Ile agonist compounds. Growth of yeast cells co-transformed with pGBK-COI1/pGAD-JAZ9 on minimal yeast media in presence of compounds of the chemical library. Molecules promoting COI1–JAZ9 interaction, and therefore mimicking JA-Ile action, were identified for their induction of yeast growth. Column 1 and 12 represent negative and positive controls; column 1 was prepared with minimal yeast medium and DMSO whereas column 12 contains minimal yeast medium supplemented with 1 μM coronatine

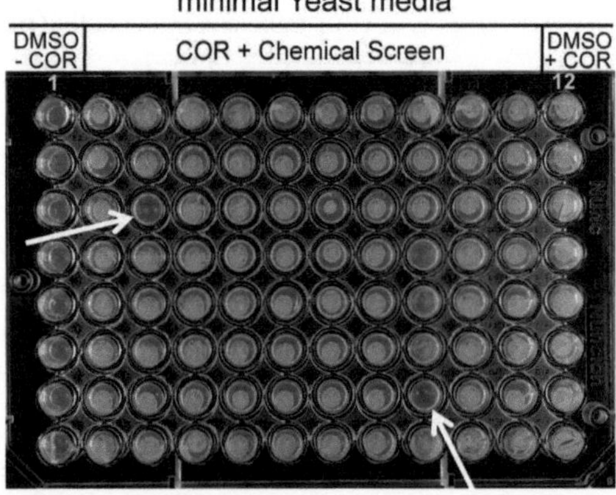

Fig. 2 Y2H screen plate for the identification of compounds showing JA-Ile antagonist effect. Growth of yeast cells co-transformed with pGBK-COI1/pGAD-JAZ9 on minimal yeast media in presence of 1 μM coronatine and compounds of the chemical library. Molecules perturbing COI1–JAZ9 interaction in presence of coronatine (indicated by *arrows*), and therefore acting as JA-Ile antagonist, were identified for the lack of yeast growth. As described in Fig. 1, column 1 and 12 represent the controls

can also be used with similar outcome. Add to the 2 ml solution the appropriate volume of the candidate compounds, previously identified in the Y2H screen, to obtain a final 100 μM solution. Use the equal amount of DMSO as control.

5. Transfer 5–8 JAZ1-GUS seedlings to each 2 ml well and incubate them for 1 h at 100 rpm at 22 °C in the same light conditions. Make sure not to damage the seedlings during the transfer (be particularly careful with the roots; *see* **Note 10**).
6. Discard the MS solution and add 1 ml of freshly prepared X-Gluc solution (*see* **Note 11**), apply vacuum to samples for 1 min to ensure homogenous uptake and store at 37 °C overnight.
7. The following day, check JAZ1-GUS degradation in roots (*see* **Note 12**). Bona fide agonist molecules should induce JAZ1-GUS degradation in absence of JA. In contrast, valid antagonistic compounds should prevent JAZ1-GUS degradation in presence of JA.

4 Notes

Practical Considerations Before Starting an Y2H Chemical Screen

The robustness and reproducibility of the Y2H protein–protein interaction are crucial for the success of a large-scale Y2H chemical screen. It is therefore essential to optimize the conditions (concentration of yeast cells, appropriate concentration of 3AT, time of incubation at 28 °C, etc.) of the Y2H protein–protein interaction assay to avoid the identification of too many false positive hits.

1. The volume added to the screen plates depends on the stock concentration of the chemical library. Generally, a final concentration between 50 and 100 μM of the compounds is employed (corresponding to 0.2 μl chemical stock solution in 100 μl yeast media in the case described here).
2. Compounds in column 1 and 12 of the 96-well format plates of all commercial chemical libraries correspond to the control, the solvent DMSO. In addition, make sure to use the eight cells of column 1 and 12 as negative and positive controls by adding 1 μM coronatine in each well of column 12 (whereas column 1 should contain only agar YSD –4 medium and DMSO) as shown in Figs. 1 and 2.
3. Freshly prepared screen plates can be stored at 4 °C in darkness for few days.
4. Prepare the yeast overnight culture using a fresh AH109 streak grown on YPDA agar plate for 2 days at 30 °C; AH109 yeast can be stored at 4 °C only for few days (avoid storage at 4 °C for a longer period).
5. Yeast culture should reach an $OD_{(600)}$ of 0.4–0.6 within 3–4 h, not longer. Avoid using overgrown yeast culture ($OD_{(600)}$ higher than 0.7).

6. Boil the carrier DNA for 5 min at 96 °C and chill immediately on ice to ensure that the DNA is single-stranded.
7. Transfer yeast cells to a sterile reservoir and use a multichannel pipette to transfer a 3 μl drop into each well of the 96-well plate. Alternatively, a liquid-handling robot can carry out this step under sterile conditions; carefully set robot to place the yeast suspension accurately on the agar medium surface.
8. Seal 96-well plate with Parafilm to avoid drying of the small volume of YSD medium.
9. Vertically grown JAZ1-GUS plants respond more homogeneously than horizontally grown seedlings to the JA treatment, possibly because they can be easily handled avoiding any mechanical stress. Be careful to avoid any mechanical pressure or wounding during seedling transfer.
10. To avoid any damage during the handling of the JAZ1-GUS plants, carefully collect with a forceps a bunch of approximately 8 seedlings around the hypocotyls as one group with their roots stuck together.
11. X-Gluc solution is light-sensitive; store it in dark conditions. X-Gluc solution can be stored at 4 °C for up to a week.
12. Roots are a more robust and better tissue to detect JA-mediated JAZ-GUS degradation than the aerial part of the plant.

Acknowledgments

The author would like to thank Roberto Solano and Isabel Monte for critical reading of the manuscript. AC is supported by the "Juan de la Cierva" and "Ramon y Cajal" programmes (MICINN). The chemical genomics screen described here was optimized in Natasha Raikhel's lab (UC Riverside) and funded by HFSO and CSIC short-term grants.

References

1. Wasternack C, Kombrink E (2010) Jasmonates: structural requirements for lipid-derived signals active in plant stress responses and development. ACS Chem Biol 5:63–77
2. Zhao Y (2010) Auxin biosynthesis and its role in plant development. Annu Rev Plant Biol 61:49–64
3. Chini A, Boter M, Solano R (2009) Plant oxylipins: COI1/JAZs/MYC2 as the core jasmonic acid-signalling module. FEBS J 276: 4682–4692
4. Lumba S, Cutler S, McCourt P (2010) Plant nuclear hormone receptors: a role for small molecules in protein-protein interactions. Annu Rev Cell Dev Biol 26:445–469
5. Chini A, Fonseca S, Fernández G, Adie B, Chico JM, Lorenzo O, García-Casado G, López-Vidriero I, Lozano FM, Ponce MR, Micol JL, Solano R (2007) The JAZ family of repressors is the missing link in jasmonate signalling. Nature 448:666–671
6. Thines B, Katsir L, Melotto M, Niu Y, Mandaokar A, Liu G, Nomura K, He SY, Howe GA, Browse J (2007) JAZ repressor proteins are targets of the SCF(COI1) complex during jasmonate signalling. Nature 448:661–665

7. Sheard LB, Tan X, Mao H, Withers J, Ben-Nissan G, Hinds TR, Kobayashi Y, Hsu FF, Sharon M, Browse J, He SY, Rizo J, Howe GA, Zheng N (2010) Jasmonate perception by inositol-phosphate-potentiated COI1-JAZ co-receptor. Nature 468:400–405
8. Fonseca S, Chini A, Hamberg M, Adie B, Porzel A, Kramell R, Miersch O, Wasternack C, Solano R (2009) (+)-7-iso-Jasmonoyl-L-isoleucine is the endogenous bioactive jasmonate. Nat Chem Biol 5:344–350
9. Dharmasiri N, Dharmasiri S, Estelle M (2005) The F-box protein TIR1 is an auxin receptor. Nature 435:441–445
10. Kepinski S, Leyser O (2005) The Arabidopsis F-box protein TIR1 is an auxin receptor. Nature 435:446–451
11. Tan X, Calderon-Villalobos LI, Sharon M, Zheng C, Robinson CV, Estelle M, Zheng N (2007) Mechanism of auxin perception by the TIR1 ubiquitin ligase. Nature 446:640–645
12. Prigge MJ, Lavy M, Ashton NW, Estelle M (2010) Physcomitrella patens auxin-resistant mutants affect conserved elements of an auxin-signaling pathway. Curr Biol 20:1907–1912
13. Park SY, Fung P, Nishimura N, Jensen DR, Fujii H, Zhao Y, Lumba S, Santiago J, Rodrigues A, Chow TF, Alfred SE, Bonetta D, Finkelstein R, Provart NJ, Desveaux D, Rodriguez PL, McCourt P, Zhu JK, Schroeder JI, Volkman BF, Cutler SR (2009) Abscisic acid inhibits type 2C protein phosphatases via the PYR/PYL family of START proteins. Science 324:1068–1071
14. Ueguchi-Tanaka M, Ashikari M, Nakajima M, Itoh H, Katoh E, Kobayashi M, Chow TY, Hsing YI, Kitano H, Yamaguchi I, Matsuoka M (2005) GIBBERELLIN INSENSITIVE DWARF1 encodes a soluble receptor for gibberellin. Nature 437:693–698
15. Murase K, Hirano Y, Sun TP, Hakoshima T (2008) Gibberellin-induced DELLA recognition by the gibberellin receptor GID1. Nature 456:459–463
16. Shimada A, Ueguchi-Tanaka M, Nakatsu T, Nakajima M, Naoe Y, Ohmiya H, Kato H, Matsuoka M (2008) Structural basis for gibberellin recognition by its receptor GID1. Nature 456:520–523
17. Chini A, Fonseca S, Chico JM, Fernández-Calvo P, Solano R (2009) The ZIM domain mediates homo- and heteromeric interactions between Arabidopsis JAZ proteins. Plant J 59:77–87
18. Katsir L, Schilmiller AL, Staswick PE, He SY, Howe GA (2008) COI1 is a critical component of a receptor for jasmonate and the bacterial virulence factor coronatine. Proc Natl Acad Sci U S A 105:7100–7105
19. Wells JA, McClendon CL (2007) Reaching for high-hanging fruit in drug discovery at protein-protein interfaces. Nature 450:1001–1009
20. Norambuena L, Raikhel NV, Hicks GR (2009) Chemical genomics approaches in plant biology. Methods Mol Biol 553:345–354

Chapter 5

High-Throughput Screening of Small-Molecule Libraries for Inducers of Plant Defense Responses

Colleen Knoth and Thomas Eulgem

Abstract

Transgenic Arabidopsis seedlings containing a pathogen-responsive reporter gene allow for convenient high-throughput screening of chemical libraries for compounds that induce plant defense responses. Candidates identified by such screens can be further tested for their ability to protect plants from pathogen-caused diseases. Using Arabidopsis defense signaling mutants, defined regulatory processes that are targeted by a given candidate molecule can be easily narrowed down. Here, we provide a detailed high-throughput screening protocol for library compounds that activate a pathogen-responsive reporter gene in liquid-grown Arabidopsis seedlings.

Key words Reporter gene, Plant defense inducer, Synthetic elicitor, High-throughput chemical screen

1 Introduction

Resistance of plants to microbial pathogens is mediated by a complex immune system that is inducible upon plant receptor-mediated recognition of a great diversity of pathogen-derived molecules, such as (poly)saccharides, proteins/peptides, or lipids. Such natural elicitors of plant immunity have been subdivided into three functional categories [1]: (1) *p*athogen/*m*icrobe-*a*ssociated *m*olecular *p*atterns (PAMPs/MAMPs), which are pathogen biomolecules (or parts of them) essential for pathogen fitness; (2) *d*anger-*a*ssociated *m*olecular *p*atters (DAMPs), which are plant molecules (or parts of them) that are released as a result of pathogen activity; (3) effectors, which are secreted by pathogens in host tissues to enhance pathogen virulence by manipulating plant processes. Furthermore, plant immune responses are controlled by a regulatory network that involves several small messenger molecules, such as reactive oxygen intermediates (ROI), salicylic acid (SA), jasmonic acid (JA), and ethylene (ET) [2, 3]. Thus, the plant immune system harbors a large number of proteins that are capable of interacting with ligands and, therefore, can likely be modulated

Glenn R. Hicks and Stéphanie Robert (eds.), *Plant Chemical Genomics: Methods and Protocols*, Methods in Molecular Biology, vol. 1056, DOI 10.1007/978-1-62703-592-7_5,

in their activity by drug-like small molecules. Consequently, a vast diversity of small molecules has been found to interfere with the plant immune system upon exogenous application to plants leading to the activation or the suppression of defense responses [4].

While both activators and repressors of plant immune responses can serve as valuable tools for the experimental dissection of regulatory processes [5–7], defense activators have also a substantial potential for the development of crop protectants [4, 8]. For example, benzo-1,2,3-thiadiazole-7-carbothioic acid *S*-methyl ester (BTH), a potent analog of SA [9], is currently on the market under the names Bion® and Actigard®. Several screens for plant defense-inducing small molecules have been successfully performed by both the academic and private sectors [4]. Often the main goal of such screens has been the identification of compounds that efficiently protect plants from pathogen-caused disease symptoms and yield loss. However, monitoring the outcome of plant–pathogen interactions in high-throughput settings is often impractical and has been successfully applied to chemical screens only in a few cases [10, 11]. Instead, a defined type of defense response that is easy to measure, such as expression of a pathogen-responsive reporter gene, can be used as a surrogate [5, 6].

We have previously described a high-throughput screen of a collection of 42,000 diversity-oriented drug-like molecules for compounds that induce the pathogen-responsive *CaBP22*$^{-333}$*::GUS* reporter gene in liquid-grown transgenic *Arabidopsis thaliana* (Arabidopsis) seedlings [6]. This screen led to the identification of 114 synthetic elicitors that reproducibly induced expression of this reporter in the absence of pathogens. Detailed analysis of one of these candidates, 3,5-dichloroanthranilic acid (DCA), showed that this compound-mediated efficient immunity of Arabidopsis against virulent strains of the oomycete *Hyaloperonospora arabidopsidis* and the bacterial pathogen *Pseudomonas syringae*. Using a set of Arabidopsis mutants with defined defects in defense signaling, we found that DCA likely interferes with regulatory processes that link SA to expression of the WRKY70 transcription factor. Below we provide a detailed protocol for our screening procedure using transgenic Arabidopsis seedlings containing a pathogen-responsive *GUS* reporter gene.

2 Materials

2.1 Material for Plant Growth and Chemical Treatment

1. Seeds from *Arabidopsis thaliana* plants stably transformed with a pathogen-responsive *promoter::GUS* translational fusion of interest described previously [6].
2. 70 % Ethanol (EtOH) and 0.05 % Triton-X-100.
3. 10 % Bleach and 0.05 % Triton-X-100.

4. 0.5× MS media.
5. Flat-bottom 96-well plates.
6. Library of compounds in DMSO.

2.2 Material for Analysis of GUS Activity

2.2.1 Stock Solutions

1. 0.5 M Na_2PO_4 (monobasic).
2. 0.5 M Na_2HPO_4 (dibasic).
3. 100 mM $K_4Fe(CN)_6$.
4. 100 mM $K_3Fe(CN)_6$.
5. Enough X-Gluc in DMF to make a 1 mg/ml final volume staining solution (for example dissolve 25 mg of X-Gluc in 1 ml DMF if making a 25 ml final volume staining solution). Prepare the solution fresh each time needed.

2.2.2 5-Bromo-4-Chloro-3-Indoyl-β-d-Glucuronide (X-Gluc) Staining Solution

1. 50 Na_2PO_4 pH 7.2 (~68.4 parts of Na2HPO4 (dibasic) with ~31.6 parts of Na_2PO_4 (monobasic) to make pH 7.2 (adjust pH by addition of monobasic or dibasic sodium phosphate as necessary).
2. 0.5 mM $K_4Fe(CN)_6$.
3. 0.5 mM $K_4Fe(CN)_6$.
4. 1 mg/ml X-Gluc.
5. Make enough staining solution that will be used in 2–3 days and store it at 4 °C. For longer storage store at −20 °C and thaw at 37 °C. Keep solution out of light.

3 Methods

3.1 Plant Growth

Homozygous Arabidopsis seedlings transformed with a pathogen-responsive promoter of interest translationally fused to the *GUS* (*uidA*) reporter gene [12] are grown in 200 μL liquid 0.5 MS media in 96-well plates (Costar) for 7 days on an orbital shaker under long day conditions (16-h light, 8-h dark, 22 °C, 100 Einstein/m^2/s) following the steps below:

1. Sterilize seeds.
   (a) Place seeds in an Eppendorf tube (1.5–2 ml) or a 15 ml polypropylene tube depending on quantity and suspected level of contamination of seeds. The more seeds there are (or the more contamination there is) the larger the volume should be. For up to 0.05 g of seeds, use 1–1.2 ml volumes for each step. For 15 ml tubes use 2–3 ml volumes for each step.
   (b) 70 % EtOH and 0.05 % Triton-X-100 for 10 min.
   (c) Remove EtOH.
   (d) 10 % Bleach and 0.05 % Triton-X-100 for 15 min.
   (e) Remove bleach and rinse five times in ½ MS.

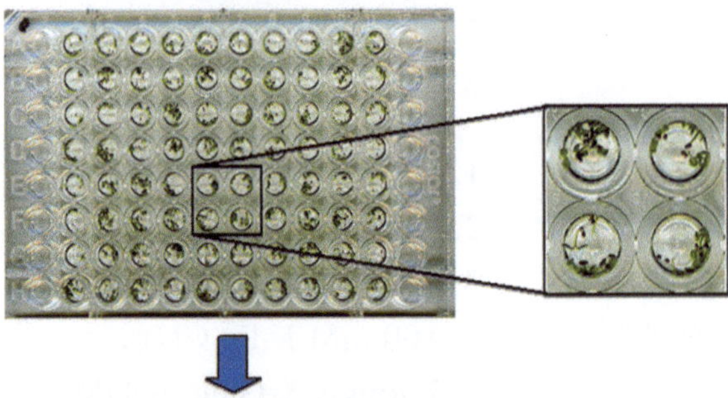

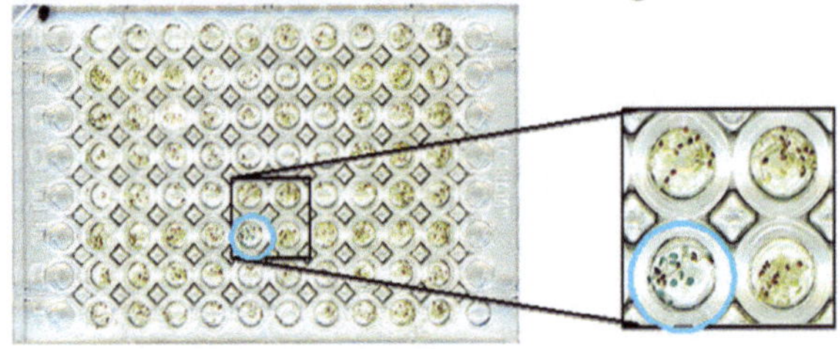

Fig. 1 Scheme illustrating screening procedure. (**a**) Example of 96-well screening plate containing 7-day-old liquid-grown *CaBP22-333 promoter::GUS* seedlings after a 24-h incubation with library compounds at concentrations ranging from 4 to 20 μM (*top*). (**b**) Screening plate after GUS histochemical staining. *Blue circle* indicates positive reaction (**a** and **b**) (adapted from [6])

2. Plate seeds in sterile 96-well plate.
   (a) Pipette 200 μl sterile ½ MS in each well (use multichannel pipettor if possible).
   (b) Pipette 3–10 seeds per well (e.g., use a P20 multichannel pipettor with the tips cut to create a wider pore, set pipettor to ~1–3 μl depending on the density of sterilized seeds in the source plate).
   (c) Add lid and seal plates with Micropore tape on its edges.
3. Grow seeds with constant rotation (150 rpm) under long day conditions (16-h days) in a growth chamber for 7 days.

3.2 Chemical Treatment

1. After 7 days, replace 0.5× MS media with 200 μL fresh 0.5× MS media.
2. Add 0.2 μL of each compound in DMSO to individual wells (recommended mode of delivery is a robotic pin tool on an automated fluid handling workstation).
3. Return treated plates to orbital shaker in growth chamber, and incubate seedlings for 24 h under long-day conditions.
4. After that, remove compound/media and replace with fresh media. Visually score health and size of seedlings. Analyze GUS activity immediately.

3.3 Analysis of GUS Activity

1. Replace media with prepared 5-bromo-4-chloro-3-indoyl-β-d-glucuronide (X-gluc) staining solution (*see* Subheading 2.2).
2. Incubate in the dark at 37 °C for 2–48 h (this should be optimized for *promoter::GUS* construct being used).
3. Clear chlorophyll with 70 % ethanol for 24 h.
4. Visually score and photographically document intensity and pattern of blue staining for each seedling (Fig. 1).

References

1. Dodds PN, Rathjen JP (2010) Plant immunity: towards an integrated view of plant-pathogen interactions. Nat Rev Genet 11:539–548
2. Nimchuk Z, Eulgem T, Holt IB, Dangl JL (2003) Recognition and response in the plant immune system. Annu Rev Genet 37:579–609
3. Glazebrook J (2005) Contrasting mechanisms of defense against biotrophic and necrotrophic pathogens. Annu Rev Phytopathol 43: 205–227
4. Schreiber KJ, Desveaux D (2008) Message in a bottle: Chemical biology of induced disease resistance in plants. The Plant Pathology Journal 24:245–268
5. Serrano M, Robatzek S, Torres M, Kombrink E, Somssich IE, Robinson M, Schulze-Lefert P (2007) Chemical interference of pathogen-associated molecular pattern-triggered immune responses in Arabidopsis reveals a potential role for fatty-acid synthase type II complex-derived lipid signals. J Biol Chem 282:6803–6811
6. Knoth C, Salus MS, Girke T, Eulgem T (2009) The synthetic elicitor 3,5-dichloroanthranilic acid induces NPR1-dependent and NPR1-independent mechanisms of disease resistance in Arabidopsis. Plant Physiol 150:333–347
7. Kim TH, Hauser F, Ha T, Xue S, Bohmer M, Nishimura N, Munemasa S, Hubbard K, Peine N, Lee BH, Lee S, Robert N, Parker JE, Schroeder JI (2011) Chemical genetics reveals negative regulation of abscisic Acid signaling by a plant immune response pathway. Curr Biol 21:990–997
8. Kessmann H, Staub T, Hofmann C, Maetzke T, Herzog J, Ward E, Uknes S, Ryals J (1994) Induction of systemic acquired resistance in plants by chemicals. Annu Rev Phytopathol 32:439–459
9. Gorlach J, Volrath S, Knauf-Beiter G, Hengy G, Beckhove U, Kogel KH, Oostendorp M, Staub T, Ward E, Kessmann H, Ryals J (1996) Benzothiadiazole, a novel class of inducers of systemic acquired resistance, activates gene expression and disease resistance in wheat. Plant Cell 8:629–643
10. Schreiber K, Ckurshumova W, Peek J, Desveaux D (2008) A high-throughput chemical screen for resistance to Pseudomonas syringae in Arabidopsis. Plant J 54:522–531
11. Schreiber KJ, Nasmith CG, Allard G, Singh J, Subramaniam R, Desveaux D (2011) Found in translation: High-throughput chemical screening in Arabidopsis thaliana identifies small molecules that reduce Fusarium head blight disease in wheat. Mol Plant Microbe Interact 24:640–648
12. Jefferson RA, Kavanagh TA, Bevan MW (1987) GUS fusions: b-glucuronidase as a sensitive and versatile gene fusion marker in higher plants. EMBO J 6:3901–3907

Chapter 6

Using a Reverse Genetics Approach to Investigate Small-Molecule Activity

Siamsa M. Doyle and Stéphanie Robert

Abstract

Chemical genomics is a highly effective approach for understanding complex and dynamic biological processes in plants. A chemical activity can be investigated by a reverse genetics strategy, for which a huge abundance and diversity of *Arabidopsis thaliana* mutants are readily available for exploitation. Here we present an approach to characterize a chemical of interest, as well as examples of studies demonstrating an effective combination of chemical genomics with reverse genetics strategies, drawn from recent literature on phytohormone signalling and auxin transport.

Key words Chemical genomics, *Arabidopsis* mutants, Reverse genetics, Phytohormone signalling, Auxin transport

1 Introduction

Chemical genomics is a powerful strategy employing small molecules as probes for dissecting biological processes [1, 2]. The many benefits this strategy presents, such as a high level of temporal, reversible control over a chemical's target protein, result in a highly effective approach towards unraveling complex and dynamic plant processes. Once an interesting molecule has been identified through chemical screening and its effects characterized, reverse genetics can then greatly assist in dissecting the precise actions of the chemical. *Arabidopsis thaliana* is an ideal tool for reverse genetics studies; a large and diverse abundance of mutants of this model species are widely and cheaply available to the research community.

A mutant plant can be defined in simple terms as a plant that has been altered in its DNA sequence, resulting in modified gene expression compared to the wild-type plant. If a gene's expression is partially suppressed due to a mutation, the plant is referred to as a knockdown or a leaky mutant. Knockout or null mutants are those in which a gene's expression is absent completely. These types of mutants are known as loss-of-function mutants,

Glenn R. Hicks and Stéphanie Robert (eds.), *Plant Chemical Genomics: Methods and Protocols*, Methods in Molecular Biology, vol. 1056, DOI 10.1007/978-1-62703-592-7_6, © Springer Science+Business Media New York 2014

whereas mutants in which a new or an enhanced protein activity has been introduced are known as gain-of-function mutants. A mutation could also result in permanently active or inactive protein function, for which the mutant is known as constitutively active or dominant negative, respectively. Different promoters can be used to control both the location and level of a protein's expression, such as overexpressor mutants, in which a strong and constitutive promoter is used, such as the cauliflower mosaic virus 35S promoter.

In a reverse genetics approach towards identifying the activities and targets of a molecule of interest, suitable mutants are tested for resistance, sensitivity, or hypersensitivity to the chemical. A mutant is termed resistant if it is less affected by the chemical than the wild-type plant. In contrast, a mutant is sensitive or hypersensitive if it responds to the chemical to the same degree as or to a greater degree than the wild-type plant, respectively.

In this chapter, we present an ideal approach towards characterizing the biological effects of a molecule isolated through chemical screening. We then outline the benefits of using *Arabidopsis* mutants in a reverse genetics strategy towards revealing a chemical's specific activities *in planta*, using examples from recently published research in phytohormone signalling and auxin transport.

2 Materials

2.1 Plant Material and Growth Medium

1. *Arabidopsis thaliana*, Columbia-0 (Col-0) ecotype, was used for chemical characterization.
2. Sodium hypochlorite (VWR International AB, Stockholm, Sweden) and concentrated hydrochloric acid (37 %) (Sigma-Aldrich GmbH, Seelze, Germany) were used for seed sterilization.
3. Solid and liquid ½ Murashige and Skoog (MS) medium were used for plant growth. The medium contains 2.2 g/L MS medium (Duchefa Biochemie, Haarlem, The Netherlands) in distilled water. Adjust the pH to 5.6 using 10 M potassium hydroxide (Eka Chemicals AB, Bohus, Sweden). For medium supplemented with 1 % sucrose, add 10 g/L sucrose (Fisher Scientific UK Ltd, Leicestershire, UK) before adjusting the pH. For solid medium only, add 5 g/L Phytagel (Sigma-Aldrich GmbH, Seelze, Germany) after adjusting the pH (*see* **Note 1**). Autoclave before pouring plates (about 50 mL medium per plate) (*see* **Note 2**).

2.2 Hardware and Software

1. Square Petri plates, aseptic, measuring 120 mm × 120 mm with height 15.8 mm (Gosselin, Hazebrouck, France) were used for plant growth.

2. Plates were imaged by high-resolution scanning with an HP Scanjet 4890 scanner (Hewlett-Packard Company, Palo Alto, California, USA).
3. Confocal microscopy was performed using a Zeiss LSM780 laser scanning microscope (Carl Zeiss AG, Oberkochen, Germany).
4. ImageJ software (National Institutes of Health, Bethesda, Maryland, USA) (http://rsbweb.nih.gov/ij/) was used for measurements on images.
5. Excel (Microsoft, Redmond, Washington, USA) and JMP (SAS, Cary, North Carolina, USA) software were used for statistical analyses of measurements.

3 Methods: Characterizing the Biological Effects of a Chemical

3.1 Seed Sterilization

1. *Arabidopsis* seeds can be surface-sterilized by treatment with chlorine gas, which eliminates the need for rinsing the seeds and allows many tubes of seeds to be sterilized simultaneously. Place open tubes of seeds upright in tube racks inside a glass container with a lid (such as a desiccator jar or other large jar), and place the container inside a fume hood.
2. Place a glass beaker containing 40 mL sodium hypochlorite inside the container beside the seeds.
3. Lift the container's lid slightly while using a syringe to rapidly add 3 mL concentrated hydrochloric acid to the beaker of sodium hypochlorite, which causes chlorine gas to be released, before quickly closing the container lid.
4. Three hours later, close all seed tube lids and remove the tubes (*see* **Note 3**).

3.2 Analysis of Chemical-Induced Physiological Phenotype

1. To observe the physiological phenotype associated with the chemical of interest, sow surface-sterilized *Arabidopsis* seeds on Petri plates of solid ½ MS medium, supplemented with various concentrations of the chemical (*see* **Notes 4** and **5**).
2. Supplement control plates with the chemical's solvent only (e.g., DMSO), adding the same volume as used for chemical treatments.
3. Store plates at 4 °C for 2 days to allow seed stratification.
4. Transfer plates to 22 °C, and grow the seedlings with the plates positioned vertically (*see* **Notes 6** and **7**).
5. Test combinations of several growth conditions (*see* **Note 8**).
6. After a certain time of growth (depending on your research interest), image the plates by high-resolution scanning and examine the images.

7. Compare seedlings grown on chemical-supplemented medium to those grown on control plates to determine the physiological phenotypic dose response to the chemical.
8. Perform 3 biological replicates, using at least 30 seedlings per treatment (*see* **Note 9**).

3.3 Analysis of Chemical-Induced Intracellular Phenotype

1. Depending on your research interest, you may wish to observe the intracellular effects associated with the chemical, such as mis-localization of fluorescent protein markers for various endomembrane compartments (see, for example, refs. 3, 4). We study chemical-induced effects on the endomembrane system according to the following methods. Grow *Arabidopsis* seedlings under the above-described conditions on control plates.
2. Immerse 5-day-old seedlings in a solution of the chemical in liquid medium (*see* **Note 10**) at a range of concentrations for a range of treatment times (we use from 0.5 to 2 h).
3. Immerse control seedlings in solvent-treated liquid medium.
4. Mount the seedlings on microscope slides in their treatment medium for microscopic observation. For our purposes, we use confocal microscopy to examine root cells in *Arabidopsis* seedlings treated with fluorescent dyes such as FM4-64 as well as transgenic *Arabidopsis* seedlings expressing fluorescent protein-tagged endomembrane-localized proteins.
5. Image the cells using the confocal microscope's software, and examine the images.
6. Compare images of cells from chemical-treated seedlings to images of control seedling cells to determine the intracellular phenotypic dose response to the chemical.
7. Perform three biological replicates, using at least ten seedlings per treatment (*see* **Note 9**).

3.4 Determination of Reversibility of the Chemical's Effects

1. It is important to determine the reversibility of the chemical's effects (*see* **Note 11**). This can be done using the plates of seedlings grown to analyze the chemical-induced physiological phenotype (*see* Subheading 3.2). After a chosen time of growth, when the chemical-induced phenotype is apparent (we chose 5 days of growth), image the plates by high-resolution scanning.
2. Transfer half of the seedlings on each plate to new plates of the same chemical-supplemented medium.
3. Transfer the remaining seedlings on each plate to chemical-free plates of the same medium.
4. Incubate all plates vertically under the original growth conditions for a chosen time (we chose 5 more days of growth).
5. Image all plates by scanning at high resolution, and examine the images.

6. Compare the images of transferred seedlings to the images of the same seedlings before transferring.
7. Compare the phenotypes of seedlings transferred to chemical-free plates to those transferred to chemical-supplemented plates.
8. Determine whether the seedlings were able to "recover" from the chemical treatment, indicating that the effects of the chemical are reversible.

3.5 Quantification and Statistical Analysis of the Chemical's Effects

1. The ImageJ freeware program is particularly useful for making measurements directly on scanned plate images and microscopy images.
2. Open an image with ImageJ, and use the *set scale* feature (using a known length such as plate width) so that all ImageJ measurements are made in real units. The scale must be reset for each new image opened.
3. On scanned plate images, lengths of plant organs and angles of growth, for example, can be measured using ImageJ and the data can be copied and pasted to an Excel spreadsheet.
4. On microscopy images, lengths, areas, and angles can be measured using ImageJ, as well as fluorescence intensity, which can be particularly useful for confocal images.
5. Once quantifications have been made, statistics should be performed to confirm any differences between control- and chemical-treated plants (we use Excel and the user-friendly JMP software from SAS for statistical analyses).

3.6 Selection of a Suitable Phenotype for Reverse Genetics

1. A chemical-induced phenotype should be selected to use for screening mutants in a reverse genetics strategy towards dissecting the specific actions of the chemical (*see* Subheading 4). This choice depends on your research interest.
2. The phenotype selected should be obvious enough so that resistant mutants can be found but not extreme enough that sensitive or hypersensitive mutants will not grow.

4 Important Applications of the Reverse Genetics Approach Towards Identifying a Chemical's Precise Activities *In Planta*

4.1 Availability of Arabidopsis Mutants

An enormous abundance of *Arabidopsis* mutants are now widely and cheaply available to the research community. Mutants can be found through The *Arabidopsis* Information Resource (TAIR), a large, unified database of information and tools available to the *Arabidopsis* research community (http://www.arabidopsis.org/). Many mutants are commercially available through several resources, including the *Arabidopsis* Biological Resource Center (ABRC), the Nottingham *Arabidopsis* Stock Centre (NASC), the INRA-Versailles

Genome Resource Centre, the RIKEN Bioresource Center (RBRC), and the SENDAI *Arabidopsis* Seed Stock Center (SASSC). Links to the websites of all these resources can be found on TAIR.

An effective use of mutants in a reverse genetics approach towards dissecting the actions of a chemical is best demonstrated using examples from published research. Several pertinent studies from the fields of phytohormone signalling and auxin transport research will now be highlighted.

4.2 Dissection of Phytohormone Signalling

Arabidopsis mutants have proven to be extremely useful as part of a reverse genetics strategy to dissect plant hormone signalling. A recent study on germination signals in smoke [5] highlights the clever use of a simple mutant study to distinguish between the roles of different plant signals. In post-fire landscapes, smoke acts as an important stimulator of seed germination for many plant species [6, 7]. Karrikins have been identified as a unique family of butenolides present in smoke [7] that trigger germination and enhance post-germination responses to light [6, 8]. Interestingly, karrikins share partial structural similarity with strigolactones [8], a family of phytohormones. Strigolactones also stimulate seed germination in some species as well as inhibit shoot branching in pea, rice, and *Arabidopsis* [9–11]. In a screen for karrikin-insensitive germination mutants, two alleles of a known strigolactone-insensitive mutant, *more axillary growth2* (*max2*), were isolated [5], suggesting that strigolactones and karrikins might share some signalling mechanisms. However, while germination of *max2* seeds is insensitive to both strigolactones and karrikins, side branching in *max2* plants is inhibited by strigolactones but insensitive to karrikins, suggesting that these two types of molecules maintain some distinct functional roles. To test this hypothesis, the pea mutants *rms4* and *rms1*, equivalent to the *Arabidopsis* mutants *max2* and *max4*, respectively, were used. MAX2 is thought to act downstream of MAX4, which is involved in strigolactone biosynthesis [11, 12]. As was found for *max2* mutants, shoot branching (bud outgrowth) is suppressed in *rms1* mutants by strigolactone but not by karrikin treatment. In contrast, shoot branching in *rms4* mutants is insensitive to both strigolactone and karrikin treatment. Thus, a simple reverse genetics strategy demonstrated that, despite sharing signalling roles in seed germination, strigolactones and karrikins also maintain some distinct functional roles.

A clever choice of mutants can also be used to distinguish whether a chemical affects biosynthesis or signalling of phytohormones. This is clearly demonstrated in a study on vesicle trafficking and brassinosteroid (BR) signalling [3]. BRs are phytohormones that promote plant growth [13]. Endosidin1 (ES1) was identified as a chemical that inhibits polar pollen tube growth [3], which is dependent on vesicle transport [14]. Treatment of dark-grown seedlings with ES1 results in a light-like phenotype including short, thickened hypocotyls. This phenotype is very similar to mutants affected in BR biosynthesis or perception. Seedlings were grown in the presence of both ES1 and the most bioactive BR, brassinolide [15], but brassinolide was not found to rescue the ES1 phenotype, suggesting that ES1 does not interfere with BR biosynthesis. It was shown that loss-of-function mutants of the BR-INSENSITIVE1 (BRI1) BR receptor, *bri1-116* [15], are resistant to ES1, implying that ES1 targets the BR signalling pathway. This was confirmed by demonstrating that ES1 inhibits the induction of a brassinolide reporter gene [3].

A reverse genetics strategy was also used to dissect the action of bikinin, a chemical targeting the BR signalling pathway [16]. BR perception by BRI1 leads to inhibition of the glycogen synthase kinase BR-INSENSITIVE2 (BIN2) resulting in expression of BR-responsive genes [13, 17]. Bikinin induces constitutive BR-like responses, including increased hypocotyl and petiole length, suggesting that bikinin is a strong activator of BR signalling [16]. Despite no structural analogy, both brassinolide and bikinin induce similar expression changes in a subset of BR target genes, suggesting that they promote BR-related responses through a common transcriptional module. To further investigate bikinin activity, the BR biosynthesis mutant, *constitutive photomorphogenesis and dwarfism* (*cpd*) [18], was used as well as two BR signalling mutants, *bri1-116* [15], and the BIN2 gain-of-function mutant *bin2-1* [19], all of which display dwarf light-grown phenotypes and de-etiolated dark-grown phenotypes. Both bikinin and brassinolide treatments were shown to rescue the *cpd* phenotype, while bikinin but not brassinolide treatment rescues the *bri1-116* phenotype, suggesting that bikinin acts downstream of the BRI1 receptor [16]. Similarly, bikinin but not brassinolide treatment rescues the *bin2-1* phenotype, suggesting that bikinin acts either at or downstream of BIN2. This straightforward mutant study showed that bikinin is a novel BR signalling activator acting downstream of BRI1, which provides new potential to unravel aspects of BR signalling that are not yet understood.

Like ES1, endosidin3 (ES3) was identified in a screen for pollen tube growth inhibitors and was also shown to alter leaf epidermal cell lobe formation [4], which is dependent on auxin, a phytohormone involved in regulating many aspects of plant development [20]. In *Arabidopsis*, lobe formation is coordinated by the counteraction of two antagonistic auxin-activated Rho-related GTPase of plants (ROP)/ROP-interacting CRIB-containing protein (RIC) pathways: the ROP2/RIC4 and ROP6/RIC1 pathways [21, 22]. In a reverse genetics strategy, *ric1-1* and *ric4-1* mutants were used to identify whether ES3 targets one of these antagonistic pathways [4]. Lobe formation in *ric1-1* is unaffected by and hence resistant to ES3, while *ric4-1* is ES3 sensitive, similar to the wild type, implying that ES3 likely targets the ROP6/RIC1 pathway. This hypothesis was confirmed by demonstrating that ES3 treatment disrupts the association of 35S::ROP6:GFP with the plasma membrane (PM), an association which is required for ROP activity, while the localization of 35S::ROP2:GFP is unaffected by ES3 [4].

A study investigating the role of the putative auxin co-receptor AUXIN-BINDING PROTEIN1 (ABP1) in auxin signalling also highlights an effective reverse genetics strategy [23]. PIN-FORMED (PIN) proteins are polar auxin efflux transporters that constitutively cycle to and from the PM [24, 25]. Auxin inhibits endocytosis, retaining PINs at the PM and thus directing its own flow [25]. PINs rapidly internalize in response to treatment with brefeldin A (BFA), a cycling inhibitor [24]. However, BFA-induced PIN internalization is reduced in the knockdown *abp1* line, implying that ABP1 promotes PIN endocytosis [23]. Furthermore, PIN internalization is not inhibited by auxin in *abp1-5* mutants, in which the auxin-binding pocket of ABP1 is defective due to a point mutation in a single allele of *ABP1* [22]. By using different *abp1* mutants, this study confirmed that auxin controls its own flow by targeting ABP1, leading to retention of PINs at the PM.

4.3 Dissection of Auxin Transport

The combination of reverse genetics and chemical genomics strategies has also proven to be extremely useful in the dissection of the auxin transport system. One pertinent example is a study that identified gravacin as a strong inhibitor of *Arabidopsis* root and shoot gravitropism [26]. In a forward genetics screen for gravacin-resistant mutants, a mutant affected in the auxin transporter *ATP-BINDING CASSETTE TYPE B19* (*ABCB19*) (formerly *PGP19*) gene was isolated. ABCB19 is known to interact with TWISTED DWARF1 (TWD1), an immunophilin-like protein that is required for both ABCB19 and ABCB1 activity [27]. The effect of gravacin on *twd1* mutants, *abcb1* mutants, and *abcb1 abcb19* double mutants was tested with respect to gravitropism [26]. While *twd1* mutants and *abcb1 abcb19-1* double mutants were found to be resistant to gravacin, *abcb1* mutants were shown to be gravacin sensitive, revealing that both ABCB19 and TWD1 are required for the effect of gravacin on gravitropism but ABCB1 is not. However, as TWD1 is required for ABCB1 activity, the fact that *abcb1* is not resistant to gravacin rules out TWD1 as a target. Thus, a clever choice of mutants in this study excluded ABCB1 and TWD1 as gravacin targets and confirmed that gravacin inhibits auxin transport by targeting ABCB19 leading to the conclusion that ABCB19 is essential for auxin transport in the gravitropic response.

The compound 2-[4-(diethylamino)-2-hydroxybenzoyl]benzoic acid (BUM) has also recently been identified as a novel auxin transport inhibitor, inducing strong phenotypes in *Arabidopsis*, including reduced root growth, reduced lateral root formation, and formation of pin-formed inflorescences [28]. These effects resemble the phenotypes induced by treatment with 1-N-naphthylphthalamic acid (NPA), a well-established auxin transport inhibitor [29]. Mutants defective in different auxin transporters were used to investigate the target of BUM [28]. While BUM treatment suppresses lateral root growth in *pin2* mutants, *abcb1* mutants are resistant to BUM treatment, excluding PIN2 and implicating ABCB1 as a potential target of BUM.

Reverse genetics strategies have also been used to distinguish between endomembrane trafficking pathways of different auxin transporters [30]. BFA inhibits GNOM, an endosome-localized ADP-ribosylation factor GDP/GTP exchange factor (ARF-GEF), which mediates the transport of PIN proteins from endosomes to the PM [24, 31]. In root protophloem cells, the auxin influx transporter AUXIN-RESISTANT1 (AUX1) is localized at the apical side of the cell [32], while the auxin efflux transporter PIN1 is localized at the opposite basal side [33], thus directing auxin transport. It was shown that unlike PIN1, PM-localized AUX1 is BFA insensitive, implying that AUX1 targeting to the PM involves BFA-insensitive ARF-GEFs [30]. However, an intracellular pool of AUX1, residing on dynamic endosomal and Golgi apparatus membranes, was shown to be BFA sensitive. To address whether the dynamics of the intracellular AUX1 pool involve GNOM, a transgenic mutant was used, in which the GNOM protein is replaced by a BFA-insensitive version, GN^{M696L}-myc [31]. In the GN^{M696L}-myc line, PIN1 trafficking was shown to be BFA insensitive; however, the intracellular pool of AUX1 is still BFA sensitive, suggesting that AUX1 trafficking is GNOM independent and rather involves a different BFA-sensitive ARF-GEF [30]. Through the use of the transgenic GN^{M696L}-myc mutant, it was shown that PIN and AUX1 endomembrane trafficking pathways are distinct, PIN trafficking being dependent on GNOM and AUX1 trafficking on a different BFA-sensitive ARF-GEF.

5 Conclusions

Combining chemical genomics with a reverse genetics strategy provides a powerful approach towards dissecting plant development. A huge abundance and diversity of *Arabidopsis* mutants are widely and cheaply available to the research community, providing great potential as useful tools for characterizing novel chemical interactors of cell signalling pathways.

6 Notes

1. For experiments involving growth of seedlings on chemical-treated medium, we recommend using Phytagel rather than plant agar to solidify the medium. From our experience, agar

often limits the effectiveness of chemicals added to the medium, resulting in wide variation of plant responses, while Phytagel never appears to cause such problems.

2. Any unused medium can be allowed to cool and solidify and stored at room temperature for use at a later time. The medium is suitable for reheating and liquifying in a microwave.
3. Directly after sterilizing seeds with chlorine gas, transfer the tubes (with closed lids) to a sterile laminar flow bench, where the tubes should be opened for 1 h to allow the gas inside the tubes to dissipate. This prolongs seed viability, as does storing the sterilized seeds at 4 °C.
4. When supplementing medium with the chemical of interest, add the chemical directly to bottles of medium after autoclaving, before pouring plates. After autoclaving, allow the medium to cool for a while to a temperature that is not very hot but not yet starting to solidify and mix well. This is important to protect the chemical from heat destruction.
5. If your chemicals are in short supply or expensive, it may be advisable to use smaller Petri dishes with lower volumes of medium to save on the amount of chemical used.
6. Growing the seedlings on vertically rather than horizontally positioned plates allows easier observation and measurements of parameters such as hypocotyl and root lengths. It is also easier to remove seedlings from vertical plates for microscopic observation without damaging the seedlings.
7. Germination of dark-grown seedlings is improved by placing the plates in light for 4 h after vernalization, before transferring to darkness, which can be done simply by wrapping the plates carefully in two layers of aluminum foil. Take care that no light whatsoever can get to the seeds as even a tiny pinpoint of light can induce a partial light-grown phenotype.
8. It is important to test several growth conditions, such as growth in 16-h light per day and growth in total darkness as well as growth on sucrose-free medium and growth on medium supplemented with 1 % sucrose. This is because many interesting phenotypes only reveal themselves under certain conditions. For example, changes in root or hypocotyl phenotype may be more obvious during light or dark growth, respectively. We have found sucrose to often strongly inhibit chemical-induced phenotypes, perhaps due to effects on plant metabolism.
9. Due to biological variation, repetition is extremely important, with at least 30 seedlings recommended per replicate for physiological phenotype observations. For cell biology observations, several cells should be observed per seedling,

so the recommended number of seedlings per replicate is lower, at 10. Biological replicates should be performed on different days.

10. Again, if your chemicals are limited or expensive, consider minimizing the volume of medium by using small tubes or multiwell plates for liquid treatments to save on the amount of chemical used.
11. Here we discuss determining the reversibility of a chemical's effects on physiological phenotype over several days. However, reversibility of a chemical's effects on intracellular phenotype can also be determined, over shorter time spans of just hours. This can be done by simply washing off the chemical treatment, using chemical-free liquid medium, before observation.

Acknowledgements

We thank the Vetenskapsrådet and VINNOVA for supporting this work.

References

1. Robert S, Raikhel NV, Hicks GR (2009) Powerful partners: Arabidopsis and chemical genomics. The Arabidopsis Book 7:e0109. http://thearabidopsisbook.org/how-to-cite/. doi:10.1199/tab.0109
2. Hicks GR, Raikhel NV (2012) Small molecules present large opportunities in plant biology. Annu Rev Plant Biol 63(1):261–282
3. Robert S, Chary SN, Drakakaki G, Li S, Yang Z, Raikhel NV, Hicks GR (2008) Endosidin1 defines a compartment involved in endocytosis of the brassinosteroid receptor BRI1 and the auxin transporters PIN2 and AUX1. Proc Natl Acad Sci U S A 105(24):8464–8469
4. Drakakaki G, Robert S, Szatmari A-M, Brown MQ, Nagawa S, Van Damme D, Leonard M, Yang Z, Girke T, Schmid SL, Russinova E, Friml J, Raikhel NV, Hicks GR (2011) Clusters of bioactive compounds target dynamic endomembrane networks in vivo. Proc Natl Acad Sci U S A 108(43):17850–17855
5. Nelson DC, Scaffidi A, Dun EA, Waters MT, Flematti GR, Dixon KW, Beveridge CA, Ghisalberti EL, Smith SM (2011) F-box protein MAX2 has dual roles in karrikin and strigolactone signaling in *Arabidopsis thaliana*. Proc Natl Acad Sci U S A 108(21):8897–8902
6. Nelson DC, Riseborough JA, Flematti GR, Stevens J, Ghisalberti EL, Dixon KW, Smith SM (2009) Karrikins discovered in smoke trigger Arabidopsis seed germination by a mechanism requiring gibberellic acid synthesis and light. Plant Physiol 149(2):863–873
7. Flematti GR, Ghisalberti EL, Dixon KW, Trengove RD (2004) A compound from smoke that promotes seed germination. Science 305(5686):977
8. Chiwocha SDS, Dixon KW, Flematti GR, Ghisalberti EL, Merritt DJ, Nelson DC, Riseborough J-AM, Smith SM, Stevens JC (2009) Karrikins: A new family of plant growth regulators in smoke. Plant Sci 177(4):252–256
9. Gomez-Roldan V, Fermas S, Brewer PB, Puech-Pagès V, Dun EA, Pillot JP, Letisse F, Matusova R, Danoun S, Portais JC, Bouwmeester H, Bécard G, Beveridge CA, Rameau C, Rochange SF (2008) Strigolactone inhibition of shoot branching. Nature 455(7210):189–194
10. Umehara M, Hanada A, Yoshida S, Akiyama K, Arite T, Takeda-Kamiya N, Magome H, Kamiya Y, Shirasu K, Yoneyama K, Kyozuka J, Yamaguchi S (2008) Inhibition of shoot branching by new terpenoid plant hormones. Nature 455(7210):195–200
11. Beveridge CA, Kyozuka J (2010) New genes in the strigolactone-related shoot branching pathway. Curr Opin Plant Biol 13(1):34–39
12. Booker J, Sieberer T, Wright W, Williamson L, Willett B, Stirnberg P, Turnbull C, Srinivasan M, Goddard P, Leyser O (2005) MAX1 encodes a cytochrome P450 family member

that acts downstream of MAX3/4 to produce a carotenoid-derived branch-inhibiting hormone. Dev Cell 8(3):443–449
13. Clouse SD (2011) Brassinosteroid signal transduction: from receptor kinase activation to transcriptional networks regulating plant development. Plant Cell 23(4):1219–1230
14. Cole RA, Fowler JE (2006) Polarized growth: maintaining focus on the tip. Curr Opin Plant Biol 9(6):579–588
15. Li J, Chory J (1997) A putative leucine-rich repeat receptor kinase involved in brassinosteroid signal transduction. Cell 90(5):929–938
16. De Rybel B, Audenaert D, Vert G, Rozhon W, Mayerhofer J, Peelman F, Coutuer S, Denayer T, Jansen L, Nguyen L, Vanhoutte I, Beemster GTS, Vleminckx K, Jonak C, Chory J, Inzé D, Russinova E, Beeckman T (2009) Chemical inhibition of a subset of *Arabidopsis thaliana* GSK3-like kinases activates brassinosteroid signaling. Chem Biol 16(6):594–604
17. Vert G, Nemhauser JL, Geldner N, Hong F, Chory J (2005) Molecular mechanisms of steroid hormone signaling in plants. Annu Rev Cell Dev Biol 21:177–201
18. Szekeres M, Németh K, Koncz-Kálmán Z, Mathur J, Kauschmann A, Altmann T, Rédei GP, Nagy F, Schell J, Koncz C (1996) Brassinosteroids rescue the deficiency of CYP90, a cytochrome P450, controlling cell elongation and de-etiolation in Arabidopsis. Cell 85(2):171–182
19. Li J, Nam KH, Vafeados D, Chory J (2001) BIN2, a new brassinosteroid-insensitive locus in Arabidopsis. Plant Physiol 127(1):14–22
20. Vanneste S, Friml J (2009) Auxin: a trigger for change in plant development. Cell 136(6): 1005–1016
21. Fu Y, Gu Y, Zheng Z, Wasteneys G, Yang Z (2005) Arabidopsis interdigitating cell growth requires two antagonistic pathways with opposing action on cell morphogenesis. Cell 120(5):687–700
22. Xu T, Wen M, Nagawa S, Fu Y, Chen J-G, Wu M-J, Perrot-Rechenmann C, Friml J, Jones AM, Yang Z (2010) Cell surface- and rho GTPase-based auxin signaling controls cellular interdigitation in Arabidopsis. Cell 143(1):99–110
23. Robert S, Kleine-Vehn J, Barbez E, Sauer M, Paciorek T, Baster P, Vanneste S, Zhang J, Simon S, Čovanová M, Hayashi K, Dhonukshe P, Yang Z, Bednarek SY, Jones AM, Luschnig C, Aniento F, Zažímalová E, Friml J (2010) ABP1 mediates auxin inhibition of clathrin-dependent endocytosis in Arabidopsis. Cell 143(1):111–121
24. Geldner N, Friml J, Stierhof Y-D, Jürgens G, Palme K (2001) Auxin transport inhibitors block PIN1 cycling and vesicle trafficking. Nature 413(6854):425–428
25. Paciorek T, Zažímalová E, Ruthardt N, Petrášek J, Stierhof Y-D, Kleine-Vehn J, Morris DA, Emans N, Jürgens G, Geldner N, Friml J (2005) Auxin inhibits endocytosis and promotes its own efflux from cells. Nature 435(7046):1251–1256
26. Rojas-Pierce M, Titapiwatanakun B, Sohn EJ, Fang F, Larive CK, Blakeslee J, Cheng Y, Cuttler S, Peer WA, Murphy AS, Raikhel NV (2007) Arabidopsis P-glycoprotein19 participates in the inhibition of gravitropism by gravacin. Chem Biol 14(12):1366–1376
27. Geisler M, Kolukisaoglu HÜ, Bouchard R, Billion K, Berger J, Saal B, Frangne N, Koncz-Kálmán Z, Koncz C, Dudler R, Blakeslee JJ, Murphy AS, Martinoia E, Schulz B (2003) TWISTED DWARF1, a unique plasma membrane-anchored immunophilin-like protein, interacts with Arabidopsis multidrug resistance-like transporters AtPGP1 and AtPGP19. Mol Biol Cell 14(10):4238–4249
28. Kim J-Y, Henrichs S, Bailly A, Vincenzetti V, Sovero V, Mancuso S, Pollmann S, Kim D, Geisler M, Nam H-G (2010) Identification of an ABCB/P-glycoprotein-specific inhibitor of auxin transport by chemical genomics. J Biol Chem 285(30):23309–23317
29. Luschnig C (2001) Auxin transport: why plants like to think BIG. Curr Biol 11(20):R831–R833
30. Kleine-Vehn J, Dhonukshe P, Swarup R, Bennett M, Friml J (2006) Subcellular trafficking of the Arabidopsis auxin influx carrier AUX1 uses a novel pathway distinct from PIN1. Plant Cell 18(11):3171–3181
31. Geldner N, Anders N, Wolters H, Keicher J, Kornberger W, Muller P, Delbarre A, Ueda T, Nakano A, Jürgens G (2003) The Arabidopsis GNOM ARF-GEF mediates endosomal recycling, auxin transport, and auxin-dependent plant growth. Cell 112(2):219–230
32. Swarup R, Friml J, Marchant A, Ljung K, Sandberg G, Palme K, Bennett M (2001) Localization of the auxin permease AUX1 suggests two functionally distinct hormone transport pathways operate in the Arabidopsis root apex. Genes Dev 15(20):2648–2653
33. Friml J, Benková E, Blilou I, Wisniewska J, Hamann T, Ljung K, Woody S, Sandberg G, Scheres B, Jürgens G, Palme K (2002) AtPIN4 mediates sink-driven auxin gradients and root patterning in Arabidopsis. Cell 108(5):661–673

Chapter 7

Investigating the Phytohormone Ethylene Response Pathway by Chemical Genetics

Lee-Chung Lin, Chiao-Mei Chueh, and Long-Chi Wang

Abstract

Conventional mutant screening in forward genetics research is indispensible to understand the biological operation behind any given phenotype. However, several issues, such as functional redundancy and lethality or sterility resulting from null mutations, frequently impede the functional characterization of genetic mutants. As an alternative approach, chemical screening with natural products or synthetic small molecules that act as conditional mutagens allows for identifying bioactive compounds as bioprobes to overcome the above-mentioned issues. Ethylene is the simplest olefin and is one of the major phytohormones playing crucial roles in plant physiology. Most of the current information on how ethylene works in plants came primarily from genetic studies of ethylene mutants identified by conventional genetic screening two decades ago. However, we lack a complete picture of functional interaction among components in the ethylene pathway and cross talk of ethylene with other phytohormones. Here, we describe our methodology for using chemical genetics to identify small molecules that interfere with the ethylene response. We set up a phenotype-based screening platform and a reporter gene-based system for verification of the hit compounds identified by chemical screening. We have successfully identified small molecules affecting the ethylene phenotype in etiolated seedlings and showed that a group of structurally similar compounds are novel inhibitors of ACC synthase, a rate-limiting enzyme in the ethylene biosynthesis pathway.

Key words Chemical genetics, Phytohormone, Ethylene, Triple response, *Arabidopsis thaliana*, ACC synthase, Inhibitor

1 Introduction

Chemical screening of small molecules as modulators in biological processes of clinically important proteins has been intensively used in drug discovery [1, 2]. Two major types of chemical screening have frequently been used to discover synthetic small molecules or natural products as effective chemical compounds with biological impacts: analyzing the activity of target proteins (target-based) or scoring specified phenotypes of cells or organisms (phenotype-based) [3, 4]. Small molecules offer the advantage of reversible, conditional and kinetic effects for functional studies in organisms in which lethality of genetically null mutants is an issue.

Glenn R. Hicks and Stéphanie Robert (eds.), *Plant Chemical Genomics: Methods and Protocols*, Methods in Molecular Biology, vol. 1056, DOI 10.1007/978-1-62703-592-7_7,

In addition, small molecules can be agonists or antagonists to a group of proteins sharing conserved functions. Thus, using small molecules may provide a solution to the issues of gene redundancy and genetic lethality [3]. However, concerns of nonspecific effects and metabolism of chemical compounds are potential disadvantages when using small molecules or drug-like compounds to investigate complex biological processes. Finally, identification of the targets of effective small molecules from phenotype-based screening strategy is challenging, especially for organisms lacking genetic information or tools [4].

Plant hormones are growth factors and small molecules that function as bioactive compounds to modulate plant physiology [5]. Combining chemical screening and genetics approaches, chemical genetics has recently been found to be an alternative approach to explore plant physiology in the reference plant *Arabidopsis thaliana* [6, 7]. Sirtinol was identified as an inhibitor of yeast SIR2, a nicotine adenine dinucleotide (NAD)-dependent histone deacetylase [8]. Interestingly, sirtinol effected the auxin response in Arabidopsis seedlings [8] and was subsequently used as a chemical tool to isolate genetic mutants resistant to sirtinol to explore its physiological role in plants [9]. In addition to sirtinol, novel inhibitors or functional analogues of auxin were identified from various collections of small molecules by reporter gene-based or phenotype-based screening strategies [10–13]. *Br*assino*p*ride (BRP), identified by using a brassinosteroid (BR)-responsive reporter gene for chemical screening, showed an inhibitory effect on the biosynthesis of BRs in Arabidopsis seedlings [14]. However, the targets of most of these small molecules identified to affect hormone response or biosynthesis remain elusive. Nevertheless, the recent exciting discovery of abscisic acid (ABA) receptors was initially achieved by a chemical genetics approach [15]. Pyrabactin was identified as a chemical inhibitor of seed germination and was used to isolate pyrabactin-resistant (*pyr*) mutants in Arabidopsis. Characterization of the *pyr* mutants and cloning of the corresponding genes led to the finding that PYR/PYL/*R*egulatory *C*omponent of *A*BA *R*eceptor (RCAR) bound ABA and function as ABA receptors [15]. Further X-ray crystallography experiments gave the molecular details of how ABA receptors perceive ligands to trigger the ABA response [15–17].

We reasoned that conventional screening for genetic mutants defective in the ethylene response was nearly saturated [18, 19]. Modified strategies to screen for additional ethylene mutants identified *weak ethylene insensitive* (*wei*) [20] and *enhanced ethylene response* (*eer*) [21] mutants and yielded new information and potentially relevant components in the ethylene response pathway. Therefore, we sought to use chemical genetics to further our understanding of ethylene hormone functions in plants. By screening a collection of 10,000 structurally diverse small molecules, we identified chemical compounds suppressing the constitutive triple response phenotype in the ethylene overproducer (*eto*) mutant, *eto1-4* [22]. Structure and activity analysis revealed that these chemical compounds contain a quinazolinone backbone and are novel uncompetitive inhibitors of ACC synthase (ACS). Therefore, we designated these small molecules *ACS in*hibitor quinazoli*nones*, acsinones. The discovery of acsinones can benefit academic research and possibly the agrochemical industry by providing effective lead chemicals for use in post-harvest management to improve quality in climacteric fruits and ornamental crops.

2 Materials

2.1 Plant, Bacteria, and DNA Materials

1. All transgenic plants and mutants are derived from the wild-type *Arabidopsis thaliana* Columbia ecotype (Col-0) and cultivated under long-day conditions (16-h light/8-h dark at 22 °C) with white light (100–150 μE m^{-2} s^{-1}).
2. Seeds of the transgenic Arabidopsis reporter line *5xEBS::LUC* (five copies of *E*IN3 *b*inding *s*equence [*EBS*] fused with luciferase gene, *LUC*) in an *eto1-4* background.
3. Seeds of ethylene mutants *eto1-4, eto2-1,* and *ctr1-1* and transgenic Arabidopsis line with overexpression of *EIN3* under the CaMV 35S promoter (*35S::EIN3*) in a Col-0 background.
4. pETDuet (Novagen, Merck KGaA, Darmstadt, Germany).
5. *E. coli* (BL21-CodonPlus, Stratagene).
6. Luria–Bertani (LB) broth (MDBio, Piscataway, NJ, USA).

2.2 Chemicals and Reagents

Chemicals and reagents are prepared by use of double deionized water (DDW) or suitable solvents, such as dimethyl sulfoxide (DMSO), sterilized by use of an autoclave (121 °C, 20 min, 15 psi [1.05 kg/cm^2]) or by 0.2-μm disc filters and store at room temperature unless otherwise indicated.

1. A chemical library (DIVERSet, ChemBridge, San Diego, CA, USA) containing 10,000 small molecules is used for chemical screening. All chemicals in the DIVERSet library are shipped in lyophilized form in dry ice and prepared by dissolving in DMSO to 10 mM. The small molecules are stored in 96-well microtiter plates at −80 °C and thawed at room temperature before use.
2. DMSO (J.T. Baker, Austin, TX, USA; Cat. no. 9224–03).
3. Aminoethoxyvinylglycine (AVG; Sigma-Aldrich, St. Louis, MO, USA; Cat. no. 359629): Dissolve 16.02 mg AVG in 10 mL distilled water for a 10-mM stock solution and store at −20 °C. The working concentration is 10 μM.

4. Silver nitrate (Sigma-Aldrich, St. Louis, MO; USA; Cat. no. S6506): Dissolve 1.7 g silver nitrate in 100 mL DDW for a 100-mM stock solution.
5. Sodium thiosulfate (Sigma-Aldrich, St. Louis, MO; USA; Cat. no. S7026): Dissolve 1.58 g sodium thiosulfate in 100 mL DDW for a 100-mM stock solution.
6. Silver thiosulfate (STS): Mix silver nitrate and sodium thiosulfate at a 1:4 molar ratio (*see* **Note 1**) and in a brown bottle at 4 °C. The working concentration is 10 μM.
7. 1-Aminocyclopropane-1-carboxylic acid (ACC; Merck, Whitehouse Station, NJ, USA; Cat. no. 149101): Dissolve 101 mg ACC in 10 mL DDW for a 100 mM ACC stock solution and store at −20 °C. The working concentration is 10 μM.
8. Triton-X 100 (AMERSCO, Solon, OH, USA; Cat. no. 0694).
9. Luciferin (d-Luciferin potassium salt, Biosynth International, Itasca, IL, USA; Cat. no. L-8220): Dissolve 318.42 mg luciferin in 10 mL DDW for a 100 mM stock solution and store in the dark at −20 °C. Use 0.01 % Triton-X 100 as diluent to prepare the working solution.
10. Isopropyl-beta-d-thiogalactoside (IPTG; Sigma-Aldrich, St. Louis, MO; USA; Cat. no. I5502): Dissolve 23.8 mg IPTG in 1 mL DDW for a 100 mM stock solution and store at −20 °C.
11. NaCl (AMERSCO, Solon, OH, USA; Cat. no. 0241).
12. Potassium phosphate buffer (0.1 M, pH 7.4): Add 802 mL of 0.1 M K_2HPO_4 and 198 mL of 0.1 M KH_2PO_4 to prepare 1 L of 0.1 M potassium phosphate buffer. K_2HPO_4 (Cat. no. 1.05104.1000) and KH_2PO_4 (Cat. no. 1.04873.1000) are available from MERCK (Whitehouse Station, NJ, USA).
13. Phosphate buffer (300 mM NaCl, 20 mM potassium phosphate buffer, pH 7.4).
14. Pyridoxal-5′-phosphate (PLP; Sigma-Aldrich, St. Louis, MO; USA; Cat. no. P9255): Dissolve 265.16 mg PLP in 10 mL DDW for a 100 mM stock solution, which appears yellowish. Keep the solution in the dark at −20 °C.
15. S-(5′-Adenosyl)-l-methionine salt (SAM; Sigma-Aldrich, St. Louis, MO; USA; Cat. no. A7007): Dissolve 434.9 mg SAM in 1 mL DDW to prepare a 1 M stock solution and store aliquots at −20 °C. Prepare working solution freshly for enzyme kinetics assay.
16. (N-[2-Hydroxyethyl]piperazine-N'-[2-Ethanesulfonic Acid]) (HEPES); USB, Cleveland, OH, USA; Cat. no. 16926).
17. Dithiothreitol (DTT, USB, Cleveland, OH, USA; Cat. no. 15397): Dissolve 3.09 g DTT in 20 mL DDW to prepare a

1-M stock and make small aliquots (0.5–1 mL) for storage at –20 °C. Use a 0.2-μm disc filter for sterilization if necessary.

18. Buffer A: 100 mM HEPES buffer containing 10 μM PLP, adjust to pH 8.0 with 10 M NaOH. Add 5 mM DTT just before use.
19. Imidazole (ACS grade, MERCK, Whitehouse Station, NJ, USA; Cat. no. 1.04716.0250).
20. Protease inhibitor tablets (Sigma-Aldrich, St. Louis, MO; USA; Cat. no. S8820).
21. Murashige and Skoog (MS) basal salt mixture (PhytoTechnology Laboratories, Shawnee Mission, KS, USA; Cat. no. M524).
22. Diethylpyrocarbonate (DEPC; Sigma-Aldrich, St. Louis, MO; USA; Cat. no. D5758)-treated water: Dissolve 200 μL DEPC in 100 mL DDW and mix well. Let it stand for at least 1 h at 37 °C or on the bench top overnight before using an autoclave to inactive traces of DEPC.
23. Chloroform–isoamyl alcohol (24:1) (AMERSCO, Solon, OH, USA; Cat. no. X205).
24. LiCl (AMERSCO, Solon, OH, USA; Cat. no. 0416): Dissolve 42.4 g LiCl in 80 mL DEPC-treated water to prepare a 10 M stock solution and adjust the final volume to 100 mL. Store at room temperature after sterilization.
25. Hexadecyltrimethylammonium bromide (CTAB; Sigma-Aldrich, St. Louis, MO; USA; Cat. no. H5882).
26. Polyvinylpyrrolidone (PVP; Sigma-Aldrich, St. Louis, MO; USA; Cat. no. P5288).
27. 1 M Tris–HCl, pH 8.0 (Invitrogen, Carlsbad, CA, USA; Cat. no. 15568–025).
28. Ethylenediaminetetraacetic acid (EDTA; Sigma-Aldrich, St. Louis, MO; USA; Cat. no. ED2SS).
29. Spermidine (Sigma-Aldrich, St. Louis, MO; USA; Cat. no. S2626).
30. Beta-mercaptoethanol (AMERSCO, Solon, OH, USA; Cat. no. M131).
31. Ampicillin (USB, Cleveland, OH, USA; Cat. no. 11259).
32. Pine tree extraction buffer. To prepare pine tree extraction buffer, mix and autoclave chemicals at the final concentrations indicated: 2 % (w/v) CTAB; 2 % (w/v) PVP; 100 mM Tris–HCl, pH 8.0; 25 mM EDTA; 2.0 M NaCl; 0.5 g/L spermidine. Add beta-mercaptoethanol to a final concentration of 2 % [v/v] just before use.

2.3 Instruments, Supply, and Software

1. Plate reader (Plate CHAMELEON™ V, Hidex, Turku, Finland).
2. Xenogen IVIS System (IVIS® Lumina, Caliper Life Sciences, Hopkinton, MA, USA).
3. Gas chromatograph (GC; HP 6890, Hewlett-Packard) equipped with a headspace sampler (HP 7694, Agilent Technologies, Santa Clara CA, USA) and a capillary column (19095P-U04, Agilent Technologies, Santa Clara CA, USA).
4. Headspace crimp-top vials (Cat. no. 5182-0838, Agilent Technologies, Santa Clara CA, USA).
5. Aluminum crimp seals with septa (Cat. no. 5183–4477) and manual crimper (Cat. no. 9301–0720, both Agilent Technologies, Santa Clara CA, USA).
6. Orbital shaker (Excella E25R, New Brunswick Scientific, Edison, NJ, USA).
7. High-pressure cell disrupter (TS 2.2 KW, Constant Systems, Low March, Daventry Northants, UK).
8. HisTrap FF columns (Cat. no. 17-5319-01, GE Healthcare Life Sciences, Pittsburgh, PA, USA).
9. AKTAprime™ system (GE Healthcare Life Sciences, Pittsburgh, PA, USA).
10. 96-well white plate (Packard Optiplate-96, Perkin Elmer, Waltham Massachusetts, USA).
11. SigmaPlot 10 (Systat Software, Chicago, IL, USA).
12. ChemOffice Ultra 11 (CambridgeSoft, Cambridge, MA, USA).
13. ImageJ [http://rsbweb.nih.gov/ij/].
14. GeneChip® Fluidics Station 450 (Affymetrix, Santa Clara, CA, USA).
15. GeneChip Scanner 7G system (Affymetrix, Santa Clara, CA, USA).
16. Genespring GX10 (Agilent Technologies, Santa Clara, CA, USA).

3 Methods

3.1 Chemical Screening Based on the Triple Response Phenotype in Etiolated Seedlings

The phenotype-based strategy involves use of etiolated Arabidopsis *eto1-4* seedlings for chemical screening, which is performed by scoring the triple response phenotype of *eto1-4* seedlings germinated in micro-titer plates harboring single, unique small molecules in individual wells. Etiolated (dark-grown) *eto1-4* seedlings overproduce ethylene and show a typical triple response phenotype of a swollen hypocotyl, short roots, and an apical hook [23]. A representative image is shown (Fig. 1). Three rounds of chemical

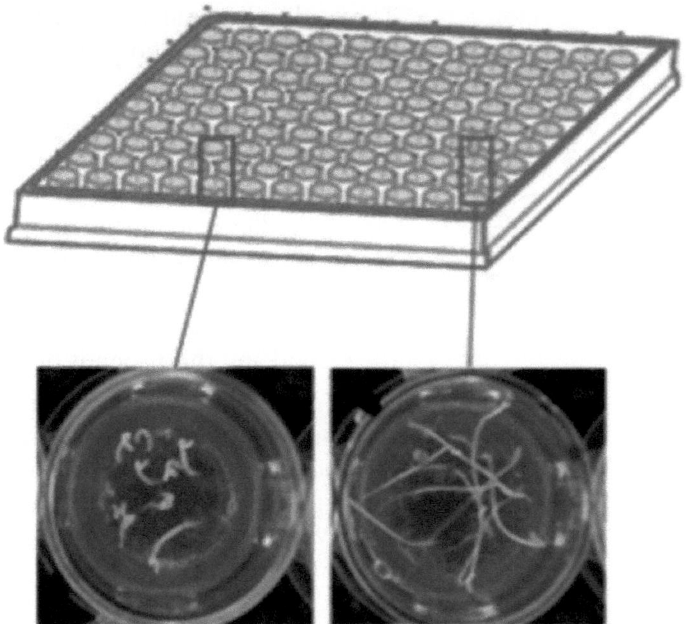

Fig. 1 A representative image of the phenotype of seedlings germinated in a 96-well microtiter plate for chemical screening. Three-day etiolated *eto1-4* seedlings show a typical triple response phenotype (*left*) that is suppressed by incubation with an effective compound (*right*)

screening are performed to identify and verify the hit compounds, called acsinones hereafter.

1. To prepare agar medium containing small molecules in 96-well microtiter plates, add 98 μL of 1/2 MS salt solution into each well by use of a multichannel pipette (RAININ, Pipet-Lite Multichannel), then mix 2 μL of 5 mM chemical compounds from the DIVERSet 96-well microtiter plates by pipetting three to five times (*see* **Note 2**). Add another 100 μL of 1.6 % 1/2 MS agar medium (*see* **Note 3**) and mix well for a final 200-μL agar medium containing 50 μM chemical compounds.
2. To prepare agar medium in 24-well plates, add 3 μL of 5 mM chemical compounds and mix well with 297 μL 1/2 MS solution, then add 300 μL of 1.6 % 1/2 MS agar medium and mix well for a final 600 μL of agar medium containing 25 μM chemical compounds (*see* **Note 4**).
3. Seeds are sterilized with 30 % (v/v) household bleach (Clorox®) for 6 min and then washed with sterile water three to four times. Sterilized seeds are sown directly into 96-well (~10 seeds/well) or 24-well (20–25 seeds/well) plates, then stratified in the dark at 4 °C for 3–4 days before germination.

The seeds are germinated in the dark at 22 °C for 3 days before scoring the triple response phenotype.

4. Use a dissection microscope to quickly inspect seedlings showing long hypocotyls and loss of apical hooks. The loss of the triple response phenotype is confirmed by comparing the wild-type and *eto1-4* seedlings germinated in the same microtiter plates. Seedlings are quantitatively measured by taking images of 10–15 seedlings under a dissection microscope then using ImageJ to measure hypocotyl length. Potential hits of small molecules obtained from the primary screening are verified by the second and third rounds of analyses.
5. Perform subsequent tests by using 24-well microtiter plates to accommodate more seeds sufficient for quantitative analyses of hypocotyl length and luciferase activity (*see* Subheading 3.3 below). A reduced concentration of candidate compounds at 25 μM is used to increase stringency of the phenotype analysis.

3.2 Measurement of Ethylene Levels in Etiolated Seedlings Germinated with Acsinones

At least 30 seedlings germinated in the dark for 3 days on the 1/2 MS agar medium supplemented with acsinones are used for quantification of ethylene levels by use of a GC with a headspace autosampler.

1. Add 1.5 mL of 1/2 MS liquid medium without (DMSO as control) and with individual 20 μM compounds to 10-mL headspace crimp-top vials and mix well with 1.5 mL of 1.9 % 1/2 MS agar medium.
2. Sow 30–40 seeds in the crimp-top vials and seal the vials with aluminum caps and septa by use of a manual crimper under the laminar flow hood to avoid contamination. Ethylene emitted by seedlings is collected from the headspace of crimp-top vials and analyzed.
3. Measure ethylene levels by use of a GC equipped with a capillary column and a headspace autosampler. Insert crimp-top vials in the vial tray of the autosampler and use ChemStation software, an instrument control for the 6890 GC, to set up programs to run analyses and collect data. N_2 is used as the carrier gas, and 1 mL gaseous sample in the headspace of crimp-top vials is automatically drawn for injection to a capillary column connected with flame ionization detector (FID) to measure ethylene levels.
4. Compared with an ethylene standard (1 parts per million [ppm]), the levels of ethylene generated by seedlings are presented as the rate of ethylene production (nmole/L/seedling/day).

3.3 Luciferase Reporter System to Verify the Effects of the Hit Compounds on Ethylene Response

Both quantification of luciferase activity (**steps 1–4**) and imaging of luminescence (**steps 5–7**) involve use of Arabidopsis seedlings germinated in multiwell plates.

1. Sow 20–25 seeds in the wells of 24-well microtiter plates containing chemical compounds and germinate in the dark for 3 days. Add 300 μL of 1/2 MS liquid medium containing 25 μM of individual compounds to each well of a 96-well white plate. Transfer 3-day-old etiolated seedlings to 96-well white plates for continued growth under light for another 3 days before measuring luciferase activity (*see* **Note 5**).
2. Transfer 9 seedlings for testing each compound and place 3 seedlings in individual wells for measuring luciferase activity in triplicate.
3. Add 1 mM luciferin to each well and mix well by pipetting up and down several times. Keep the 96-well white plate in the dark for 5 min at room temperature before placing it in the microplate reader (Plate CHAMELEON™ V) for analysis of luciferase activity.
4. Use "luminescence" as the detection method in the control panel of CHAMELEON™ V. Select "direct" measurement mode and specify the sample range on the microplate. Luminescence intensity in selected wells is shown on the screen and can be exported to Microsoft Excel for further analysis.
5. To view the luminescence of seedlings, germinate ~50 seeds in 9-cm diameter Petri dishes in the dark at 22 °C for 3 days, then transfer to a growth room to continue growing under white light for 3 more days before acquiring images of the luminescence of seedlings. In some experiments, 6-day-old etiolated seedlings are used to acquire images.
6. Apply 2 mM luciferin evenly on seedlings by use of a mini mist sprayer and keep the plates in the dark for 5 min before images are acquired by use of the Xenogen IVIS System. Representative images are shown (Fig. 2).
7. Choose "luminescent" as the image mode in the control panel of the IVIS System software. Adjust the height of the sample platform and the exposure time; click the "acquire" button to capture live images of seedlings.

3.4 Phenotype Analysis of Ethylene Mutants and Transgenic Plants to Determine the Effect of Acsinones on the Ethylene Pathway

1. Prepare growth medium in 9-cm diameter Petri dishes by adding 400 μL of 10 mM chemical stock solution to 400 mL 1/2 MS agar medium for approximately 20 plates.
2. Use DMSO to replace acsinones as a control. Supplements including AVG, STS, and ACC are prepared as described previously and are used at 10 μM. Agar medium plates are stored at 4 °C and should be used within 2 weeks; plates containing STS should be wrapped with aluminum to avoid the light.

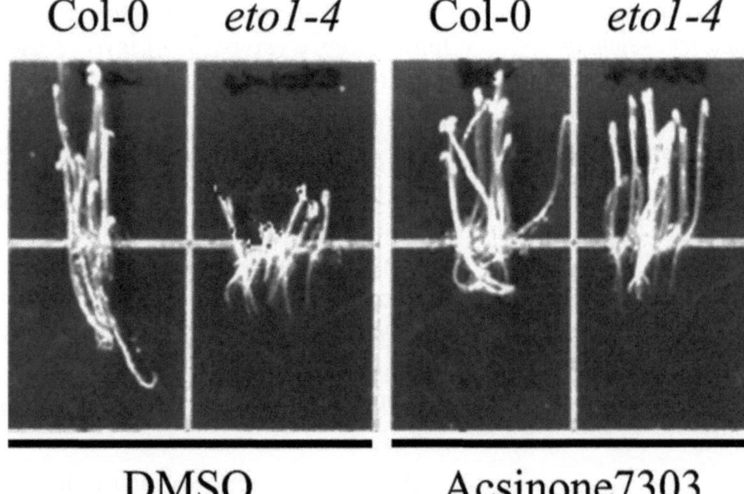

Fig. 2 Luminescent images of etiolated Arabidopsis seedlings. Wild-type *Arabidopsis* (ecotype Columbia, Col-0) and a *5xEBS::LUC* transgenic line (in *eto1-4* background) were germinated on 1/2 MS agar medium in the dark without (*left panel*, DMSO) and with (*right panel*) acsinone7303. Luminescent images were obtained by use of a Xenogen IVIS System from 6-day-old etiolated seedlings after spraying 2 mM luciferin directly on the plants. The triple response phenotype and luminescence in *eto1-4* seedlings are both suppressed by acsinone7303 (*right panel*)

3. For quantitative phenotype analyses, use a fine-point tweezer to carefully transfer at least 25 seedlings from each plate and place them between two transparent slides to avoid desiccation of seedlings. Put a ruler on the side of transparency slides and take pictures. Use ImageJ software to measure the hypocotyl length of seedlings for comparing the chemical effect on the hypocotyl phenotype.

3.5 Enzyme Kinetics Assays to Determine the Biochemical Properties of an Acsinone (see Note 6)

1. To generate an expression construct, the full-length cDNA of Arabidopsis ACS5 (At5g65800) is cloned into pETDuet (Novagen) to generate pETDuet-6His-ACS5 for expression of recombinant protein in *E. coli* (BL21-CodonPlus, Stratagene).
2. Inoculate a single colony into 50 mL LB broth supplemented with antibiotic (ampicillin, 100 μg/mL) for growth overnight in an orbital shaker set at 200 rpm and 37 °C. On the following day, 50 mL of culture is transferred to a fresh 500 mL of LB broth medium for continued growth under the same condition until OD_{600} 0.6.

3. Add 0.4 mM IPTG to induce protein expression at 16 °C for 18 h and harvest cells by centrifugation at 6,000 × *g* for 10 min at 4 °C.
4. Dissolve 6.8 g imidazole to 100 mL to prepare a 1 M stock solution for use in the phosphate buffer. Wash and resuspend the cell pellet in 50 mL phosphate buffer containing 20 mM imidazole and a cocktail of protease inhibitors.
5. Disrupt cells by use of a continuous high-pressure cell disrupter with 30 KPsi at 4 °C and rinse the sample cup twice with 50 mL phosphate buffer containing 20 mM imidazole to collect residual cells.
6. Centrifuge the final 150-mL suspension at 10,000 × *g* for 30 min at 4 °C and keep the supernatant.
7. Load the supernatant into a 5 mL HisTrap FF column by the AKTAprime™ system and then wash stepwise with 100 mL buffer A (5 mM DTT, 10 μM PLP, 100 mM HEPES buffer, pH 8.0) containing 20 mM imidazole, then 100 mL buffer A containing 100 mM imidazole.
8. Elute the bound protein with a 0.1–1 M gradient of imidazole in 25 mL buffer A and store the purified proteins in aliquots at -80 °C.
9. In vitro ACS activity assay is performed as described [24, 25] with minor modification. Mix purified recombinant ACS5 protein (1 μg) in 2 mL buffer A containing 10 μM PLP and 250 μM SAM with different concentrations of acsinones or DMSO (as a control) in 20-mL crimp-top GC vials for enzymatic reaction for 30 min at 25 °C.
10. Add 100 μL $HgCl_2$ (20 mM) and 100 μL NaOH–bleach (1:1 with saturated NaOH and household bleach) and immediately seal the crimp-top GC vials with caps and septa to stop reactions and to oxidize ACC to ethylene (*see* **Note 7**), respectively, which continues on ice for at least 10 min before using gas chromatography for analysis of ethylene levels.
11. To measure the ethylene level by GC, follow **step 3** in Subheading 3.2 above.
12. Use various concentrations of SAM as a substrate to determine the apparent kinetic parameters (K_m and V_{max}) of recombinant ACS5 in the enzyme kinetic assay.
13. Both AVG and a hit compound (acsinone7303) are used with 0.01 and 0.05 μM in the enzyme kinetics assay to determine the inhibition constant (K_i) and apparent kinetic parameters (K_m and V_{max}).

14. The enzyme kinetics assay is performed by adding different concentrations of SAM (80, 100, 150, 200, 300, and 400 μM) in 20-mL crimp-top GC vials containing 10 μM PLP in 2 mL buffer A then mixing inhibitors or DMSO and recombinant ACS5 (1.6 μg) concurrently to initiate the enzyme reaction for 30 min at 25 °C.
15. Terminate the reaction by adding $HgCl_2$ and NaOH–bleach as described in **step 10**. Gas chromatography is used to quantify the levels of ethylene chemically converted from ACC.
16. Statistical analysis of data from enzyme kinetic assays and preparation of Lineweaver–Burk plots involve use of the add-on Enzyme Kinetic module of SigmaPlot.

3.6 Transcriptome Analysis of Differential Gene Expression Patterns Affected by Acsinones and AVG

1. Sow 500 seeds per plate in 9-cm-diameter Petri dishes with and without acsinones and AVG. At least 2,000 seeds for each chemical treatment are required.
2. Seeds are germinated in the dark for 3 days before aerial tissues are collected and ground to a fine powder with liquid nitrogen. Weigh the ground tissue powder to isolate total RNA by the pine tree method described below (**steps 3–6**) [26].
3. Pre-warm 5 mL extraction buffer to 65 °C in a water bath. Add ground tissue powder to the extraction buffer and mix thoroughly by inverting or vortexing centrifuge tubes.
4. Extract twice with an equal volume of chloroform–isoamyl alcohol by vortexing, and save the supernatant after separating aqueous and organic phases at room temperature by centrifugation at 12,000 × *g* for 10 min. A longer centrifugation may be required if the two phases do not separate well.
5. Add 1/4 volume of 10 M LiCl to the supernatant and mix well. Total RNA is precipitated by centrifugation at 12,000 × *g* for at least 20 min at 4 °C.
6. Dissolve the pellet with an appropriate volume of DEPC-treated water (50–100 μL), and measure RNA concentration by spectrophotometry with OD_{260} spectrum.
7. Ten micrograms of total RNA are used for cDNA synthesis and labeling with biotin by in vitro transcription, then fragmentation according to the manufacturer's protocol (GeneChip Expression Analysis Technical Manual rev5, Affymetrix).
8. Labeled RNA samples are hybridized to GeneChip ATH1 at 45 °C for 16.5 h. The hybridized chips are washed with Fluidic Station 450 and then scanned by use of the Affymetrix GeneChip Scanner 7G. Data are analyzed by use of Agilent GeneSpring GX software. Methods and variables in the data analysis are described below.

9. Use the MAS5 method for probe summarization, with normalization to the all-sample median. Genes with expression > 100 and with present or marginal calls in duplicate experiments are selected for subsequent data analysis.
10. Genes with differential expression (twofold cut-off) in *eto1* with and without AVG or acsinones are identified as genes co-regulated by AVG and acsinones. Use one-way ANOVA to examine the statistical significance ($p < 0.05$) of the expression of these genes.

4 Notes

1. Although silver nitrate also blocks the action of ethylene, its in vivo transport efficiency is not as good as that of STS solution [27]. Therefore, STS solution is used as an ethylene signal-blocking agent. Preparation of STS follows the mixing order: add silver nitrate into sodium thiosulfate and mix quickly by gently shaking the bottle while adding silver nitrate. We recommend preparing a fresh working solution of STS (10 μM) before use. The STS stock solution (10 mM) should be stored in a brown bottle at 4° to avoid light for no more than 2 weeks.
2. Place the DIVERSet chemicals in 96-well plates on a bench to thaw before use. It takes approximately 30 min to 1 h to completely thaw DIVERSet chemicals after being removed from a −80 °C freezer. Use a multiple pipette to make aliquots of small molecules from the original to a working 96-well plate under the laminar flow hood to avoid contamination of the original chemical library set.
3. To achieve a consistent triple response phenotype in etiolated seedlings during chemical screening, we recommend not including sucrose in the 1/2 MS agar medium.
4. To avoid solidification of 1.6 % agar medium before mixing with liquid MS medium containing small molecules, use a hot plate to keep the temperature of the agar medium between 50 and 55 °C during preparation.
5. We found decreased luciferase activity derived from the EBS promoter after seedlings continued to grow under white light, which is probably due to loss of EIN3 protein. The optimal condition for our quantitative luciferase activity assays is to use seedlings with 3 days of etiolated growth and another 3 days under light.
6. Because acsinones suppress the ethylene overproduction in etiolated *eto1-4* seedlings, these hit compounds may interfere with ethylene biosynthesis. To test this hypothesis, we used a

bacterial expression system to generate recombinant ACS enzyme for the enzyme activity assay.

7. The NaOH–bleach is a highly oxidative reagent to oxidize ACC to ethylene gas. To avoid loss of ethylene gas, seal the crimp-top vials as quickly as possible. Alternatively, use a 1-mL syringe with a needle to inject NaOH–bleach into capped crimp-top vials.

Acknowledgments

This work was supported by grants from National Science Council (NSC972311B001003) and Development Program of Industrialization for Agricultural Biotechnology (99S0030088 and 100S0030016).

References

1. Cong F, Cheung AK, Huang SM (2012) Chemical genetics-based target identification in drug discovery. Annu Rev Pharmacol Toxicol 52:57–78
2. Knight ZA, Garrison JL, Chan K, King DS, Shokat KM (2007) A remodelled protease that cleaves phosphotyrosine substrates. J Am Chem Soc 129:11672–11673
3. O'Connor CJ, Laraia L, Spring DR (2011) Chemical genetics. Chem Soc Rev 40:4332–4345
4. Stockwell BR (2000) Frontiers in chemical genetics. Trends Biotechnol 18:449–455
5. McCourt P, Desveaux D (2010) Plant chemical genetics. New Phytol 185:15–26
6. Osawa Y, Lau M, Lowe ER (2007) Plant-derived small molecule inhibitors of Neuronal NO-Synthase: Potential effects on protein degradation. Plant Signal Behav 2:129–130
7. Blackwell HE, Zhao Y (2003) Chemical genetic approaches to plant biology. Plant Physiol 133:448–455
8. Grozinger CM, Chao ED, Blackwell HE, Moazed D, Schreiber SL (2001) Identification of a class of small molecule inhibitors of the sirtuin family of NAD-dependent deacetylases by phenotypic screening. J Biol Chem 276:38837–38843
9. Zhao Y, Dai X, Blackwell HE, Schreiber SL, Chory J (2003) SIR1, an upstream component in auxin signaling identified by chemical genetics. Science 301:1107–1110
10. Hayashi K, Jones AM, Ogino K, Yamazoe A, Oono Y, Inoguchi M, Kondo H, Nozaki H, Yokonolide B (2003) A novel inhibitor of auxin action, blocks degradation of AUX/IAA factors. J Biol Chem 278:23797–23806
11. Armstrong JI, Yuan S, Dale JM, Tanner VN, Theologis A (2004) Identification of inhibitors of auxin transcriptional activation by means of chemical genetics in Arabidopsis. Proc Natl Acad Sci U S A 101:14978–14983
12. Christian M, Hannah WB, Lüthen H, Jones AM (2008) Identification of auxins by a chemical genomics approach. J Exp Bot 59:2757–2767
13. Savaldi-Goldstein S, Baiga TJ, Pojer F, Dabi T, Butterfield C, Parry G, Santner A, Dharmasiri N, Tao Y, Estelle M, Noel JP, Chory J (2008) New auxin analogs with growth-promoting effects in intact plants reveal a chemical strategy to improve hormone delivery. Proc Natl Acad Sci U S A 105:15190–15195
14. Gendron JM, Haque A, Gendron N, Chang T, Asami T, Wang ZY (2008) Chemical genetic dissection of brassinosteroid-ethylene interaction. Mol Plant 1:368–379
15. Park SY, Fung P, Nishimura N, Jensen DR, Fujii H, Zhao Y, Lumba S, Santiago J, Rodrigues A, Chow TF, Alfred SE, Bonetta D, Finkelstein R, Provart NJ, Desveaux D, Rodriguez PL, McCourt P, Zhu JK, Schroeder JI, Volkman BF, Cutler SR (2009) Abscisic acid inhibits type 2C protein phosphatases via the PYR/PYL family of START proteins. Science 324:1068–1071
16. Nishimura N, Hitomi K, Arvai AS, Rambo RP, Hitomi C, Cutler SR, Schroeder JI, Getzoff ED (2009) Structural mechanism of abscisic acid binding and signaling by dimeric PYR1. Science 326:1373–1379

17. Peterson FC, Burgie ES, Park SY, Jensen DR, Weiner JJ, Bingman CA, Chang CE, Cutler SR, Phillips GN Jr, Volkman BF (2010) Structural basis for selective activation of ABA receptors. Nat Struct Mol Biol 17:1109–1113
18. Hua J, Sakai H, Nourizadeh S, Chen QG, Bleecker AB, Ecker JR, Meyerowitz EM (1998) EIN4 and ERS2 are members of the putative ethylene receptor gene family in Arabidopsis. Plant Cell 10:1321–1332
19. Stepanova AN, Alonso JM (2009) Ethylene signaling and response: Where different regulatory modules meet. Curr Opin Plant Biol 12:548–555
20. Alonso JM, Stepanova AN, Solano R, Wisman E, Ferrari S, Ausubel FM, Ecker JR (2003) Five components of the ethylene-response pathway identified in a screen for weak ethylene-insensitive mutants in Arabidopsis. Proc Natl Acad Sci U S A 100:2992–2997
21. Larsen PB, Chang C (2001) The Arabidopsis eer1 mutant has enhanced ethylene responses in the hypocotyl and stem. Plant Physiol 125:1061–1073
22. Lin LC, Hsu JH, Wang LC (2010) Identification of novel inhibitors of 1-aminocyclopropane-1-carboxylic acid synthase by chemical screening in Arabidopsis thaliana. J Biol Chem 285:33445–33456
23. Guzmán P, Ecker JR (1990) Exploiting the triple response of Arabidopsis to identify ethylene-related mutants. Plant Cell 2:513–523
24. Lizada MC, Yang SF (1979) A simple and sensitive assay for 1-aminocyclopropane-1-carboxylic acid. Anal Biochem 100: 140–145
25. Yamagami T, Tsuchisaka A, Yamada K, Haddon WF, Harden LA, Theologis A (2003) Biochemical diversity among the 1-aminocyclopropane-1-carboxylate synthase isozymes encoded by the Arabidopsis gene family. J Biol Chem 278:49102–49112
26. Chang S, Puryear J, Cairney J (1993) A simple and efficient method for isolating RNA from pine trees. Plant Mol Biol 11:693–699
27. Veen H (1983) Silver thiosulphate: An experimental tool in plant science. Sci Hortic 20: 211–214

Chapter 8

Screening for Inhibitors of Chloroplast Galactolipid Synthesis Acting in Membrano and in Planta

Laurence Boudière and Eric Maréchal

Abstract

The knowledge of the membrane lipid metabolism in photosynthetic cells is expected to benefit from the availability of inhibitors acting at the level of specific enzymes like MGD1 (E.C. 2.4.1.46) that catalyzes the synthesis of monogalactosyldiacylglycerol (MGDG) in chloroplasts. MGDG is a major lipid of photosynthetic membrane, interacting with photosystems. It is the precursor of digalactosyldiacylglycerol that serves as a phospholipid surrogate when plants are deprived of phosphate, and it is a source of polyunsaturated fatty acids for jasmonic acid syntheses. MGD1 is activated by phosphatidic acid and thus a coupling point between phospholipid and galactolipid metabolisms. Here we describe a method to screen for inhibitors of MGD1 assayed in liposomes. Selected compounds can therefore reach the core of the biological membranes in which the target sits. We then describe a secondary screen to evaluate the efficiency of developed compounds at the whole plant level. Major issues raised by the screening of inhibitors acting on membrane proteins are discussed and can be useful for similar targets.

Key words Chloroplast, Galactolipids, Monogalactosyldiacylglycerol, High throughput enzymatic screening, Galvestine-1

1 Introduction

The search for compounds acting on enzymes synthesizing membrane lipids faces multiple issues. Firstly, molecules must be hydrophobic enough to reach the core of biological membranes where the target sits, and be hydrophilic enough to circulate in the medium without precipitating: the library of compounds should therefore be selected with caution to avoid the screening of mostly hydrophilic molecules. The assay should also be developed so that the target is inhibited in a membrane environment, i.e., membrane vesicles or liposomes. Secondly, the effect should be measured at the level of the plant phenotype. In contrast with other biological processes occurring within the cell, like the emission of a signal or the dynamic modification of a polymer of the cytoskeleton, the turnover of membrane lipids is very slow: long-term effects are

Glenn R. Hicks and Stéphanie Robert (eds.), *Plant Chemical Genomics: Methods and Protocols*, Methods in Molecular Biology, vol. 1056, DOI 10.1007/978-1-62703-592-7_8,

therefore expected and plants should be analyzed after days of treatment. A secondary screen should therefore be performed in planta to evaluate if the molecule can be absorbed by the root, circulate in the xylem, and have an effect after 1–2 weeks. Here, we detail the primary in vitro screening of inhibitors of *Arabidopsis* UDP-galactose:1,2-*sn*-diacylglycerol galactosyltransferase 1 (E.C. 2.4.1.46) also called MGD1, based on an assay specifically developed in liposomes [1, 2]. This enzyme catalyzes the following reaction in the inner membrane of the envelope of chloroplasts:

1,2-*sn*-diacylglycerol + UDP-galactose → monogalactosyldiacylglycerol + UDP

MGD1 has two substrates. The first one is a hydrophilic nucleotide-sugar, for which numerous nonspecific competitors are known, like UDP. The other substrate is diacylglycerol, for which few competitors have been described, like ceramides or phorbol-esters, which are inactive when tested on MGD1. Monogalactosyldiacylglycerol (MGDG) is the major galactolipid of chloroplast membranes, including photosynthetic membranes. This lipid interacts with photosystems and other plastid proteins. It is the precursor of digalacosyldiacylglycerol (DGDG), which is exported to different compartments of plant cells during phosphate deprivation, acting as a surrogate for phosphatidylcholine. MGDG is also a donor of unsaturated fatty acids for the production of oxylipins including jasmonic acids in the response of plants to various stresses. MGD1 is activated by phosphatidic acid [3, 4], generated in response to various changes of the environment, like a depletion of phosphate in the soil, and is therefore considered as a key enzyme coupling phospholipid syntheses in endomembranes and galactolipid syntheses in the plastids. For all these reasons, the availability of a specific inhibitor of MGD1 would be useful for chemical genetic studies and help advancing our knowledge of the complex glycerolipid system in various physiological contexts [5]. This procedure has been successfully used to identify galvestine-1 [2].

2 Materials

The complete screening consists in four steps that should be performed independently, with a necessary validation experiment at the end of each step. The first step consists in the preparation of bacteria (*Escherichia coli*) expressing sufficient proportions of the screening target, i.e., the monogalactosyldiacylglycerol 1 (MGD1) enzyme from *Arabidopsis thaliana*. It should be validated by a SDS Polyacrylamide gel and an enzymatic assay. The second step consists in the enrichment of the bacteria membranes containing MGD1 in diacylglycerol, one of the two substrates of the enzyme. It should be validated by a kinetic study of the enzymatic catalysis showing that diacylglycerol is in excess compared to MGD1.

The third step is the automatic screening of a library of compounds in order to identify inhibitors. It should be validated, firstly by the statistical evaluation of the screen based on the *Z'* value, and secondly by an estimate of IC50s measured in same conditions as those used for the screen, thus confirming the level of the inhibitory activity of the selected compounds. At this stage, inhibitors acting in vitro (=in membrano) are thus selected. The last step consists of the screening of MGD1 inhibitors acting at the whole plant level (=in planta). It should be validated by a phenotypic analysis of the treated plant, and in particular the lipidomic profile showing a change in MGDG content. This last step could be ignored if one is uniquely interested in the development of inhibitors acting in vitro.

2.1 Laboratory Environment, Specific Materials, and Instruments

1. Laboratory environment and specific instruments: Molecular biology and bacterial growth require a microbiology laboratory environment. A screening platform complying with the automatic manipulation of 96-well microplates is necessary, including devices for the pipetting of samples, the preparation of reaction mixtures and molecules, incubations at room temperature, washing of media, and the filtration of microplates with filter-bottom wells. Such an automatic platform can be assembled with modules purchased from Tecan, Packard, or other suppliers. The screening is based on a radioactivity assay. A specific laboratory environment and practices complying with radioactivity manipulation are requested. A scintillation counter (Beckman, PerkinElmer, or equivalent) is necessary. *Arabidopsis* growth on sterile agar medium requires a microbiology workstation for sterile manipulations and greenhouses or growth chambers (white light 130 $\mu E/m^2/s$; 26 °C) for plant cultivation under controlled illumination.
2. Glass tubes, liquid scintillation counting vials and 96-well and 48-well microplates: Glass tubes (5 mL) with hermetic plastic caps are necessary whenever chloroform is used. Scintillation vials are purchased based on the recommendations of the radioactivity counter supplier. Two types of 96-well microplates are required for the screening: classical microplates for the storage of compounds and filter-bottom well microplates (1.2 μm fluoride polyvinylidene, PVDF; Millipore) for the screening assay. 48-well Cellstar-type microplates, with working volumes of at least 1 mL and transparent coverlids, used for *Arabidopsis* growth, are purchased from Greiner or other supplier.

2.2 Buffers, Detergents, Organic Solvents, Enzyme Substrates, Control Inhibitors

1. Vehicle for screened compounds: All compounds are solubilized in dimethylsulfoxide (DMSO), called therefore the "vehicle" for tested compounds (*see* **Note 1**). The same volume of DMSO shall be added in "vehicle control" experiments. For in vivo (=in planta) screening, Tween-20 and/or ethanol can be used as vehicles (*see* **Note 2**).

2. Components for buffers, detergent solutions, enzyme substrates, control inhibitors and solvents: 3-(*N*-morpholino) propanesulfonic acid (MOPS), dithiothreitol (DTT), 3[(3-cholamidopropyl)-dimethylammonio]-1-propanesulfonate (CHAPS), solvents (glycerol, chloroform, and methanol), lipids (phosphatidylglycerol), substrates (*sn-1*,2-dioleoyl-glycerol, DAG, and uridine diphospho galactose, UDP-Galactose), unspecific inhibitors (uridine diphosphate, UDP and *N*-ethylmaleimide, NEM) are purchased from Sigma-Aldrich or another supplier (*see* **Note 3**). Radiolabelled [^{14}C]-UDP-galactose (11 GBq/mmol) is purchased from Dupont–New England Nuclear, PerkinElmer Life Sciences, or another supplier (*see* **Note 4**).
3. Preparation of concentrated detergent: The Critical Micellar Concentration of CHAPS (CMC) being around 3–5 mM, it should be provided in excess to help mix enzymes, added diacylglycerol and other lipids. Using stock solutions of detergent and buffers, prepare 1 mL of CHAPS 85 mM, 0.7 M MOPS, pH 7.8 and 14 mM DTT. Store at −20 °C until use (*see* **Note 5**).
4. Preparation of UDP-Galactose substrate solution: The UDP-Galactose substrate is provided as a mixture of unlabelled and radioactive substrates, mixed so that the specific radioactivity is low-enough to avoid excessive use of radioactivity and sufficient to allow a measurable transfer of galactose from the UDP-Galactose substrate to the diacylglycerol substrate. Ten milligrams of unlabelled UDP-galactose is dissolved in 820 μL MOPS 50 mM, pH 7.8, and then mixed with 820 μL de [^{14}C]-UDP-galactose. A 0.5 μL fraction is transferred into a scintillation vial, completed with 10 mL scintillation for the calculation of the specific activity in dpm/μmol or Bq/μmol. The solution is stored at −20 °C until use.
5. Preparation of control inhibitors' solutions: 23.6 mg UDP is mixed with 655 μL H_2O to obtain a solution of 9×10^{-2} M. 4.64 mg NEM is mixed with 1,237 μL H_2O to obtain a solution of 3×10^{-2} M. Solutions are stored at −20 °C until use. Desired final concentrations of inhibitors and DMSO are obtained by dilution with water.

2.3 Vectors, Biological Strains and Components for the Functional Expression of Arabidopsis thaliana MGD1 in Escherichia coli Membranes

1. Vector and bacteria strain: The vector used for the expression of MGD1 has been described previously [6] and is available upon request. The sequence corresponds to the portion of the *MGD1* gene coding for the mature form of the protein (i.e., without the 105-amino acid transit peptide), cloned in the Nde1-BamH1 site of the pET-Y3a plasmid (*see* **Note 6**). *Escherichia coli* strain used for MGD1 expression is BL21 (DE3) pLys. This strain allows the expression of genes under the control of a T7 promoter. This strain also contains the

pLysS plasmid, carrying the T7 lysozyme gene, lowering the background expression level of target genes. Induction of expression is induced by isopropyl-β-d-thiogalactopyranoside (IPTG).

2. Components for bacteria cultures and induction of recombinant MGD1 expression: Luria Broth (LB), carbenicillin, and IPTG are purchased from Sigma-Aldrich or other supplier.
3. Growth media: All media used for bacteria growth are sterilized by autoclaving and stored at room temperature. LB liquid medium is prepared by solubilizing 10 g Bacto-tryptone, 5 g yeast extract, and 10 g NaCl in 800 mL deionized H_2O, followed by an adjustment of pH to 7.5 with NaOH and the addition of deionized H_2O to reach a final volume of 1 L. LB-agar is prepared by adding 15 g of agar powder in 1 L LB liquid medium. It is sterilized by autoclaving, and poured into Petri dish plates, following conventional procedures for microbiology.
4. IPTG 1 M solution: Dissolve 2.38 g of IPTG in 10 mL (final volume) deionized H_2O, filter-sterilize with a 0.22 μm syringe filter, and store in 1 mL aliquots at −20 °C until use.

2.4 Components for the Preparation of Liposomes Sequestering MGD1 and Its Lipophilic Diacylglycerol Substrate

1. Enzyme for DAG enrichment of liposomes: *Bacillus cereus* phospholipase C (PLC) Grade I, 2,000 U/mL is purchased from Sigma-Aldrich (*see* **Note 7**). The enzyme stock solution in 3.2 M ammonium sulfate is stored at -20 °C (*see* **Note 8**). 100 μL aliquot fractions are stored at 4 °C until use.
2. Components for liposome treatment and purification: Ethylenediaminetetraacetic acid (EDTA) and Percoll are purchased from Sigma-Aldrich or another supplier.
3. Argon stream: The quality control consists in an assay of the activity of MGD1. This assay is based on the transfer of galactose from a radiolabelled UDP-galactose donor to a DAG acceptor, forming a radiolabelled MGDG which is extracted by solvents and dried under a stream of Argon (*see* **Note 9**). This evaporation step should be performed under a fume hood.

2.5 Components for the High Throughput Screening of MGD1 Inhibitors Acting In Membrano

1. Library of small molecules, formatted in 96-well microplates: The library to be screened consists of a selected series of molecules dissolved in 100 % DMSO, in order to have a final concentration of 10 μM of compounds in reaction medium, and a final proportion of 3 % DMSO. Each molecule solution is dispensed in a well of a 96-well microplate, with the help of a Tecan workstation, or any other multi-pipetting automatic device.
2. Liposome suspension, reaction buffers and washing media: The screening is based on large numbers of assays run in parallel. All media should therefore be prepared in batches so that

the complete screening can be performed in a single large-scale experiment, using a unique source of materials. The total volumes of all samples and buffers should be calculated based on the volumes used in one assay (one well), adding a 10–20 % extra for pipetting errors and void volumes.

2.6 Biological Material and Components for the Secondary Screening of MGD1 Inhibitors Acting In Planta

1. *Arabidopsis* seeds: *Arabidopsis* seeds used for the secondary screen are obtained from a homogenous batch, corresponding to a wild type genotype of the *Columbia* ecotype (*see* **Note 10**). The seeds are stored between 10 and 15 °C in a drying room or cabinet. Germination must be tested prior to utilization and seeds should be used only if the germination is >90 %.
2. Components for *Arabidopsis* growth in sterile conditions: Murashige and Skoog (MS) powder is purchased from Sigma and stored at room temperature (*see* **Note 11**). DMSO or Tween-20 used to dissolve candidate inhibitors are purchased from Sigma-Aldrich or another supplier.
3. Solid MS medium: 2.2 g of Murashige and Skoog (MS) powder, 2.5 g sucrose, and 7.5 g agar are dissolved in 500 mL deionized water and sterilized by autoclaving.

3 Methods

3.1 Recombinant Expression of an Active and Membrane Embedded form of Arabidopsis thaliana MGD1 in Escherichia coli

1. Induction of recombinant MGD1 expression in *Escherichia coli*: Streak BL21 (DE3) *Escherichia coli*, previously transformed with the pET-Y3a plasmid containing the *Arabidopsis MGD1* sequence, on LB agar plates containing 100 μg/mL carbenicillin. Incubate plates overnight at 37 °C. Only recombinant bacteria can grow in presence of the antibiotic. Pick up a single colony of *E. coli*, inoculate a 10 mL LB medium in presence of 100 μg/mL carbenicillin and grow cells overnight with shaking at 37 °C. 5 mL of this primary culture are transferred in 400 mL LB medium and grown at 37 °C with regular control of the absorbance at 600 nm (A_{600}). When A_{600} reaches 0.4–0.5, induce the expression of MGD1 by adding 0.4 mM IPTG, and incubate for 3 h at 28 °C with vigorous shaking (*see* **Note 12**).
2. Harvesting and storage of bacteria expressing MGD1: After induction by IPTG, transfer culture vials in ice and pellet bacteria by centrifugation for 15 min at 6,000 ×*g*. Discard the supernatant. Up to 45 mL of induced bacteria can be harvested. Transfer 100 μL of the pellet in a 1.5 mL microtube for control analyses (analyses of the MGD1 induction by SDS PAGE, using 10–20 μL of the bacteria pellet, and of the MGD1 activity, using 1–5 μL of the bacteria pellet) and store this aliquot fraction at -20 °C until use. Immediately store the harvested bacteria at −80 °C until use.

3.2 Preparation of Liposomes Containing Both MGD1 and Its Lipophilic Diacylglycerol Substrate

1. Enrichment of bacteria membranes with diacylglycerol (*see* **Note 13**): Mix one volume (45 mL) of induced bacteria with an equal volume (45 mL) of MOPS 10 mM, pH 7.8, glycerol 10 %, DTT 1 mM. Add 16 U of phospholipase C from *Bacillus cereus* per mL of reaction buffer and hydrolyze bacterial phospholipids, mainly phosphatidylethanolamine, by incubation for 3 h, at room temperature, under slow horizontal rotation (Glas-Col tissue culture rotator, 5 rpm). PLC reaction is stopped by transfer on ice and addition of 10 mM EDTA. The suspension is then stored at 80 °C until use.
2. Preparation and purification of liposomes containing both MGD1 and diacylglycerol: Harvest the DAG-enriched liposomes by centrifugation at 15,000 × *g* for 15 min, dilute the pellet with the same volume of MOPS 10 mM, pH 7.8, glycerol 10 %, DTT 1 mM and pass the obtained suspension through a chilled pressure cell (French press; SLM Instruments) at 85 MPa. PLC-treated bacteria are thus broken, leading to the production of inside-out liposomes enriched in active MGD1 and diacylglycerol, mixed with broken cell materials and aggregates of unfolded and inactive MGD1 polypeptides in inclusion bodies. To purify liposomes from bacterial contaminants and high-density inclusion bodies, the suspension is divided in 30-mL fractions, loaded gently on top of 35 % Percoll cushions (45 mL of 35 % Percoll in the same buffer) and centrifuged at 6,000 × *g* for 15 min. After centrifugation, the purified liposomes make a broad band at the upper surface of the Percoll cushion. Collect this fraction with a pipette, and dilute it ten times in MOPS 10 mM, pH 7.8. Wash the liposomes by centrifugation for 15 min, at 6,000 × *g*. Remove the supernatant and resuspend the pellet in 13 mL MOPS 10 mM, pH 7.8. An aliquot fraction of 100 μL is collected for determination of the protein content using a standard protein assay procedure, and for controlling the MGDG synthase activity in quality control experiments. The suspended liposomes are stored at −80 °C until use.
3. Quality validation of liposome samples before utilization in large-scale screening: The liposomes should contain sufficient proportions of diacylglycerol so that the Michaelis–Menten conditions are filled, i.e., an excess of substrate compared to the enzyme. Upon addition of a non-limiting concentration of UDP-galactose, the measured production of MGDG should therefore vary linearly with the quantity of sample, reflecting that *the enzyme is the limiting factor* of the system. Prepare enough 5-mL glass tubes to test the activity of MGDG synthesis without any control inhibitors, and in presence of 25 μM, 50 μM, 100 μM UDP and 50 μM NEM in duplicate (=10 tubes in total). A fraction of liposomes corresponding to 50 μg of

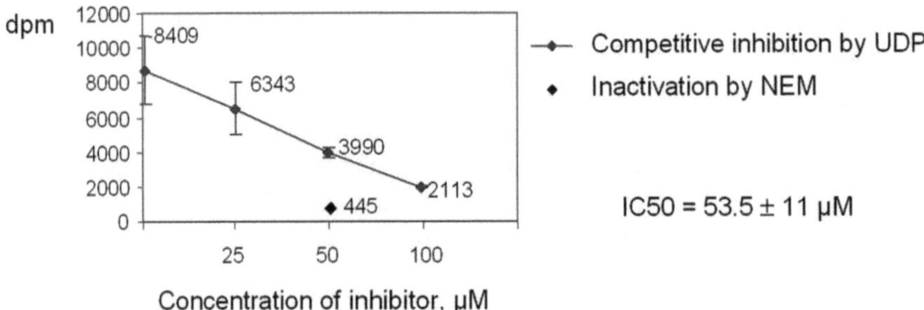

Fig. 1 Example of an evaluation of the quality of liposomes containing both MGD1 and DAG, prepared for a subsequent large-scale screening. Note that UDP is a competitive inhibitor docking to the UDP-galactose binding site, whereas NEM is a specific reagent of cysteine residues, inactivating MGD1 by the formation of irreversible covalent bonds with sulfhydryl groups. An IC50 is measured with UDP

proteins (around 3 μL of the liposome suspension) is transferred inside each glass tube and completed stepwise (*see* **Note 14**) with 75 μL KCl 1 M, 75 μL KH_2PO_4 1 M, pH 7.8, 11 μL of concentrated detergent, a volume of UDP or NEM corresponding to the desired final concentration, and a volume of water so that the final volume is 300 μL. Prepare a timer to monitor the reaction time. Reaction is started ($t=0$) by addition of 10 μL d'UDP-galactose (10 mM, 37 Bq/μmol). Reaction is stopped after 20 min by addition of 1.5 mL of a mixture of chloroform–methanol (1:2, v/v) (*see* **Note 15**). A monophasic solution is obtained. After 10 min of incubation, 0.5 mL of chloroform and 0.6 mL of H_2O are added. Tubes are capped, mixed using a vortex, and decanted until a clear biphasic system is obtained (*see* **Note 16**). The lower phase is collected using a Pasteur pipette, transferred into a scintillation vial and dried under Argon. Dried lipids are then suspended in 10 mL scintillation medium and radioactivity is measured using a scintillation counter following the manufacturer instruction for recording ^{14}C radioactivity. Up to 8,000–10,000 dpm (disintegrations per minute) can be measured without any inhibitor, corresponding to 12–15 nmol galactose transferred from UDP-Gal to DAG. Identical tests performed in presence of 25, 50 or 100 μM UDP or of 50 μM NEM must show an inhibition of the incorporation of radioactivity in MGDG. An example of a quality evaluation (and validation) is given in Fig. 1.

3.3 High Throughput Screening of MGD1 Inhibitors Acting In Membrano

1. Preparation of liposome batch: The quantity of liposome fraction to perform the complete screening must be calculated and the procedure for the preparation of liposomes should be repeated if necessary. Quality controls should be performed for each preparation of liposomes. All suspensions should then be mixed and homogenized prior to the screening.

The stock solution of liposomes should be adjusted to 1–10 μg protein/μL in MOPS 50 mM, pH 7.8, glycerol 50 %. Homogenization is carried out using a potter to avoid the formation of aggregates.

2. Protocol (per well): The reaction volume per point is 300 μL + 10 μL of UDP-galactose when starting the enzymatic reaction. Preincubate the filter-bottom microplate with 100 μL water for 10 min and filtrate the water out using a Packard 96-well plate filtration device. Program the platform workstation so that the following components are subsequently added in each well: 75 μL KCl 1 M, 75 μL KH_2PO_4 1 M, 9 μL of compounds in DMSO, 91 μL H_2O, and 50 μL liposomes 1 μg protein/μL. Incubate for 20 min at 22 °C (*see* **Note 17**). Add 10 μL UDP-galactose 10 mM, 2.64 Bq/μmol. Incubate for 20 min at 22 °C. Filtrate the well using a Packard 96-well plate filtration device and discard filtered liquids in an appropriate radioactivity waste. Wash three times by filtering with 300 μL water and discard filtered liquids in an appropriate radioactivity waste. Dry plates and add 50 μL Microscint 0 in each well. Count the radioactivity in each well using a TopCount from Packard or an equivalent device. Plot results using a classical spreadsheet or specific software.

3. Microplate design for screening: Microplates are designed so that the tested library is compared with the effect of the control vehicle (DMSO) and control inhibitors, (i.e., UDP and NEM), as shown in Fig. 2 (*see* **Note 18**).

4. Statistical validity of the screening: A Z' factor for each plate is calculated based on the formula: 1-((standard deviation positive sample values + standard deviation negative sample values)/(mean positive sample values − mean negative sample values)). This Z' factor is used for assessing the validity of an assay. When it scores between 0.5 and 1, it indicates an excellent screening assay. The average for Z' factor for the MGD1 screen should be higher than 0.7.

5. Microplate design for in vitro IC50 evaluation: The determination of the IC50 depends on the effect of the compounds measured in the primary screening at 10 μM. Plates should be designed to allow an identical measure of the effect on liposomes, with a range of concentrations of selected compounds from the nanomolar range to the millimolar range if necessary, keeping a 3 % final proportion of DMSO. Caution should be taken to avoid compound precipitation at high concentrations. IC50s should be determined in triplicates. In these conditions IC50s allow an evaluation of the performances of the inhibitors in membrano. Based on the quality of the screened library, expected IC50s should be in the 1–100 μM range.

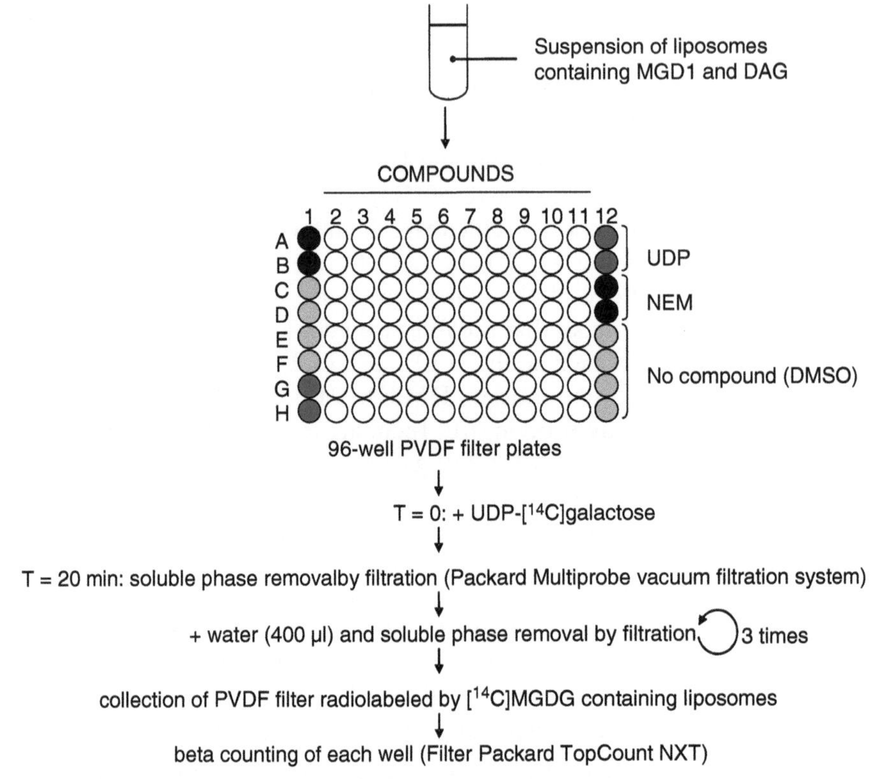

Fig. 2 Design of the microplate for the screening. The screen is performed in suspensions of *Escherichia coli* (BL21 strain) membrane liposomes, containing the mature form of MGD1 and enriched in diacylglycerol (DAG) by treatment with non specific PLC from *Bacillus cereus*. A homogenized suspension of liposomes is used for the screening (1–10 mg protein/mL). Each plate is designed to test 80 compounds (10 μM; columns 2–11). Positions A1, B1, C12, and D12 contain 10 mM *N*-ethylmaleimide (NEM), a cysteine reagent inactivating MGD1 by covalent bond formation. Positions G1, H1, A12, and B12 contain 3 mM UDP, a product of MGDG synthesis known to inhibit the enzyme. Positions C1, D1, E1, F1, E12, F12, G12, and H12 are used as vehicle controls with no compound. Reaction is started by addition of UDP-[^{14}C]galactose (0.35 mM; 2.64 Bq/μmol) and stopped by filtering out the soluble phase (Packard Multiprobe vacuum filtration system). The filters containing labelled membranes are washed three times with water (400 μL) and radioactivity is determined by beta counting (Packard Topcount NXT)

3.4 Secondary Screening of MGD1 Inhibitors Acting In Planta

1. Preparation of mini-greenhouses: The 24 central wells of 48-well plates (Cellstar) are filled with 400 μL of solid agar medium (Murashige and Skoog growth medium; sucrose 0.5 % m/v; dimethylsulfoxide (DMSO) 1 %; agar 0.8 %; pH 5.7) and variable concentrations of added compounds (*see* **Note 19**). Wells at the periphery of the plate are filled with 500 μL sterile water to ensure a humid atmosphere within the plate. The plates are prepared in advance in a sterile microbiology workstation.
2. Protocol (per well): Prior to screen, seeds of *Arabidopsis thaliana*, ecotype *Columbia*, are sterilized by incubation in 2 mL of Barychlore 10 %, v/v, Triton-X100 0.5 %, w/v in ethanol 90 %.

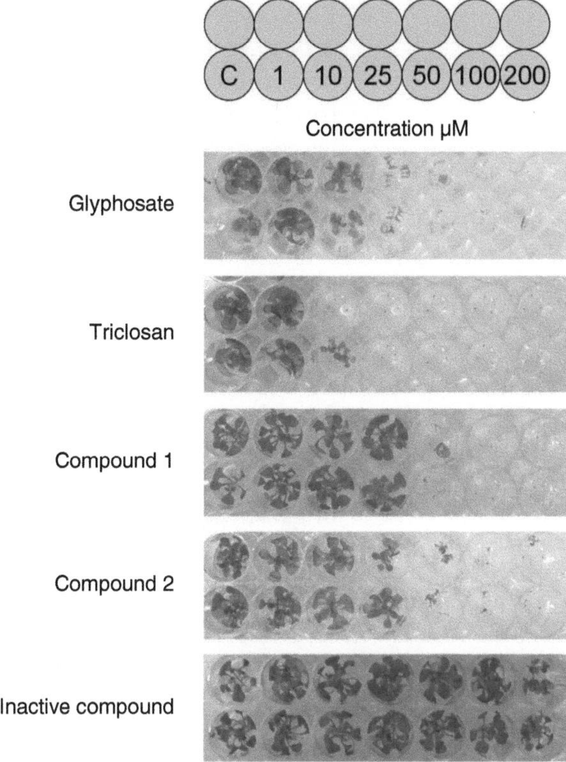

Fig. 3 Design of 48-cell microplates for secondary screening. Compounds are tested at concentrations varying from 0 to 500 μM. Three seeds of *Arabidopsis* are sown in each well and growth is measured after 15–20 days. Positive controls include herbicide (glyphosate) or antibiotic (triclosan) molecules

They are then sown in the 24 central wells of 48-well plates (Cellstar), containing 400 μL of solid agar medium and the desired concentration of the screened compound. Three seeds are sown per well. Adjust the coverlid, wrap the microplate in foil and incubate at 4 °C. After 2 days of stratification at 4 °C in order to stimulate and synchronize seed germination, foil is removed and the plate is placed in a growth chamber (white light 130 μE/m^2/s; 16 h photoperiod; 26 °C).

3. Design of the plates and readout: An herbicide is a compound that can affect the germination, slow down the growth at low concentration, and being rapidly lethal at high doses. The cultivation of *Arabidopsis* seeds in sterile conditions in Cellstar 48-well microplates is performed with a range of concentrations of compounds from 0 to 500 μM and 10 % DMSO (Fig. 3). Positive controls include herbicides, e.g., glyphosate (*see* **Note 20**) or antibiotics, e.g., triclosan (*see* **Note 21**). Negative controls include the vehicle alone or any compound that does not impair *Arabidopsis* growth.

4. Phenotype analyses: Phenotypes are analyzed at different scales. At lethal doses, germination can simply be inhibited, or germination can be initiated and stopped after the first root has broken the seed teguments. At lower doses, a plantlet can develop and die within days or weeks. Growth, level of development and lethality are recorded and compared to positive and negative controls. A convention is defined to assess an IC50 for plant development, based on the observed phenotype, e.g., the concentration of compound corresponding to a plant with half the leaf size or rosette diameter, or number of fully developed leaves, compared to untreated plants. The IC50 should be determined in the first 2 weeks after germination of control plants. In the case of MGDG inhibition, phenotype also includes a chlorosis of leaves, a decrease in chlorophyll content, a specific decrease of galactolipids, a decrease in chloroplast size, and a decrease of thylakoid stacks within chloroplasts. Any of these features can be determined using appropriate biochemical, lipidomic, or bioimaging methods.

4 Notes

1. A fresh bottle of DMSO is used to suspend molecules. Whereas all media should be sterilized using an autoclave in all experiments, tested compounds should never be heated. Storage of compounds in pure DMSO should prevent most contamination by bacteria that could interfere with bioactivity assays. When diluting this solution in an aqueous buffer, sterilization should be achieved by filtration on a 0.22 μm Millipore sterile filter, using a disposable sterile syringe of appropriate volume.
2. DMSO might not allow the accurate solubilization of compounds and it might also activate phospholipases D (PLDs) in planta, interfering with the evaluation of the bioactivity of tested compounds. Tween-20 and/or ethanol might be used as alternative.
3. NEM is sensitive to light and should be stored wrapped in foil.
4. In the absence of any fluorescent-based assay for the activity of galactosyltransferases, the MGD1 enzymatic assay relies on the use of radiolabelled substrates. Here, we suggest the utilization of [^{14}C]-labelled galactose, transferred from the UDP-galactose donor to the diacylglycerol acceptor, but other labelling could be used, like [^{3}H]-labelled galactose. The labelling of the diacylglycerol substrate would require the development of a different assay, since in the presented

conditions, the monogalactosyldiacylglycerol product is separated only from the UDP-galactose substrate and not from diacylglycerol.

5. This solution should be rapidly stored at −20 °C after each utilization to prevent the degradation of DTT.
6. The pET-Y3a plasmid is a modified version of the pET-3a plasmid developed by Novagen, comprising the ArgU rare codon gene, allowing a higher level of expression of plant genes in *E. coli*. Any other plasmid/bacteria system optimized for the expression of plant sequences, or synthetic gene with optimized codons can be used as an alternative.
7. The PLC is used to hydrolyze the phosphoglycerolipids of the *Escherichia coli* membranes, in which the recombinant MGD1 is embedded, thus enriching the system in DAG. It is therefore essential to select a nonspecific PLC or a PLC with the broadest spectrum of substrates and avoid those that are specific of a lipid class like phosphoinositide PLCs. Because the production of enzymes by commercial suppliers is sometimes interrupted, it is advised to test the enzyme prior to large-scale use, and ascertain that at least phosphatidylethanolamine is hydrolyzed.
8. When stored at 4 °C, a decrease in PLC activity of approximately 10 % occurs within 6 months.
9. This enzymatic assay based on a solvent extraction of monogalactosyldiacylglycerol produced by the action of MGD1 cannot be used for the high throughput screening, which prevents the use of solvents.
10. Other ecotypes or genetic backgrounds can be used. Comparison with Columbia seed run in parallel should be performed for comparisons of results.
11. When a MS vial is unsealed, the powder hydrates rapidly and should therefore be used within 10 days.
12. MGD1 is a membrane protein. Part of it is therefore incorporated in the membrane of the bacteria, whereas another portion accumulates in inactive form as inclusion bodies. In the bacterial membrane, MGD1 leads to the synthesis of monogalactosyldiacylglycerol, a lipid usually absent from *E. coli*. The bacterial cells elongate and the division is altered. Growth of the bacteria, as measured by OD600, is therefore slowed down. It is therefore expected that the induction by IPTG triggers an arrest of the growth when MGD1 is produced in an active form.
13. Bacteria membranes are naturally poor in diacylglycerol, but are rich in phosphoglycerolipids, phosphatidylethanolamine being the most abundant. Since the assay requires that

Michaelis–Menten conditions are filled, both substrates of the enzyme, i.e., diacylglycerol and UDP-galactose should be in excess compared to the enzyme itself. UDP-galactose will be added to start the reaction. Exogenous diacylglycerol cannot be provided without adding detergent for its proper solubilization in the medium. Treatment of bacterial membranes with a nonspecific phospholipase C (from *Bacillus cereus*) allows the hydrolysis of endogenous phospholipids that are then converted into diacylglycerol, within the membrane in which MGD1 is embedded, and therefore at the vicinity of the enzyme active site.

14. The stepwise addition of components is important to ensure that the salts, phosphate buffer and concentrated detergent break the liposome structure, allowing the solubilization of lipids, prior to their extraction by solvents.
15. In these proportions of water, chloroform, and methanol, the mixture is monophasic. This unique phase helps therefore the mixture of the organic solvent with liposomes, and the dissolution of lipids. The addition of chloroform will induce the separation of the organic phase from the water phase, and the subsequent purification of lipids in the lower phase.
16. Decantation can be improved by centrifugation for 10 min at low speed (700–1,000 rpm), with cautious adjustment of caps on top of the glass-tubes, to avoid leaks of radioactive liquid.
17. The incubation allows liposomes to be stabilized and homogenized in the reaction medium. The incorporation of hydrophobic inhibitors within the liposome membranes at the vicinity of the MGD1 active site is therefore made possible.
18. The positions of control vehicle (DMSO) and inhibitors (UDP and NEM) on the right and left sides of the plate are changed as shown in the Figure to avoid or limit positional biases within plates.
19. Mix compounds in 1–2 mL MS-agar in a sterile plastic tube and then dispense 400 μL in the desired well of the microplate. Because compounds might be temperature-sensitive, compounds should be mixed in MS-agar between 50 and 60 °C and rapidly dispensed in the microplate.
20. Glyphosate, (*N*-(phosphonométhyl)glycine), is an inhibitor of 5-enolpyruvoyl-shikimate-3-phosphate synthase (EPSPS). It is a widely used herbicide and can serve as a positive control for toxicity.
21. Triclosan (5-chloro-2-(2,4-dichlorophenoxy)phenol) is an antibiotic with a broad spectrum of targets including the plastid fatty acid synthesis. This molecule can serve as a generic control for the impairment of acyl-lipid metabolism.

Acknowledgements

The authors wish to thank Stéphane Miras, Hélène Hardré, Amélie Zoppé, Aymeric Roccia, and Maryse Block for some technical developments. E.M. is supported by ANR-05-EMPB-017-01, ANR-06-MDCA-014, and ANR-10-BLAN-1524-01 (ReGal) grants from Agence Nationale de la Recherche, and L.B. is supported by Conseil Régional Rhône-Alpes (PhD program, Cluster 9).

References

1. Nishiyama Y, Hardré-Liénard H, Miras S et al (2003) Refolding from denatured inclusion bodies, purification to homogeneity and simplified assay of MGDG synthases from land plants. Protein Expr Purif 31:79–87
2. Botté C, Deligny M, Roccia A et al (2011) Chemical inhibitors of monogalactosyldiacylglycerol synthases in Arabidopsis thaliana. Nat Chem Biol 7:834–842
3. Dubots E, Audry M, Yamaryo Y et al (2010) Activation by phosphatidic acid of the MGD1 monogalactosyldiacylglycerol synthase in leaves. J Biol Chem 285:6003–6011
4. Dubots E, Botté C, Boudière L et al (2012) Role of phosphatidic acid in plant galactolipid synthesis. Biochimie 94:86–93
5. Boudière L, Botté CY, Saidani N et al (2012) Galvestine-1, a novel chemical probe for the study of the glycerolipid homeostasis system in plant cells. Mol BioSyst. doi:10.1039/C2MB25067E
6. Awai K, Maréchal E, Block MA et al (2001) Two types of MGDG synthase genes, found widely in both 16:3 and 18:3 plants, differentially mediate galactolipid syntheses in photosynthetic and nonphotosynthetic tissues in Arabidopsis thaliana. Proc Natl Acad Sci U S A 98:10960–10965

Chapter 9

Forward Chemical Screening of Small RNA Pathways

Yifan Lii and Hailing Jin

Abstract

RNA silencing is a mechanism of gene expression regulation mediated by short noncoding RNAs called small RNAs. Small RNAs can suppress gene expression transcriptionally or posttranscriptionally by base pairing to their targets. In plants, they are involved in a diverse range of pathways and processes. Forward genetic screens have led to the identification of many components in small RNA biogenesis and functional pathways. However, it cannot identify essential or functionally redundant genes. Forward chemical screens can overcome these limitations by targeting multiple redundant proteins within a family and by giving the user temporal control of the application of the chemical. Here, we describe a method to quickly screen chemicals that perturb small RNA pathways using *Arabidopsis thaliana* reporter lines in a 96-well format.

Key words Small RNAs, Gene silencing, *Arabidopsis thaliana*

1 Introduction

Small RNAs (sRNAs) are short, 20–30 nucleotide, noncoding RNAs that regulate gene expression by base pairing to their targets. They can mediate transcriptional gene silencing (by directing cytosine methylation and histone modifications) or posttranscriptional gene silencing (by directing mRNA degradation and/or translational inhibition) [1]. In plants, small RNAs are involved in a range of pathways and processes including development, genome maintenance, genomic imprinting, and biotic and abiotic stress responses [2–7].

There are two categories of plant sRNAs: microRNAs (miRNAs) and short-interfering RNAs (siRNAs). Although their precursors and biogenesis pathways differ, they share many commonalities. Biogenesis begins at the formation of double-stranded RNA (dsRNA) that can be derived from endogenous transcripts producing stem-loop structures, inverted repeats, natural antisense transcripts, RNA-dependent RNA polymerase-produced transcripts and viral RNA transcripts. RNase-III ribonuclease

Glenn R. Hicks and Stéphanie Robert (eds.), *Plant Chemical Genomics: Methods and Protocols*, Methods in Molecular Biology, vol. 1056, DOI 10.1007/978-1-62703-592-7_9,

Dicer-like (DCL) proteins dice dsRNA into short transcripts. After dicing, Hua Enhancerl (HEN1) methylates the ends of sRNAs, thereby stabilizing them. Some sRNAs are retained in the nucleus for chromatin modification involving Argonaute (AGO) proteins and DNA-dependent plant specific RNA Polymerase IV (Pol IV) and Pol V, while others are exported by Hasty into the cytoplasm for posttranscriptional gene silencing. In the latter case, sRNA is loaded onto AGO, which is the major component of the RNA-induced silencing complex (RISC). The sRNA guides RISC to its complementary target mRNA, leading to the silencing of the gene through mRNA cleavage or translation inhibition. *Arabidopsis thaliana* has four DCL proteins (DCL1-4) and ten AGO proteins (AGO1-10), each functioning in a specific pathway but some are also able to act redundantly in place of another [1].

Traditionally, forward genetic screens are used to identify components in silencing pathways. However, genetic screens are limited because of loss-of-function lethality or the lack of phenotype due to gene redundancy. To circumvent these limitations and complement genetic screens, chemicals can be used to perturb gene functions. Chemicals can be added after seed germination to prevent embryo lethality. Also, chemicals can be removed to see if their effects are permanent. Chemicals may be able to target the conserved domains of all proteins of a family with redundant function [8, 9]. We present a general forward chemical screening method to discover new components involved in the biogenesis or function of a sRNA using *Arabidopsis thaliana* reporter lines. Luciferase and green fluorescent protein (GFP) are described here, but other reporter genes can also be used. The reporter gene is fused to the known target of the miRNA or siRNA of interest and a constitutive promoter. Under normal conditions, the reporter is silenced by the miRNA or siRNA, but a small compound that perturbs a component of the silencing pathway will result in the recovery of reporter expression. To quickly screen for recovery of reporter expression using a small compound library, we use a 96-well format with one chemical per well. Chemicals perturbing silencing may target known or novel proteins involved in sRNA pathways.

2 Materials

2.1 Reagents and Equipment

1. Bleach.
2. Triton X-100.
3. *Arabidopsis thaliana* reporter line.
4. Potassium hydroxide (KOH).

5. Murashige and Skoog basal salt mixture (MS).
6. Gamborg B5 Vitamin Mix (bio-WORLD).
7. Sucrose.
8. 2-(*N*-morpholino)ethanesulfonic acid (MES).
9. Granulated agar.
10. Chemical library, 96-well format, stored in −20 °C freezer.
11. Dimethyl sulfoxide (DMSO).
12. D-luciferin, potassium salt.
13. Solution reservoir (sterile).
14. 96-Well culture plates.
15. 1.5 ml microcentrifuge tube.
16. 200 μl pipet tips, cut about 5 mm from tip and autoclaved.
17. Razor blade (sterile).
18. 1 ml disposable plastic transfer pipet.
19. ½ in. 3 M™ Microspore medical tape.
20. 50 ml plastic centrifuge tube.
21. Spray nozzle (short) or Preval spray gun with dip tube.
22. Growth chamber: 12-h light–12-h dark photoperiod, 27 °C ± 1 °C, 80 % humidity.
23. Luminescent dark box imager.
24. Handheld UV lamp (UVP).
25. Digital camera and tripod.

2.2 Solutions

1. Sterilization solution: 40 % bleach, 0.1 % Triton X-100 in sterile ddH_2O.
2. 1,000× Gamborg B5 Vitamin Mix: 11.2 % (w/v) vitamin mix, filter-sterilize. Store at −20 °C.
3. 0.5× MS media: 0.22 % (w/v) MS, 1 % (w/v) sucrose, 2.6 mM MES, pH 5.7; autoclave, once cooled, add 0.5 % (v/v) 1,000× Gamborg B5 Vitamin Mix. Store at room temperature or 4 °C for longer storage.
4. 0.5× MS agar: 0.22 % (w/v) MS, 1 % (w/v) sucrose, 2,6 mM MES, pH 5.7, 12 % (w/v) agar; autoclave, once cooled, add 0.5 % (v/v) 1,000× Gamborg B5 Vitamin Mix. Store at room temperature or 4 °C for longer storage.
5. Agar solution: 0.1 % agar in ddH_2O, autoclave.
6. Luciferin solution: 10 mM D-luciferin in 0.05 % Triton X-100. Should be made fresh.

3 Methods

3.1 Seed Sterilization

1. Aliquot seeds into labeled microcentrifuge tubes.
2. From now on, the tubes should only be opened in a flow hood. Add 1 ml of the sterilization solution.
3. Pulse vortex for 20 s. Make sure that there are no clumps present.
4. Incubate for 5 min.
5. Spin for 20 s at 3,380 × *g*.
6. Decant or pour off the liquid (be careful not to lose any seeds) and add 1 ml sterile ddH_2O.
7. Pulse vortex for 10 s. Make sure that there are no clumps present.
8. Spin for 20 s at 3,380 × *g*.
9. Repeat **steps 5–6** six more times for a total of seven washes. After the last wash, add 1 ml sterile ddH_2O to the tubes and stratify for 2 day at 4 °C.

3.2 Plate Preparation

1. Remove chemical plates from freezer to thaw. Wipe off the ice and water from the surface. Place in a fume hood for 30 min.
2. Spin down the plates in a centrifuge for 60 s at 1,000 rpm (*see* **Note 1**).
3. This step and the following steps should be done only in a laminar flow hood. All materials used should also be pre-sterilized (e.g., tips, media, tubes, reservoir). Using an 8-channel multichannel pipettor, transfer each chemical from the chemical plate to the labeled 96-well plate. The final concentration should be 25 μM (*see* **Note 2**).
4. Add the same volume of DMSO as the chemical to each well in columns 1 and 12 (*see* **Note 3**).
5. Pour 0.5× MS media into a reservoir and add 98 μl into each well (*see* **Note 4**).
6. Melt 0.5× MS agar in the microwave. Pour the agar into a reservoir and add 100 μl to each well (*see* **Note 5**).
7. Leave the plates uncovered in the laminar flow hood until the agar is solidified.
8. After the agar solidifies, prepare the seeds to add to the plates. Carefully pour or pipet out water from the tubes from the last sterilization step. Add 0.1 % agar to the tubes. The agar will help suspend the seeds.

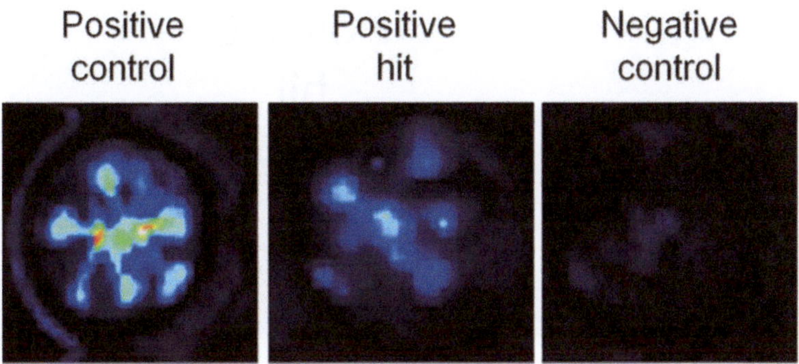

Fig. 1 CCD images of 2-week-old seedlings with a siRNA target site fused to a luciferase reporter gene. The positive control had a genetic mutation in the silencing pathway and was treated with DMSO. The positive hit was treated with a chemical dissolved in DMSO. The negative control was treated with DMSO

9. Using a sterile razor blade, cut off the narrow tip of a plastic transfer pipet. Place a sterile 200 μl pipet tip, cut about 5 mm from the tip, on the end of the transfer pipet. Use this to add 2–3 seeds to each well. Column 1 and 12 can be used for positive or negative controls.
10. Wrap the plates with surgical tape twice around and add ~1 in. pieces to the edges of the plates. This prevents the media from drying out.
11. Place the plates in a growth chamber for 2 weeks.

3.3 Screening for Reporter Expression

3.3.1 Luciferase Screening

1. Make fresh luciferin solution in a 50 ml centrifuge tube. Each plate will need 5 ml solution.
2. In a fume hood, use a spray nozzle or a Preval spray can to spray the plants with the fresh luciferin solution. Coat the plants evenly. Re-cover the plates.
3. Place the plates in the dark for at least 5 min before imaging. Take pictures of the plates in a luminescent dark box imager with an attached charge coupled device (CCD) camera chilled to −70 °C.
4. Putative hits are seedlings that show recovered luciferase expression (Fig. 1).

3.3.2 GFP Screening

1. In a dark room, set up a camera and tripod to take pictures of plates. Preset the camera shutter speed to 8 s.
2. With the lights off, shine UV light onto the plates for 8 s during the exposure time (*see* **Note 6**).
3. Putative hits are seedlings that show recovered GFP expression (Fig. 2).

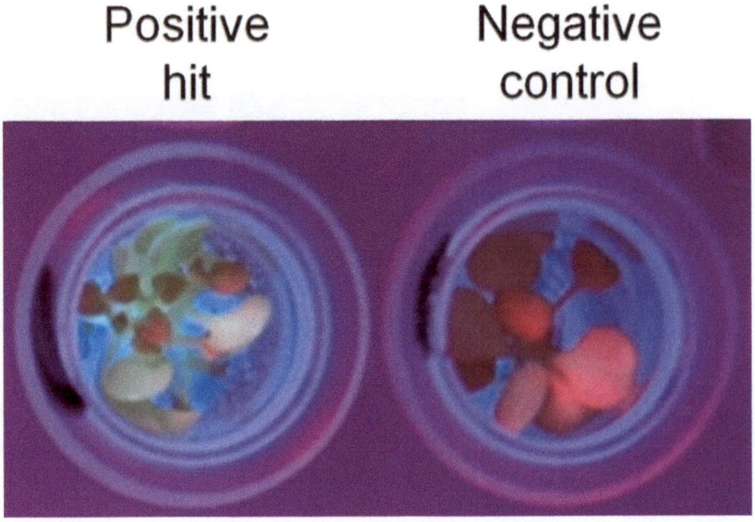

Fig. 2 Images of 10-day-old seedlings with a siRNA target site fused to a GFP reporter gene. The positive hit was treated with a chemical dissolved in DMSO. The negative control was treated with a compound dissolved in DMSO that does not affect the silencing pathway. Chlorophyll fluoresces red under UV light

4 Notes

1. Refer to the chemical plate manufacturer handling instructions for more information. The plates should be covered with a foil seal, which should be sealed tightly after every use.
2. Make sure not to splatter or cross-contaminate any of the chemicals. A higher concentration of chemical can be used but no higher than 100 μM.
3. Since most chemicals are stored in DMSO, DMSO should be added to the control wells. Columns 1 and 12 in chemical plates are empty and can be used for positive and negative controls.
4. Pipet the media against the side of the wells to avoid splashes. If the tips touch the chemicals, replace them. Avoid any cross-contamination of the chemicals in the wells.
5. This step must be done quickly since the agar will solidify. If the agar begins to solidify in the reservoir, pour new agar into a fresh reservoir. If the agar begins to solidify in the tips, replace them.
6. Be sure to rotate the UV lamp to ensure even distribution of the light. Less exposed regions of the plate will appear darker if this is not done.

Acknowledgements

This work was funded by the NSF-IGERT in Chemical Genomics, DGE0504249 to Y.L., a NIH R01 GM093008, and a NSF Career Award MCB-0642843 to H. J.

References

1. Vazquez F (2006) Arabidopsis endogenous small RNAs: highways and byways. Trends Plant Sci 11(9):460–468
2. Chen X (2012) Small RNAs in development—insights from plants. Curr Opin Genet Dev 22: 361–367
3. Law J, Jacobsen S (2010) Establishing, maintaining and modifying DNA methylation patterns in plants and animals. Nat Genet 11: 204–220
4. Sunkar R, Zhu J (2004) Novel and stress-regulated MicroRNAs and other small RNAs from arabidopsis. Plant Cell 16: 2001–2019
5. Katiyar-Agarwal S, Jin H (2010) Role of small RNAs in host-microbe interactions. Annu Rev Phytopathol 48:225–246
6. Padmanabhan C et al (2009) Host small RNAs are big contributors to plant innate immunity. Curr Opin Plant Biol 12:465–472
7. Ding S (2010) RNA-based anti-viral immunity. Nat Immunol 10:632–644
8. Robert S et al (2009) Powerful partners: arabidopsis and chemical genomics. Arabidopsis Book 7:e0109. doi:10.1199/tab.0109
9. Tóth R, van der Hoorn R (2010) Emerging principles in plant chemical genetics. Trends Plant Sci 15:81–88

Chapter 10

Identification and Use of Fluorescent Dyes for Plant Cell Wall Imaging Using High-Throughput Screening

Charles T. Anderson and Andrew Carroll

Abstract

Plant cell walls define cell shape during development and are composed of interlaced carbohydrate and protein networks. Fluorescent dyes have long been used to label plant cell walls, enabling optical microscopy-based interrogation of cell wall structure and composition. However, the specific cell wall components to which these dyes bind are often poorly defined. The availability of fluorescent compound libraries provides the potential to screen for and identify new fluorescent compounds that interact with specific plant cell wall components, enabling the study of cell wall architecture in intact, living tissues. Here, we describe a technique for screening fluorescent compound libraries for enhanced fluorescence upon interaction with plant cell walls, a secondary screening method to identify which cell wall components interact with a given dye, and a protocol for staining and observing Arabidopsis seedlings using a fluorescent cell wall-labeling dye. These methods have the potential to be applied to screening for differences in cell wall structure and composition among genetically diverse plant varieties or species.

Key words Fluorescent, Plant cell wall, Chemical library, In vivo imaging, Carbohydrate, Cellulose, Hemicellulose, Pectin, High-throughput

1 Introduction

Plants use carbohydrates to transfer and store energy and to build rigid cell walls [1] that encase and protect the cell, but these walls can also undergo controlled deformation to allow for cell growth [2]. Unlike proteins, carbohydrates cannot be labeled with genetically encoded fluorescent tags, and other methods are required for their in situ detection. Plant cell walls present the additional challenge of having low porosity [3] that might prevent the penetration of larger detection probes, such as lectins and antibodies [4]. Thus, small fluorescent molecules that recognize specific cell wall polymers are highly desirable for microscopic studies of intact cell wall architecture and dynamics [5]. Because many fluorescent dyes contain ring moieties [6], they represent good potential candidates for interaction with cell wall carbohydrates [7].

Glenn R. Hicks and Stéphanie Robert (eds.), *Plant Chemical Genomics: Methods and Protocols*, Methods in Molecular Biology, vol. 1056, DOI 10.1007/978-1-62703-592-7_10, © Springer Science+Business Media New York 2014

Recently, combinatorial libraries of fluorescent compounds have become available [6]. However, screening these libraries for interaction with cell walls in intact plants is laborious, and it can be difficult to distinguish cell wall binding from plasma membrane or cytoplasmic labeling at the low magnifications often used for screening. The methods described below include screening of fluorescent compounds for interaction with powdered plant cell walls, secondary screening for interaction with isolated cell wall components, and imaging of plant seedlings using fluorescent dyes.

2 Materials

Prepare all solutions using nanopure water.

2.1 Primary Screening Materials

1. Dried plant tissue (*see* **Note 1**).
2. Fluorescent compound library in 96-well plate format (*see* **Note 2**).
3. 2 mL screw-cap polypropylene tubes (Sarstedt).
4. 3 mm diameter steel milling balls (Retsch).
5. Bead beater or ball mill (e.g., Retsch MM 400) suitable for 2 mL polypropylene tubes.
6. 50 mL conical tubes.
7. 12-Channel pipettes (1–10 μL, 50–300 μL), multichannel dispensing reservoirs (Matrix Technologies).
8. Black polystyrene 96-well plates (e.g., Nunc).
9. Anhydrous dimethyl sulfoxide (DMSO).
10. ½ MS liquid: add 2.2 g Murashige and Skoog salts (Caisson Laboratories), 0.6 g MES (3-(*N*-morpholino)propanesulfonic acid, Research Organics) to 975 mL nanopure water, adjust pH to 5.6 with 1 M KOH, autoclave, add 25 mL 40 % (w/v) filter-sterilized sucrose solution.
11. Paradigm plate reader (Beckman) with multiple wavelength cassettes and Multimode Analysis software.

2.2 Secondary Screening Materials

1. 1 mg/mL solutions/suspensions of isolated cell wall components (*see* Table 1) in ½ MS liquid.

2.3 Plant Growth and In Vivo Imaging Materials

1. *Arabidopsis thaliana*, ecotype Columbia seeds.
2. Sterilization solution: 30 % (v/v) bleach in nanopure water, 0.01 % sodium dodecyl sulfate (SDS).
3. 1.5 mL microcentrifuge tubes.
4. Autoclaved nanopure water and 0.15 % agar (w/v) in nanopure water.

Table 1
Isolated cell wall components for secondary screening of cell wall-interacting fluorescent compounds

Carbohydrate	Description	Source
Cellulose	Semicrystalline; multiple β-1,4-linked glucose chains	Avicel; Sigma-Aldrich
Cellobiose	Dimer of β-1,4-linked glucose	Seikagaku Corporation
Cellotriose	Trimer of β-1,4-linked glucose	Seikagaku Corporation
Cellotetrose	Tetramer of β-1,4-linked glucose	Seikagaku Corporation
Cellopentose	Pentamer of β-1,4-linked glucose	Seikagaku Corporation
Cellohexose	Hexamer of β-1,4-linked glucose	Seikagaku Corporation
Xyloglucan	Hemicellulose; β-1,4-linked glucose backbone with neutral sugar side chains	Tamarind seed; Megazyme
Isoprimeverose	Hemicellulose; α-xylose-1,6-glucose dimer	Megazyme
Xylan	Hemicellulose; β-1,4-linked xylose chain	Larchwood; Sigma-Aldrich
Arabinoxylan	Hemicellulose; β-1,4-xylose backbone with arabinose side chains	Rye; Megazyme
Mannan	Hemicellulose; β-1,4-linked mannose chain	Borohydride reduced; Megazyme
Glucomannan	Hemicellulose; β-1-4-linked mannose and glucose with β-1,6-linked branches	Konjac; Megazyme
Galactomannan	Hemicellulose; β-1,4-linked mannose backbone with α-1,6-linked galactose side groups	Guar; Megazyme
Rhamnogalacturonan	Pectin; repeating galacturonic acid-α-1,2-rhamnose-α-1,4-backbone with branched side chains	Soybean pectic fiber; Megazyme
Pectic galactan	Pectin; α-1,4-linked galactose	Lupin; Megazyme
Arabinan	Pectin; α-1,5-linked arabinose	Sugar beet; Megazyme
Arabinogalactan	Pectin; branched arabinose and galactose	Larch; Megazyme
Starch	α-1,4-linked glucan with α-1,6-linked branches	Potato; Sigma-Aldrich
Curdlan	β-1,3-linked glucose	Sigma-Aldrich
Lichenan	β-1,3/1,4-linked glucose	Icelandic moss; Megazyme

5. ½ MS plates: add 2.2 g Murashige and Skoog salts (Caisson Labs), 0.6 g MES (3-(*N*-morpholino)propanesulfonic acid, Research Organics), 8 g agar-agar (Research Organics) to 975 mL nanopure water, adjust pH to 5.6 with 1 M KOH, autoclave, add 25 mL filter-sterilized 40 % (w/v) sucrose in

nanopure water, pour into 10 cm square polystyrene plates (Simport) in a laminar flow hood, store at 4 °C.

6. Plant growth chamber (e.g., Percival) set at 22 °C, 67 % relative humidity, 24 h light or greenhouse.
7. 2 mL microcentrifuge tubes.
8. 25 × 75 mm glass slides (Gold Seal).
9. #1.5 24 × 50 mm cover glass (Fisher).
10. Fine forceps (e.g., Dumont #3).
11. Vacuum grease (Beckman) loaded into a 10 mL syringe with a blunt 21-gauge needle attached.
12. Fluorescence microscope (epifluorescence, laser scanning confocal, or spinning disk confocal) with a 100× 1.4 NA oil immersion objective.

3 Methods

Perform all procedures at room temperature unless indicated.

3.1 Screening of Fluorescent Compound Libraries for Enhanced Fluorescence Upon Interaction with Plant Cell Walls

1. Break dried plant tissue into <5 mm pieces by hand or with a blade and add 150–200 mg tissue (*see* **Note 3**) to a 2 mL screw cap tube containing two 3 mm steel milling balls.
2. Mill tissue in a bead beater or Retsch mill for 5 min. No obvious chunks of tissue should remain. Remove milling balls using a magnet, weigh powdered cell wall material, and add to a 50 mL conical tube.
3. Suspend powdered cell wall material in ½ MS liquid at a concentration of 10 mg/mL; vortex to break up any clumps.
4. From a 96-well plate containing fluorescent compound library members dissolved at 1 mM (*see* **Note 4**) in anhydrous DMSO, add 1 μL of each compound to two black 96-well plates using a 12-channel pipette. Some wells should contain DMSO only as a blank.
5. Pour 12 mL of suspended cell wall material into a multichannel dispensing reservoir and add 100 μL of material to each well of one of the black 96-well plates containing the fluorescent compounds using a 50–300 μL 12-channel pipette, pipetting up and down to mix the compounds with the cell walls. Repeat this procedure for the other plate using ½ MS liquid without cell wall material as a negative control.
6. Cover both plates and incubate for 5 min, mix by gently tapping side of plate, and measure fluorescence values for each sample at multiple wavelengths using a Paradigm plate reader.

7. Calculate the mean fluorescence value at each wavelength for the DMSO-containing wells in the plates with and without cell wall material; these values represent background fluorescence levels for each plate.
8. Subtract the background fluorescence values for each plate from the fluorescence values for each sample to obtain the corrected fluorescence values for each compound with and without plant cell walls. Compounds for which fluorescence is enhanced upon interaction with plant cell walls should display a large ratio of (corrected fluorescence with cell walls)/(corrected fluorescence with cell walls) (*see* **Note 5**).

3.2 Secondary Screening to Identify Specific Cell Wall Components that Interact with Fluorescent Dyes

1. Add 1 μL of a 1 mM solution (*see* **Note 6**) of a dye that displays enhanced fluorescence in the presence of cell wall material to 84 wells of a black 96-well plate.
2. Add 100 μL of each carbohydrate listed in Table 1 to four wells of the plate (*see* **Note 7**); add ½ MS liquid with DMSO alone to four wells as a negative control.
3. Measure fluorescence at one or more wavelengths for the plate using a Paradigm plate reader; calculate the mean fluorescence values for each carbohydrate and normalize to the negative control values to determine whether any carbohydrates cause enhanced fluorescence upon interaction with the compound (*see* **Note 8**).

3.3 In Vivo Imaging of Fluorescently Labeled Plant Cell Walls in Arabidopsis Seedlings

1. Sterilize seeds by adding ~50 μL seeds and 500 μL sterilization solution to a 1.5 mL microcentrifuge tube, vortex, incubate 20 min; wash seeds 4× with sterile water, suspend in 1 mL sterile 0.15 % agar, and incubate at 4 °C in the dark for 3–7 days.
2. In a laminar flow hood, transfer seeds in two rows (one row at the top and one row halfway down) onto ½ MS plates using a 200 μL pipette with a cutoff pipette tip and place vertically in a growth chamber or a greenhouse for 5 days.
3. Make a 10 μM solution of fluorescent dye in ½ MS liquid in a 2 mL microcentrifuge tube.
4. Using fine forceps, transfer several 5-day-old seedlings from a ½ MS plate to the dye solution; incubate in the dark for 30 min (*see* **Note 9**).
5. Add four very small (~0.5 mm diameter) drops of vacuum grease to the corners of a 24 × 50 mm rectangle on a microscope slide.
6. Using fine forceps, transfer a single seedling from the dye solution to a 2 mL microcentrifuge tube containing 1.8 mL water to remove excess dye, then transfer the seedling to the slide

with the cotyledons extending beyond the edge of the vacuum grease rectangle. Add 80 μL nanopure water to the seedling and cover with a 24 × 50 mm cover glass.

7. Observe the seedling root (*see* **Note 10**) on an epifluorescence or confocal microscope, comparing any fluorescence signals to control seedlings incubated in ½ MS liquid lacking the dye.

4 Notes

1. Although dried plant material consists mainly of cell walls, other cellular components are present in low amounts. Plant material from the species to be used for microscopy studies should be used, since plant cell wall composition can vary between species. If interspecies comparisons are being performed, samples from identical tissues and developmental stages should be used. Tissue should be completely dry to enable effective ball milling, and can be dried in a 60 °C oven if air-drying is not effective.
2. Fluorescent compound libraries can be diversity-oriented [6] or preselected for interaction with cell walls [8]. It might also be possible to identify compounds that fluoresce only in the presence of plant cell walls by screening general diversity-oriented libraries.
3. Do not add more than 200 mg tissue to tube, as this will inhibit milling. For each 96-well plate, approximately 110 mg of powdered material will be needed.
4. Fluorescent compound concentrations in the 1–10 μM range can be used; however, we recommend using 10 μM to minimize the risk of false negative results due to the high autofluorescence of cell wall material at some wavelengths. Alternatively, cell wall suspensions can be pelleted by centrifugation after dye incubation and the fluorescence of the supernatants can be measured; any depletion of fluorescence relative to the dye solution in ½ MS liquid alone represents potential interaction of the dye with cell wall material.
5. Some compounds might display shifts in excitation or emission wavelengths upon interaction with plant cell wall components. These can be identified by plotting the fluorescence profile across multiple wavelengths for each compound with and without cell walls and screening for compounds for which the two profiles differ.
6. If the dye is soluble in an aqueous solution, the dye can be dissolved directly in ½ MS liquid, but if this is not the case, a stock solution in DMSO can be diluted into ½ MS liquid.

7. Like crude cell walls, many cell wall components are insoluble in water, and fluorescence measurements must be made quickly after mixing samples to avoid settling. Measurements of at least four replicates for each wall component in each experiment are recommended to reduce the chance of incorrect measurements due to heterogeneity in the suspensions.
8. To allow for clearly interpretable results, only dyes that display enhanced fluorescence in the presence of one or a few related components should be used for imaging.
9. If cellular dynamics are to be observed, the effects of the dye on seedling growth should be tested by making ½ MS plates containing a range of concentrations of the dye and comparing seedling root elongation after 5 days on these plates to that on control plates.
10. Seedling roots are amenable to dye labeling due to the lack of a cuticle, which can prevent dye penetration in aerial organs; for labeling aerial organs, the cuticle must be removed, e.g., by dissolving in DMSO, 80 % ethanol, or other organic solvents, but this method precludes in vivo imaging.

References

1. Somerville C et al (2004) Toward a systems approach to understanding plant cell walls. Science 306(5705):2206–2211
2. Cosgrove DJ (2005) Growth of the plant cell wall. Nat Rev Mol Cell Biol 6(11):850–861
3. Chesson A, Gardner PT, Wood TJ (1997) Cell wall porosity and available surface area of wheat straw and wheat grain fractions. J Sci Food Agric 75(3):289–295
4. Marcus SE et al (2010) Restricted access of proteins to mannan polysaccharides in intact plant cell walls. Plant J 64(2):191–203
5. Anderson CT, Carroll A, Akhmetova L, Somerville C (2010) Real-time imaging of cellulose reorientation during cell wall expansion in Arabidopsis roots. Plant Physiol 152(2): 787–796
6. Kang NY, Ha HH, Yun SW, Yu YH, Chang YT (2011) Diversity-driven chemical probe development for biomolecules: beyond hypothesis-driven approach. Chem Soc Rev 40(7):3613–3626
7. Smith MM, McCully ME (1978) Enhancing aniline blue fluorescent staining of cell wall structures. Stain Technol 53(2):79–85
8. Hoch HC, Galvani CD, Szarowski DH, Turner JN (2005) Two new fluorescent dyes applicable for visualization of fungal cell walls. Mycologia 97(3):580–588

Chapter 11

High-Throughput Identification of Chemical Endomembrane Cycling Disruptors Utilizing Tobacco Pollen

Michelle Q. Brown, Nolan Ung, Natasha V. Raikhel, and Glenn R. Hicks

Abstract

Endomembrane cycling processes in plants remain mostly intractable through classical genetic interrogation. Chemical disruption of these processes provides an opportunity to slow or inhibit these processes for study. Tobacco pollen, which is dependent upon endomembrane cycling for tube growth, provides a plant system that is amenable to high-throughput screening of chemical disruptors. We describe here the process that allowed the identification of over 360 endomembrane cycling disruptors.

Key words Endomembrane, Chemical biology, Tobacco pollen

1 Introduction

Plant endomembrane cycling is at the nexus of many important developmental and environmental responses. Given the critical nature of endomembrane cycling, many genes that participate in these processes are essential or functionally redundant making full dissection of endomembrane cycling processes in plants nearly impossible using classical genetic techniques. The use of chemical disruptors provides a way to inhibit or negatively modulate protein function (*see review* [1]). High-throughput cellular screening in plants remains problematic due to the different tissue types found in plants and the variability of growth rates. However, tobacco pollen offers a unicellular, plant system that is amenable to high-throughput screening. Tobacco pollen tubes, which germinate in vitro and are easily imaged at a low magnification, display highly polarized growth determined by endomembrane cycling and cellular oscillations involving calcium gradients and actin organization [2–4]. Utilizing tobacco pollen in a high-throughput assay with commercially available chemical libraries and selected activity libraries, we have identified 360 endomembrane disrupting compounds, named them Endosidins, and clustered them based on

Glenn R. Hicks and Stéphanie Robert (eds.), *Plant Chemical Genomics: Methods and Protocols*, Methods in Molecular Biology, vol. 1056, DOI 10.1007/978-1-62703-592-7_11, © Springer Science+Business Media New York 2014

their phenotypic fingerprint, meaning that compounds that caused similar phenotypes on a variety of subcellular markers would be grouped together in a matrix format [5, 6]. Here we describe a method to screening for bioactive chemicals utilizing pollen germinated in vitro as a system. This approach is very straightforward and robust and can be applied to many different types of screens provided that the chemically induced phenotypes include inhibition of pollen tube growth or other scorable phenotypes in pollen such as altered pollen tube morphology or mislocalization of fluorescent protein markers.

2 Materials

2.1 Pollen Germination Media (GM)

For 100 mL of GM, start with 80 mL of deionized water and add 18 g of sucrose (18 %) and mix until dissolved. Then add 1 mL of 1 % boric acid solution (0.01 %), 0.5 mL of 1 mM calcium chloride solution, 0.5 mL of 1 mM calcium nitrate solution, and 0.1 mL of 1 M magnesium sulfate solution. Thoroughly mix and adjust pH to 6.5–7.5 with 1 N KOH. Then bring volume to 100 mL with deionized water, filter-sterilize, and store at 4 °C (*see* **Note 1**).

2.2 Preparation of Chemical Plates

Twenty microliters of germination media was added to each well of a multiwell plate (ThermoFisher 12-566-280 (384 well) or ThermoFisher 12-566-607(96 well)). A Biomek FX fluid handling robot was used to distribute 0.2 μL of each library compound into individual wells, with the exception of the end wells for each row of the plate which were reserved for controls. Finished plates were sealed (ThermoFisher AB-0559) and stored at −20 °C to be used within a week of creation (*see* **Note 2**).

3 Methods

1. On the day of the assay, collect eight anthers in dehiscence from flowers of *Nicotiana tabacum* and place in a 50 mL Falcon tube. Anthers can be easily collected using forceps. After the anthers are collected, add 8 mL of germination media and vortex the anthers and GM thoroughly. Solutions will become cloudy with pollen. For each plate harvest eight anthers/8 mL of GM. All anthers can be combined into one tube so that the average density of pollen is consistent for each plate in the experiment. Up to 40 anthers in 40 mL of GM can be combined into a single 50 mL tube. At dehiscence, anthers with viable pollen will have a whitish yellow appearance and appear "fluffy." When anthers are older and less viable they will become more strongly yellow in appearance. This pollen will still germinate but at much lower frequency than fresh anthers.

It is also important to plan ahead because it takes several months for tobacco plants in the greenhouse to begin to flower in sufficient quantity for screening. As the plants flower, they can be continually be pruned to produce flowers over an extended period of time.

2. Pipet all liquid solution (disregarding any solid material) into a multichannel reservoir (Biologix 25-0051).
3. Using a multichannel pipet, pipet 20 μL of the pollen solution into each well of a thawed, centrifuged plate of compounds (*see* Subheading 2.2), giving a total volume of 40 μL per well (*see* **Notes 3** and **4**).
4. The plate is then placed on an orbital shaker with gentle agitation (~70 rpm) under a black cloth for 3 h. At about 3 h pollen will have germinated and pollen tubes will be of moderate length for scoring.
5. The plate is then immediately visualized by microscopy for the phenotype of interest (*see* **Notes 5**–7).

4 Notes

1. If germination media becomes cloudy during storage, simply re-filter-sterilize and store again at 4 °C. If germination appears to be reduced, make fresh GM or replace every 6 months.
2. Be sure to centrifuge plates after sealing but before freezing to keep all the solution together in the well. It's also important to keep the plates level while freezing to prevent contamination between wells.
3. After the plate is thoroughly thawed, be sure to centrifuge to keep all of the solution in the well to prevent cross-contamination between wells.
4. Be sure to pipet up and down to mix pollen solution before drawing the 20 μL for each row to help maintain an even distribution of pollen because the pollen tends to sink to the bottom of the reservoir.
5. The stock concentration of our libraries was 5 mg/mL in 100 % DMSO. For compounds averaging about 400 MW, this is a concentration of about 12.5 mM. For our screen 0.2 μL of stock into a final volume of 40 μL resulting in a screening concentration of 50–100 μM for each compound. These values are approximate because each chemical library will contain compounds that have a range of masses.
6. If it will take longer than 1 h before visualization of the plate can occur, store the plate at 4 °C until it can be visualized. This will prevent overgrowth and collapse of pollen tubes.

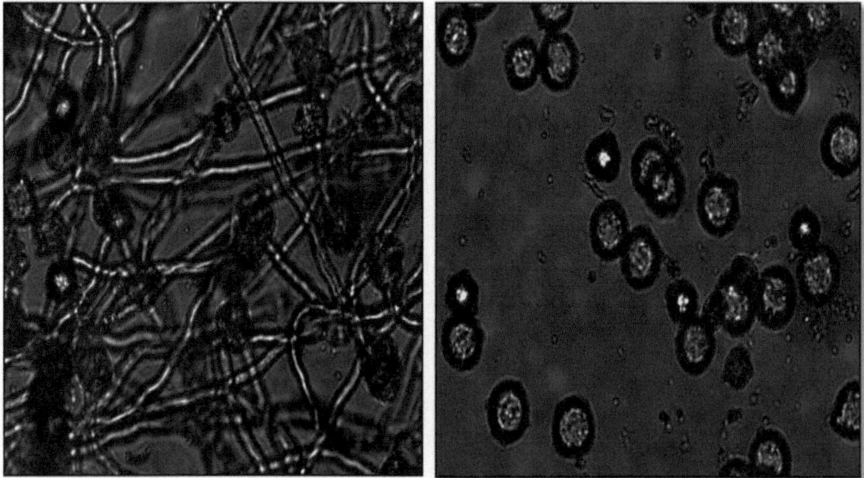

Fig. 1 The effect of a chemical inhibitor on pollen germination and tube growth (*right panel*) is strikingly distinct from normal pollen germination and tube growth (*left panel*)

7. We were able to visualize our plates using the BD Pathway inverted confocal microscope which had the capability to automatically focus and image each well, allowing a plate to be imaged in approximately 45 min. These images were archived automatically then reviewed manually for growth differences between normal pollen germination and germination in the presence of specific small molecules. Strong inhibition of germination or pollen growth was the most common chemical-induced phenotype (Fig. 1). However, other phenotypes are also possible such as pollen tube tip swelling, wavy pollen tubes, and other morphological changes.

References

1. Hicks GR, Raikhel NV (2009) Opportunities and challenges in plant chemical biology. Nat Chem Biol 5(5):268–272
2. Zonia L, Munik T (2008) Vesicle trafficking dynamics and visualization of zones of exocytosis and endocytosis in tobacco pollen tubes. J Exp Bot 59(4):861–873
3. Feijo J, Sainhas J, Holdaway-Clarke T, Cordeiro M, Kunkel J, Hepler P (2001) Cellular oscillations and the regulation of growth: the pollen tube paradigm. BioEssays 23:86–94
4. Cole R, Fowler J (2006) Polarized growth: maintaining focus on the tip. Curr Opin Plant Biol 9:579–588
5. Robert S, Chary S, Drakakaki G, Li S, Yang Z, Raikhel NV, Hicks GR (2008) Endosidin1 defines a compartment involved in endocytosis of the brassinosteroid receptor BRI1 and the auxin transporters PIN2 and AUX1. PNAS 105(24):8464–8469
6. Drakakaki G, Robert S, Szatmari A, Brown MQ, Nagawa S, Van Damme D, Leonard M, Yang Z, Girke T, Schmid SL, Russinova E, Friml J, Raikhel NV, Hicks GR (2011) Clusters of bioactive compounds target dynamic endomembrane networks in vivo. PNAS 108(43):17850–17855

Chapter 12

Plant Chemical Genomics: Gravity Sensing and Response

Marci Surpin

Abstract

The gene families that encode the vesicle trafficking machinery in plants are highly expanded compared to those from protists and animals. As such, classical genetic screens for mutants with lesions in these genes are fraught with issues of redundancy and lethality. A chemical genomics approach can, in theory, circumvent these issues because inhibitory or stimulatory molecules may be applied at any point in development at sublethal concentrations. This chapter describes the protocols for a chemical genomics screen designed to identify components of the plant cell vesicle trafficking machinery. A two-tiered screen was designed where the primary screen assayed for chemicals that modified the gravitropic response, a process that in plant cells is intimately tied to vesicle trafficking; the secondary screen employed fluorescent marker lines that were treated with gravitropic inhibitors or inducers to assay for changes in endomembrane system morphology. We thus identified four compounds by which we can further explore the relationship between gravitropic signal transduction and vesicle trafficking.

Key words Chemical genomics, Gravitropism, Plant endomembrane system, Screening, Multi-tiered screen, Auxin, Structure–activity relationship (SAR)

1 Introduction

The study of plant vesicle and protein trafficking has traditionally been accomplished via cell biology and protein biochemistry methodologies. The ascendance of *Arabidopsis thaliana* as a plant model genetic system, and the subsequent publication of an annotated genome, suggested that it might be feasible to carry out genetic screens for trafficking mutants. However, it has become apparent that gene families that encode proteins that comprise the plant cell trafficking machinery are greatly expanded with respect to yeast and mammalian counterparts [1]. Thus, genetic screens for plant trafficking mutants tend to be confounded by redundancy, or perversely, lethality [2]. This reality has made it difficult to assign physiological functions to many annotated genes, and therefore alternative approaches are required to dissect plant vesicle trafficking pathways.

Glenn R. Hicks and Stéphanie Robert (eds.), *Plant Chemical Genomics: Methods and Protocols*, Methods in Molecular Biology, vol. 1056, DOI 10.1007/978-1-62703-592-7_12, © Springer Science+Business Media New York 2014

Chemical genomics approaches offer an alternative to classical genetic screens in that they do not employ permanent alterations of the genome. Instead, small molecules are used to inhibit or modify the action of either individual proteins or a family of proteins [3]. The power of this approach is that it is tunable; that is, increasing applications of a small molecule will result in stronger effects, and also that small molecules may be applied at any point during an organism's development. Thus, it is possible to selectively inhibit the desired protein or class of proteins under the environmental and/or developmental conditions of one's choosing.

To screen thousands of molecules in a primary chemical genomic screen, it is critical to design an assay that permits facile chemotyping and scoring. In the screen described in this chapter, we were interested in identifying small molecules that would inhibit a specific trafficking pathway. It has been clear for some time that there is a connection between protein trafficking and gravitropic signal transduction. For example, a classical genetic screen for agravitropic mutants uncovered lesions in genes related to protein trafficking [4, 5], and conversely, many protein trafficking mutants exhibit agravitropic phenotypes [6].

Therefore, in order to identify trafficking components involved in a specific physiological pathway, we designed a primary chemical genomics assay to screen for small molecules that would modify the gravitropic response in Arabidopsis seedlings. We then chose a subset of chemicals and screened for alterations in endomembrane system morphology using scanning confocal laser microscopy. This tiered approach allowed us to visually screen a large number of chemical compounds (in this case, 10,000) using a relatively quick, "yes/no," phenotypic assay, and then focus on a smaller number of compounds in more painstaking and time-consuming assays [7]. It is not uncommon in plant chemical genomics studies to carry out the primary assay in a surrogate system, such as yeast or tobacco pollen, and then transition to Arabidopsis in the secondary assay [8]. The advantage of such an approach is the ability to screen larger numbers of compounds in the primary screen; however, the obvious disadvantages are that some compounds that are active in the primary tier will not translate to the second tier, and conversely, some compounds that may be active in Arabidopsis can be missed in a primary screen that is carried out in another system.

This chapter describes the specific protocols of a screen to identify vesicle trafficking components of the plant gravitropic response pathway. We screened 10,000 compounds and identified 199 inhibitors and 20 enhancers [7]. We then tested 69 of these compounds in a secondary screen where we generated gravitropic dose–response curves to (a) confirm the primary hit, and (b) evaluate the efficacy of the individual compounds. We then analyzed compounds based on structure and the severity of induced chemotypes, and chose a subset of 34 confirmed hits for tertiary screening, where we treated fluorescent marker lines for various endomembrane compartments with the selected compounds. The fluorescent lines were examined using laser scanning confocal microscopy for changes in endomembrane system morphology. Through this protocol we identified four compounds: 5850247/gravacin, which inhibits auxin transport activity through binding of the ATP-binding cassette transporter PGP19 [9] (*see* **Note 1**); 5271050, a phenazine derivative that induces the aggregation of ER-derived bodies in hypocotyls [10]; 5403629, an auxin-like compound; and 6220480, a compound that induces aggregate formation in roots.

2 Materials

2.1 Plant Materials

1. Plant materials: *Arabidopsis thaliana*, Columbia-0 ecotype, was used for the primary and secondary screens. Various fluorescent marker lines were used for imaging experiments in the tertiary screen.
2. Media: prepare media as described in Weigel and Glazebrook: ½× Murashige and Skoog salts (Plant Media, Dublin, OH), pH 5.8, no sucrose, with 0.6 % Phytoagar (Plant Media, Dublin, OH) (*see* **Note 2**).

2.2 Chemical Library

Chemical libraries are available from a number of different sources. The screen described in this chapter used a collection of 10,000 synthetic compounds that was purchased from Chembridge (San Diego). There are additional commercial sources and also various academic centers devoted to the synthesis of diverse compounds (*see* **Note 3**).

1. Chemical library: DiverSetE 10 k small molecule collection (Chembridge, San Diego, CA), dispersed into 125 98-well plates (80 compounds per plate, 16 blanks).
2. Solvent: 100 % DMSO (Fisher Scientific) added to final concentration of 0.5 %.

2.3 Hardware, Software, and Consumables

An important consideration in setting up a chemical screen is that the handling of materials and treatment of specimens is consistent throughout the screen. It is therefore a good investment of time and resources to optimize the primary assay as much as possible before commencing with the chemical screen. The development of clear, stepwise protocols and the use of automation will help minimize variability over the course of the screen.

1. Tissue culture plates: 24 well (Corning, #3337).
2. Tissue culture plates: 4 well rectangular (Nunc, #267061).
3. Liquid handling system, such as the Bio-Tek Precision XP Liquid handling system (Bio-Tek Instruments, Winooski, VT).

4. Pipet tips for liquid handling system.
5. Repeat pipettor, such as the Gilson Distriman (#F164001, Middleton, WI).
6. Tips for repeat pipettor (e.g., Distritips: Mini ST 1250 μL, F164140; and Maxi ST 12.5 mL, F164120; Gilson, Middleton, WI).
7. Flat-bed scanner (e.g., Epson Model 2450, Long Beach, CA).
8. ImageJ quantification software (National Institutes of Health, Bethesda, MD; http://imagej.nih.gov/ij/).
9. Excel (Microsoft Corp., Seattle, WA).

3 Methods

3.1 Primary Screen for Compounds that Inhibit the Gravitropic Response

The overwhelming advantage of carrying out chemical screens in plants is that it is possible to screen for whole-organism chemotypes. Most chemical screens in animal systems are cell or biochemically based. While these types of screens are amenable to miniaturization and thus higher throughput, the transition from a culture dish well to organism is not trivial. In whole-plant assays, provided a given hit is confirmed, it is already known that the compound in question induces an observable effect in the intact organism. However, the ability to screen in the whole plant also requires larger formats and thus lower throughput. Consequently, while screens in animal systems can often be carried out in 384- or 1,536-well formats, the upper limit for plant screens is often the 96-well format, and it is not unusual for plant screens to be carried out in 24-well formats. However, the lower throughput is not without its compensations. It is often possible to screen multiple plants in a single well in the 24-well format, which can increase confidence in the robustness of a single hit.

1. If using a liquid-handling robot, it may be necessary to program it to convert 96-well library plates to 24-well assay plates. This is carried out in an iterative fashion, where a single 96-well plate yields four 24-well plates (*see* **Note 4**).
2. Chemical library stock plates are typically 10 mg/mL in 100 % DMSO. It is advisable to make multiple working plates in order to minimize the number of freeze–thaw cycles to which the master plates are subjected. The library used in this screen was diluted into working plates 1:5 in autoclaved distilled deionized water (ddH_2O), or 2 mg/mL in 20 % DMSO. The final optimal concentration of DMSO in the assay plates is 0.5 % or less; however, Arabidopsis will tolerate up to 1 % DMSO.
3. Two days before making assay plates, surface sterilize the appropriate number of Arabidopsis seed. There should be approximately 8–12 seeds per well. Suspend the sterilized seed

in 0.1 % Phytoagar, wrap the tube in aluminum foil, and store at 4 °C for stratification.

4. To make the assay plates, add 40 μL of chemicals from the working plates to the 24-well plates. Be sure to include the same amount of 20 % DMSO in the control wells. In a sterile hood, place the plate on a paper towel, and using a repeat pipettor, add 360 μL of molten ½× MS media with 0.6 % Phytoagar. While adding the media, swirl the plate using a circular motion, so as to mix the chemicals with the media. Allow the media to solidify.
5. When the media is solidified, disperse the sterile stratified Arabidopsis seed, at least eight seeds per well, using a repeat pipettor. The seeds should be dispersed toward the bottoms of the wells, in order to ensure they will have sufficient room to accommodate the elongating hypocotyls, and that they will also have enough room to bend.
6. When the Phytoagar in which the seeds are suspended dries down, seal the plates with surgical tape and place them horizontally in a plant growth chamber for 4–8 h in order to photoconvert the phytochromes and ensure germination. In groups of four (corresponding to a single 96-well working plate), wrap the plates in two layers of aluminum foil. Be sure to arrange all the plates in the same orientation, and note the location of well A1. This will be the top left-hand corner for all assay plates. Label with the date and time and mark the "up" orientation. Place back in the plant growth chamber, this time in a vertical orientation.
7. Allow the seeds to germinate and grow for 48–72 h. It is helpful to create a control plate with no chemicals, but the same concentration of DMSO in the media. Sow seed and wrap and treat the same as the assay plates, and use this plate to monitor the progress of the seedlings without unwrapping the assay plates.
8. When the hypocotyls of the seedlings elongate to approximately 5–8 mm (as monitored using the control plate), turn the plates 90°. Be sure to mark the date and time and the new orientation. Also turn the control plate. Incubate the plates for another 48 h.
9. When the seedlings on the control plate have turned 90°, unwrap a single group of plates, remove one plate, and then loosely place the foil back on the remaining plates, keeping them in the dark and same orientation while you are scoring. Small amounts of light and handling while scoring will not adversely affect results.
10. Score the plates by visual inspection for wells in which the seedlings did not turn, have a noticeably smaller angle of bending, or an exaggerated angle of bending. These wells are scored

as "hits." Scan the plates in groups of four, always in the same arrangement of A, B, C and D plates.

11. Keep track of hits using Excel files. It is important to note the following information: 24-well plate number and location, 96-well plate number and location, chemical identity and characteristics of the chemotype (e.g., inhibitor or enhancer).
12. For hit confirmation: create a sub-library plate by cherry-picking chemical hits and assembling them into a 96-well plate or plates. Prepare making assay plates, sow seed, and perform the gravitropic assay as previously described. Note in the Excel file which hits are "Confirmed."

3.2 Secondary Screen to Confirm Primary Screen Hits and Carry Out Preliminary Characterization

Potential hits are evaluated at this stage by expanding the analysis of compounds that gave a response in the primary screen.

1. Once confirmed, order 5 or 10 mg quantities of desired compounds for secondary screening and characterization. At this point, compounds should be solubilized in 100 % DMSO on a molar basis, using enough solvent to allow testing over a range of concentrations (e.g., 1–100 μM) and ensuring that the DMSO can be diluted to 1 %. It is also important at this stage to confirm that the structure of the resupply compound matches that in the library plate; this is accomplished by mass spectrometry and/or NMR analysis.
2. Set up dose–response curves in 24-well culture plates. Each row, containing six wells, is to evaluate one compound; therefore, four compounds can be evaluated per plate. Set up wells as follows: 0, 1, 5, 10, 50, and 100 μM, a range that we find empirically to typically yield activity, being sure to hold the amount of DMSO constant in every well. Sow seed and assay as described previously. Prioritize compounds based on dose response and categorization. For example, in the described screen, we subdivided the chemical hits into four groups: enhancers and inhibitors, and unique and auxin-like. Because we were interested in discovering novel chemicals that affected trafficking, we decided to de-emphasize the auxin-like chemicals and focus on chemicals that we classified as "unique." (*See* **Note 5**).
3. To further quantify the gravitropic response, set up another dose–response experiment, using four-well culture dishes. Each well accommodates 6 mL of media. Set up individual wells containing the desired amounts of the selected compounds. It is helpful to refer back to the dose–response experiment to determine which concentrations to use. For example, if 100 μM compound severely hampers growth, there is no need to include that concentration in quantification of gravitropism. Use at least two wells per single concentration. Positioning the plates along the short axis, draw a line 1/3 of the way from the bottom of the well. Sow approximately 20 sterilized and stratified seed along the line, and light-treat the plates.

Treat the plates as previously described. In order to quantify the bending angles, use the ImageJ program to open the plate scans, and then use the angle tool to measure hypocotyl bending [11]. Express the data as the average bending angle per treatment. It is sometimes more illuminating to express the data as a histogram than an average.

4. It is also useful to determine the tissue specificity of the chemical hits, that is, whether a hit is specific for modification of hypocotyl or root bending. Repeat the previous experiment, but leave enough room at the bottom of the plate to allow the roots to bend. Measure as before, using ImageJ.
5. On the basis of chemotype and dose response, assign values via a ranking system (e.g., 5 = best, 1 = worst).

3.3 Tertiary Screen to Identify Molecules that Modify Endomembrane System Morphology

1. Depending on the number of hits that pass secondary screening (i.e., demonstrate a consistent observable chemotype and dose-responsiveness), select another subset for tertiary screening. To screen for compounds that modify the morphology of the endomembrane system, sow Arabidopsis GFP:δTIP seeds on the selected chemicals and examine the resultant seedlings for endomembrane system defects using a scanning laser confocal microscope. The GFP:δTIP marker is specific for the tonoplast membrane [12]. Possible chemotypes include changes in vacuole morphology and re-localization of GFP fluorescence to other organelles, such as the endoplasmic reticulum.
2. In order to begin to determine the site of action of the compound, it is useful to test fluorescent marker lines that carry constructs specific for different compartments within the cell, i.e., endoplasmic reticulum, Golgi apparatus, etc. [13].
3. Because the gravitropic response is ultimately mediated by the action of auxins, it is necessary to determine whether hits are acting as auxins or auxin transport inhibitors. This is accomplished by treatment of the Arabidopsis DR5::GUS line [7]. The DR5::GUS transcriptional reporter is responsive to auxin, thus, seedlings treated with hit compounds will stain blue in a GUS assay, indicating that the compound has auxin-like properties. Include IAA-treated seedlings as positive controls, and DMSO-treated seedlings as negative controls.
4. Alternatively, it is also important to determine whether a given compound may be acting as an auxin transport inhibitor. In this case, 7-day old DR5::GUS seedlings are incubated with compounds for 14 h. A media plug that contains IAA is placed at the bottom of the root tips for 5 h and then stained. TIBA, an auxin transport inhibitor, is used as a negative control. The roots of seedlings treated with compounds that do not act as auxin transport inhibitors will turn completely blue in the GUS staining assay, and seedlings treated with compounds that inhibit auxin transport will be stained only at the root tips.

3.4 Structure–Activity Relationship (SAR) Studies

1. Further characterization of lead compounds includes structure–activity relationship, or SAR, studies in order to determine which domains or moieties of the compounds of interest are necessary for activity. It is also possible to identify molecules that have higher potency, although this may require the de novo synthesis of new molecules. The first steps in conducting SAR studies are to (1) identify commercially available substituents and (2) carry out an online search for structurally similar molecules, using a searchable engine such as ChemMine [14] (*see* **Note 6**). After substituent and structurally similar molecules are procured, they are tested using the previously described phenotypic assays; that is, gravitropic response and endomembrane system morphology using fluorescent marker lines.
2. Another approach to SAR determination is through de novo synthesis of custom molecules. One technique is iodine scanning. This is a variation of the methyl scanning approach, in which all available positions on a molecule of interest are derivatized with methyl groups to determine points of protein-molecule interactions [15]. The advantage of using iodine groups instead of methyl groups is that in addition to having a similar van der Waals radius, iodine may be further derivatized to create affinity matrices for possible target isolation studies [10]. A derivatized atom that is determined not to interfere with target protein binding may be biotinylated using readily available reagents.

4 Notes

1. Target identification for gravacin and chemical 5271050 was carried out via genetic chemical resistance screens. The resistance screen for 5271050 highlighted an important issue in such genetic screens: namely, that of ecotype sensitivity. The primary chemical screen was carried out in the Columbia ecotype. Out-crossing of putative 5271050-resistant mutants to the Landsberg *erecta* ecotype made clear that Landsberg was not as sensitive to the chemical. A survey of additional ecotypes identified a range of activities, ranging from fully sensitive (i.e., equivalent to Columbia) to partially sensitive (Landsberg) to completely insensitive ecotypes. Therefore, when embarking on chemical resistance or hypersensitivity target identification screens, it is of utmost importance to determine that the ecotype to be used for outcrossing is as sensitive as the parent ecotype, otherwise, it will be impossible to identify F_2 recombinants that are true mutants (Surpin and Raikhel, unpublished results).
2. We did not add sucrose to the media. The liquid handling apparatus was not situated in a sterile hood. We opened sterile plates, added chemicals, and then added media and seeds in a

laminar airflow hood. Excluding sucrose from the media kept contamination to a minimum.

3. There are a number of sources for synthetic chemical libraries [16]. Another option is to screen using natural products, though it is important to keep in mind that such molecules can often have high molecular weights and multiple chiral centers, which may later confound resupply, resynthesis, and SAR studies.
4. In order to create four 24-well assay plates from one 96-well library plate, it is necessary to split up one box of 96 pipette tips. Transfer every other row to an empty box, so one box has tips in the first, third, fifth and seventh rows, and the other in the second, fourth, sixth, and eighth rows. The first transfer, using the box with tips in the first row, will transfer compounds from rows A, C, E, and G of the library plate. Columns 1, 3, 5, 7, 9, and 11 will be transferred to Plate A, and columns 2, 4, 6, 8, 10, and 12 to Plate B. The second transfer uses the other tip box, with tips in the second row, etc., to fill plates C and D. Plate C receives compounds from library plate rows B, D, F, and H, and columns 1, 3, 5, 7, 9, and 11. Library compounds from the same rows and columns 2, 4, 6, 8, 10, and 12 are transferred to Plate D.
5. It is worth noting that the particular library used for this screen had a number of compounds that were synthesized using a phenoxyacetate building block, thus increasing the incidence of compounds with auxin-like activities.
6. ChemMine has a number of useful tools for analysis and clustering. Other search engines include PubChem (http://pubchem.ncbi.nlm.nih.gov/) [17] and SciFinder (www.cas.org/products/scifindr/sfweb).

Acknowledgments

The work described in this chapter was funded by grants from the National Aeronautics and Space Administration (NNA04CC73G) to Glenn R. Hicks and the National Science Foundation (MCB-0515963) to Natasha V. Raikhel. Special thanks are due to Dr. Paul Silverman and Dr. Clay Carter for helpful comments.

References

1. Sanderfoot AA, Assaad FF, Raikhel NV (2000) The Arabidopsis genome. An abundance of soluble N-ethylmaleimide-sensitive factor adaptor protein receptors. Plant Physiol 124:1558–1569
2. Rojo E, Gillmor CS, Kovaleva V, Somerville CR, Raikhel NV (2001) VACUOLELESS1 is an essential gene required for vacuole formation and morphogenesis in Arabidopsis. Dev Cell 1: 303–310
3. Blackwell HE, Zhao Y (2003) Chemical genetic approaches to plant biology. Plant Physiol 133:448–455
4. Kato T, Morita MT, Tasaka M (2002) Role of endodermal cell vacuoles in shoot gravitropism. J Plant Growth Regul 21:113–119

5. Kato T, Morita MT, Fukaki H, Yamauchi Y, Uehara M, Niihama M, Tasaka M (2002) SGR2, a phospholipase-like protein, and ZIG/SGR4, a SNARE, are involved in the shoot gravitropism of Arabidopsis. Plant Cell 14:33–46
6. Schumacher K, Vafeados D, McCarthy M, Sze H, Wilkins T, Chory J (1999) The Arabidopsis det3 mutant reveals a central role for the vacuolar H(+)-ATPase in plant growth and development. Genes Dev 13:3259–3270
7. Surpin M, Rojas-Pierce M, Carter C, Hicks GR, Vasquez J, Raikhel NV (2005) The power of chemical genomics to study the link between endomembrane system components and the gravitropic response. Proc Natl Acad Sci U S A 102:4902–4907
8. Zouhar J, Hicks GR, Raikhel NV (2004) Sorting inhibitors (Sortins): chemical compounds to study vacuolar sorting in Arabidopsis. Proc Natl Acad Sci U S A 101:9497–9501
9. Rojas-Pierce M, Titapiwatanakun B, Sohn EJ, Fang F, Larive CK, Blakeslee J, Cheng Y, Cutler SR, Cuttler S, Peer WA, Murphy AS, Raikhel NV (2007) Arabidopsis P-glycoprotein19 participates in the inhibition of gravitropism by gravacin. Chem Biol 14:1366–1376
10. Surpin M, Zou Y, Xiong C, Raikhel NV, Pirrung MC (2010) Iodine scanning of a phenazine inhibitor of vacuolar sorting. Bioorg Med Chem Lett 20:1496–1499
11. Abramoff MD, Magelhaes PJ, Ram SJ (2004) Image processing with ImageJ. Biophotonics Int 11:36–42
12. Cutler SR, Ehrhardt DW, Griffitts JS, Somerville CR (2000) Random GFP::cDNA fusions enable visualization of subcellular structures in cells of Arabidopsis at a high frequency. Proc Natl Acad Sci U S A 97:3718–3723
13. Robert S, Chary SN, Drakakaki G, Li S, Yang Z, Raikhel NV, Hicks GR (2008) Endosidin1 defines a compartment involved in endocytosis of the brassinosteroid receptor BRI1 and the auxin transporters PIN2 and AUX1. Proc Natl Acad Sci U S A 105:8464–8469
14. Girke T, Cheng LC, Raikhel N (2005) ChemMine. A compound mining database for chemical genomics. Plant Physiol 138:573–577
15. Pirrung MC, Liu Y, Deng L, Halstead DK, Li Z, May JF, Wedel M, Austin DA, Webster NJ (2005) Methyl scanning: total synthesis of demethylasterriquinone B1 and derivatives for identification of sites of interaction with and isolation of its receptor(s). J Am Chem Soc 127:4609–4624
16. Robert S, Raikhel NV, Hicks GR (2009) Powerful partners: Arabidopsis and chemical genomics. Arabidopsis Book 7:e0109
17. Bolton EE, Wang Y, Thiessen PA, Bryant SH (2008) PubChem: integrated platform of small molecules and biological activities. In: Annual reports in computational chemistry, vol 4, American Chemical Society, Washington, DC

Chapter 13

Screening Chemical Libraries for Compounds That Affect Protein Sorting to the Yeast Vacuole

Jan Zouhar

Abstract

Protein trafficking to the yeast vacuole has been extensively studied using a series of deletion mutants. In these genetic screens, mis-targeted vacuolar cargo proteins were used as phenotype markers. Here we describe a similar approach employing pharmacological effects of diverse chemical compounds to mimic molecular phenotypes caused by conventional genetic mutations in protein trafficking genes.

Key words Chemical genomics, Yeast vacuole, Protein trafficking

1 Introduction

Newly synthesized proteins destined for the yeast vacuole utilize two major trafficking pathways (reviewed in [1]). The principal one is referred to as the carboxypeptidase Y (CPY) pathway and involves vesicular trafficking of cargo proteins from the *trans*-Golgi network to the prevacuolar compartment (PVC), from where some trafficking machinery can be recycled back to the TGN. The second pathway is known after its principal cargo as the alkaline phosphatase (ALP) pathway and uses a different cargo-sorting signal and a vesicle coat [2]. The ALP trafficking pathway bypasses the PVC, which also allows for restrictive localization of tonoplast proteins, such as Vam3p [3].

A recent genome-wide genetic screen resulted in identification of 146 mutants in the CPY pathway, which are classified as *vacuolar protein sorting* (*vps*) mutants [4]. The approach consisted of screening nearly 5,000 homozygous diploid gene deletion strains for secretion of CPY into the medium by using a colony-blotting assay [5]. In wild-type yeast cells, vacuolar cargo is efficiently sorted to the vacuole and therefore is not mis-targeted. However, in the *vps* mutants, CPY sorting to the vacuole is hindered and different levels of the protein are secreted.

Glenn R. Hicks and Stéphanie Robert (eds.), *Plant Chemical Genomics: Methods and Protocols*, Methods in Molecular Biology, vol. 1056, DOI 10.1007/978-1-62703-592-7_13,

Here, we present a chemical genomics approach to identify chemical compounds that mimic the *vps* phenotype. Using this assay to screen a small library of 5,000 compounds, we identified 14 structurally diverse chemicals, termed sorting inhibitors or Sortins that trigger secretion of CPY into the liquid medium [6]. Importantly, this observed phenotype was strictly dose-dependent, demonstrating a unique ability of chemical genomics to easily induce phenotypes in a manner that is difficult to achieve using conventional genetics [6]. While approximately 31 % of all yeast genes have definite homologs among mammals [7], this number was found to be significantly higher for trafficking machinery genes [4]. In accordance with this observation, two Sortin drugs were also found to interfere with vacuolar protein sorting in Arabidopsis [6]. This demonstrated that high-throughput chemical genomics screens using a simple unicellular eukaryote could provide a series of compounds that could be used to study similar cellular processes in higher eukaryotes. In addition, many protein trafficking genes in higher eukaryotes underwent gene duplications leading to their functional specialization [8] and an identified chemical compound may ideally target all members of a gene family.

2 Materials

2.1 Yeast Strain

The *Saccharomyces cerevisiae* INVSc1 strain (Invitrogen, Carlsbad, CA) is a fast growing diploid strain that does not secrete carboxypeptidase Y (CPY) under normal conditions. Therefore it can be successfully used as a reference strain (*see* **Note 1**).

2.2 Chemicals and Consumables

1. YPD medium: 1 % yeast extract (BD Biosciences, Franklin Lakes, NJ), 2 % bactopeptone (BD Biosciences), and 2 % glucose in MilliQ-grade water.
2. DIVERSet, a library of diverse chemical compounds in DMSO (ChemBridge, San Diego, CA) (*see* **Note 2**).
3. V-shaped 96-well polypropylene plates (Greiner Bio-One, Longwood, FL).
4. Breathe-Easy air-permeable films (Diversified Biotech, Dedham, MA).
5. Polystyrene lids for 96-well plates, high profile (Greiner Bio-One).

2.3 Immunoassay Components

1. PBS buffer (Bio-Rad, Hercules, CA) (*see* **Note 3**).
2. PBS buffer containing 0.5 % of Tween-20 (PBST).
3. Blocking solution: 5 % nonfat milk in PBST.

4. Primary antibody: monoclonal anti-CPY antibody (Molecular Probes, Eugene, OR) in 1 % nonfat milk in PBST.
5. Secondary antibody: Goat Anti-Mouse IgG Alkaline Phosphatase antibody (Sigma-Aldrich, St. Louis, MO) (*see* **Note 3**) in 1 % nonfat milk in PBST.
6. Colorimetric alkaline phosphatase substrate reagent kit (Bio-Rad, Hercules, CA) (*see* **Note 3**).
7. Immobilon-NC HAHY nitrocellulose membrane (Millipore, Billerica, MA).
8. Bio-Dot dot-blot apparatus (Bio-Rad, Hercules, CA) (*see* **Note 4**).

3 Methods

3.1 Growth Conditions

1. Inoculate and grow a starting yeast culture in YPD medium at 28 °C with constant shaking for 24 h.
2. Dilute the culture 1,000 times in fresh YPD media and plate in 96-well plates at 100 μL per well.
3. Supplement the culture with chemical compounds to a final concentration of 10 mg/L. The final concentration of DMSO in media should be 1 %. The control cells are treated with DMSO only (*see* **Note 5**).
4. Seal the plates with air-permeable films and cover them with lids. Incubate the cells for 48 h at 28 °C in the dark with constant shaking (*see* **Note 6**).

3.2 Dot-Blotting Analysis

1. After the drug treatment, centrifuge 96-well plates at 2,000 × *g*.
2. Assemble the dot-blot apparatus according to the manufacturer's instructions.
3. Transfer 60 μL of the growth media onto a pre-wet nitrocellulose membrane by using a dot-blot apparatus (*see* **Note 7**).
4. Wash the membrane briefly with MilliQ-grade water and block with 5 % milk in PBST for 1 h at room temperature.
5. Incubate the membrane with monoclonal anti-CPY antibody for 1 h at room temperature.
6. Wash the membrane five times in PBST buffer for 5–10 min.
7. Incubate the membrane with secondary antibody for 1 h at room temperature.
8. Wash the membrane five times in PBST buffer for 5–10 min.
9. Develop the membrane using colorimetric substrates for alkaline phosphatase, according to the manufacturer's instructions (*see* **Note 8**).

4 Notes

1. In principle, many yeast strain genotypes can be used as reference strains for the CPY secretion assay. However, it is strongly recommended to test a CPY secretion background using the same conditions and reagents.
2. Any diverse compound library similar to DIVERSet can be utilized; however for initial screenings, high diversity and quality drug-like and lead-like compounds are required.
3. This can be substituted.
4. Any dot-blot or slot-blot apparatus compatible with the 96-well plate format can be used.
5. The toxicity of compounds in diverse libraries is mostly unknown. For your safety, use cabinets with vertical airflow pattern during handling of concentrated compound libraries.
6. The chemical libraries may contain compounds that are light sensitive and may yield degradation products that hamper correct analysis of the induced phenotype. Treating yeast cells in the dark may avoid these complications.
7. Using a multichannel pipet is recommended.
8. To evaluate correctly the secretion phenotype caused by the exposure to the compound, it may be useful to include some known deletion *vps* strains with different levels of CPY secretion [4].

Acknowledgements

The author is supported by the Ramón y Cajal Programme by the Spanish Ministerio de Ciencia e Innovación (MICINN).

References

1. Bowers K, Stevens TH (2005) Protein transport from the late Golgi to the vacuole in the yeast Saccharomyces cerevisiae. Biochim Biophys Acta 1744(3):438–454
2. Vowels JJ, Payne GS (1998) A dileucine-like sorting signal directs transport into an AP-3-dependent, clathrin-independent pathway to the yeast vacuole. EMBO J 17(9): 2482–2493
3. Darsow T, Burd CG, Emr SD (1998) Acidic dileucine motif essential for AP-3-dependent sorting and restriction of the functional specificity of the Vam3p vacuolar t-SNARE. J Cell Biol 142(4):913–922
4. Bonangelino CJ, Chavez EM, Bonifacino JS (2002) Genomic screen for vacuolar protein sorting genes in Saccharomyces cerevisiae. Mol Biol Cell 13(7):2486–2501
5. Roberts CJ, Raymond CK, Yamashiro CT, Stevens TH (1991) Methods for studying the yeast vacuole. Methods Enzymol 194:644–661
6. Zouhar J, Hicks GR, Raikhel NV (2004) Sorting inhibitors (Sortins): chemical compounds to study vacuolar sorting in Arabidopsis. Proc Natl Acad Sci U S A 101(25):9497–9501
7. Botstein D, Chervitz SA, Cherry JM (1997) Yeast as a model organism. Science 277(5330):1259–1260
8. Sanderfoot AA, Pilgrim M, Adam L, Raikhel NV (2001) Disruption of individual members of Arabidopsis syntaxin gene families indicates each has essential functions. Plant Cell 13(3):659–666

Part III

Cheminformatics

Chapter 14

The Use of Multidrug Approach to Uncover New Players of the Endomembrane System Trafficking Machinery

Daniela Urbina, Patricio Pérez-Henríquez, and Lorena Norambuena

Abstract

Chemical Biology is a strong tool to perform experimental procedures to study the Endomembrane System (ES) in plant biology. In the last few years, several bioactive compounds and their effects upon protein trafficking as well as organelle distribution, identity, and size in plants and yeast have been characterized. Today, several of these chemical tools are widely used to perform mutant screens and establish the trafficking pathway of a given cellular component. This chapter is a guideline to perform multidrug approaches to study the endomembrane system in plant cells. This type of approach is a powerful and useful strategy to thoroughly determine the trafficking of a specific protein as well as to perform mutant screens based on phenotypes produced by drug treatments. On the other hand, a multidrug approach can address the characterization of a new bioactive molecule and find its cellular pathway target. Overall, this approach can unravel mechanisms and identify new players in endomembrane trafficking.

Key words Chemical biology, Endomembrane, Endocytosis, Secretory route

1 Introduction

In plants, as in animals, embryonic development, cell division, and growth and development of organs are enormously dependent on subcellular vesicular trafficking and the identity of cell membrane domains. Thus, the endomembrane system, which is responsible for vesicular trafficking and cell polarity, plays a relevant role in determining cell types and physiological functions of organs and tissues. Over the years, the role of genes involved in the endomembrane system has been mainly studied by genetic approaches using gain-of-function and loss-of-function mutants. This strategy has been very productive; however, it is limited when essential genes are the subject of study in which case mutations generate embryonic or gametophytic lethality. This has been the case for many plant endomembrane system genes [1–3]. Another problem that arises is the gene redundancy that has been reported for the endomembrane system. Proteins belonging to a given family are able to

Glenn R. Hicks and Stéphanie Robert (eds.), *Plant Chemical Genomics: Methods and Protocols*, Methods in Molecular Biology, vol. 1056, DOI 10.1007/978-1-62703-592-7_14, © Springer Science+Business Media New York 2014

partially or completely compensate for the absence of a family member, masking the effects of the mutation [4, 5]. In order to overcome these limitations for the study of the endomembrane system, the use of low molecular weight molecules capable of altering specific cellular pathways via alteration of protein activity, has come to the forefront. The chemical biology approach for studying the plant endomembrane system has aimed to dissect protein trafficking pathways and organelle targeting as well as define particular proteins as drug molecular targets. For studying the endomembrane system, the advantages of chemical biology are enormous. Since bioactive chemicals can perturb multiple functions within a cell or among cells of the same organism, they allow us to observe specifically a particular function. Due to the pivotal role of protein trafficking in development, generating temporary and tunable disturbances at different stages is a tremendous benefit. Several bioactive chemicals have been useful to effect understanding biological processes within endomembrane system (Fig. 1a and Table 1). In addition, reversibility feature of chemical effects is useful to specifically assign the observed phenotypes to the chemically induced disturbance. With tunable and reversible chemical tools, multidrug approaches can be designed in order to assess further knowledge upon endomembrane system trafficking pathways.

The endomembrane system is roughly divided in two main processes that are intimately connected, the secretory and endocytic pathways [6]. These processes involve several different protein trafficking routes through the endomembrane compartments as shown (Fig. 1a) [7, 8]. Protein trafficking has to be accurate and reliable in order to ensure the optimal cellular functioning. Therefore, a clear understanding is crucial in plant cell and molecular biology of plants.

Several chemicals have been characterized that alter secretory and endocytic pathways, as well as the morphology of organelles and the dynamics of vesicular trafficking within endomembrane system (Fig. 1a) [8]. For example, brefeldin A (BFA) is a well-characterized drug in terms of molecular targets and cellular perturbation. BFA is able to cause retraction of the Golgi cisternae to the endoplasmic reticulum [9]. This drug also prevents protein

Fig. 1 (continued) question [*2*]. Plants are treated with different known drugs (KD) and in combination with the unknown drug (UD) of interest, accordingly with the used protocol [*3*]. Analyze any subcellular phenotype that is caused with the KD and that might be altered with the UD [*4*]. Considering all the gathered information could be integrated to conclude. *Experiments combining drugs can be performed along with a protein synthesis inhibitor such as cycloheximide (CHX) (*see* **Note 2**). This approach is helpful to exclusively analyze the endocytic route. Conversely, endocytosis inhibitors, such as TyrA23, can be used to isolate the secretory route in order to be examined. **In this case, a PM/endosome protein GFP marker (PIN2:GFP) localization is followed with confocal microscopy. The UD antagonizes KD effects [*4*]. Choosing between protocol option 1 or option 2 can be helpful to discriminate whether unknown drug is acting upstream or downstream the known drug action site

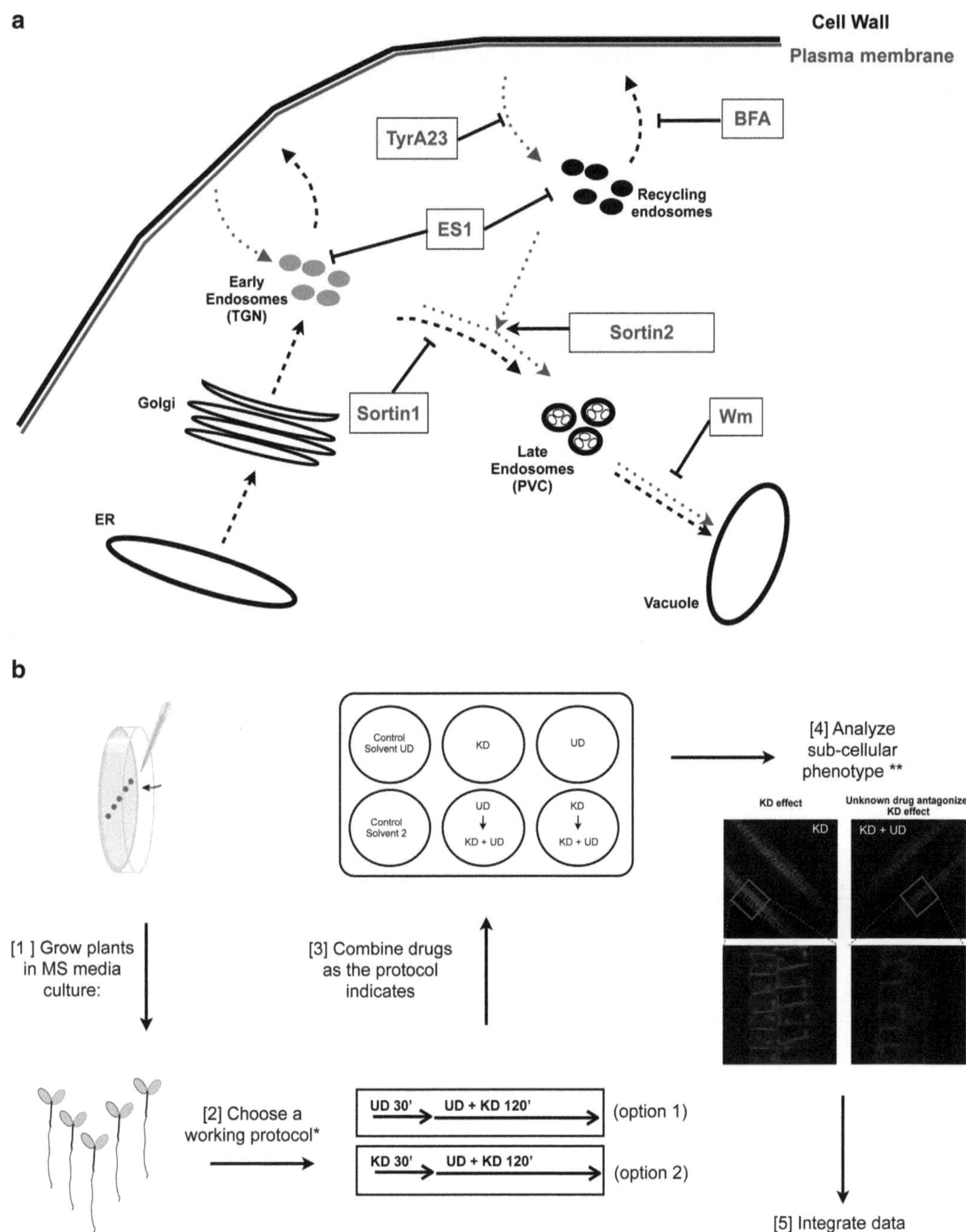

Fig. 1 (**a**) Schematic representation of the main endomembrane system components. Some subcellular components have been omitted for better understanding. *ER* endoplasmic reticulum, *TGN trans*-Golgi network, *PVC* pre-vacuolar compartment, *BFA* Brefeldin A, *Wm* Wortmannin, *TyrA23* Tyrphostin A23, *ES1* Endosidin1. Arrows indicate the direction of protein trafficking within endomembrane system. Pointed gray arrows represent the endocytic trafficking and segmented black arrows indicate the secretory route. Chemical compounds that alter specific protein trafficking are displayed in boxes pointed at the trafficking step in which they are involved. (**b**) Working Flow diagram for chemical biology research in plant cell biology. Seedlings are grown [*1*] in vertical plates in order to select and manipulate them. Different treatment protocols can be used to answer a specific

Table 1
Bioactive molecules alter endomembrane system in plant

Name	Cellular target/process involved	Effects within the endomembrane system	References
Brefeldin A	Exocytosis/vesicle fusion	Inhibits exocytosis, induces BFA bodies	[1]
Endosidin1	Endosome	Induces a SYP61/VHA-a1 endocytic bodies (ES1 bodies)	[32]
Wortmannin	Vacuolar trafficking	Inhibits PVC to vacuole trafficking	[33, 34]
Sortin1	Vacuolar trafficking	Inhibits TGN to PVC trafficking	[35]
Sortin2	Endocytosis and vacuolar trafficking	Accelerates endocytosis toward the vacuole	[24, 36]
Concanamycin A	Endosome acidification	Inhibits trafficking from TGN	[37]
Tyrphostin A23	Endocytosis/clathrin-coated vesicle formation	Inhibits cargo recruitment from PM or TGN	[13, 38]
Dynasore	Endocytosis/vesicle fission	Inhibits endocytosis	[39, 40]
Cycloheximide	Ribosomes/protein synthesis	Isolates endocytic from secretory pathway	[41]
NAA	Endocytosis	Inhibits endocytosis	[42]
TIBA	Cytoskeleton dynamics/ endocytic cycling	Inhibits vesicle trafficking	[43, 44]
NPA	Auxin influx carrier/ endocytic cycling	Inhibits vesicle trafficking	[45]
Latrunculin B	Actin cytoskeleton disruption	Reduces endomembrane system/ endosome dynamics	[46–48]
Oryzalin	Microtubule cytoskeleton disruption	Reduces endomembrane system/ endosome dynamics	[49]

recycling from early endosomes to the plasma membrane resulting in well-described BFA bodies [10]. Many vacuole proteins follow a route through the late endosome/pre-vacuolar compartment (PVC)/multi-vesicular body (MBV) to their final destination. This route can be altered by wortmannin [11]. This drug produces vacuolation of the PVC/MBV at the cellular level by inactivating a PI3 kinase, thus inhibiting PVC-to-vacuole protein trafficking [11, 12]. On the other hand, tyrphostin A23 (TyrA) is a drug that blocks endocytosis and is predicted, by molecular modeling studies, to bind to the tyrosine-binding cleft of the μ2 adapter which is crucial for YXXØ internalization motif recognition [13]. An alternative mechanism for blocking endocytosis is by drugs that alter the actin cytoskeleton such as latrunculin B or cytochalasin D [14].

Destabilization of actin filaments prevents membrane invagination, thus preventing clathrin-associated endocytosis [15]. These drugs together or in combination have helped to dissect protein trafficking pathways, such as the recycling of plasma membrane proteins and the targeting of vacuole proteins [16]. A chemical biology strategy combined with specific organelle markers or molecular probes such as the endocytic marker FM4-64 may become a powerful tool to map trafficking routes, perform screening of mutants, and characterize the subcellular location and cell processes of new proteins in the endomembrane system [17].

In this chapter, we detail experimental strategies to perform the characterization of an endomembrane system pathway which a given protein follows or to find cellular target of uncharacterized bioactive molecules. On the other hand, this method can be useful to describe, by mapping, novel routes within the endomembrane system. Furthermore, this strategy allows the identification of phenotypes that may provide the basis for screens to identify drug-resistant or hypersensitive mutants. Such mutants can reveal new proteins that function in the endomembrane system. Through the use of classical cell biology tools such as molecular dyes and fluorescent protein markers, in combination with endomembrane system-altering drugs, it is possible to dissect the endomembrane system machinery.

2 Materials

2.1 Plant Growth

1. The solid culture media for making plates (SMS Media) is 0.5× Murashige and Skoog (MS) media (MS basal media w/vitamins, Phyto Technology Laboratories #M519) pH 5.7 containing 2 % sucrose and 0.8 % Agar Plant TC (Phyto Technology Laboratories # A296).
2. *Arabidopsis thaliana* seeds are sterilized and stratified in darkness for 48 h at 4 °C prior plating. Germination and growth is performed at 22 °C under 61.6 μmol/m^2/s cool white fluorescent light (Philips TLD 30W/33) and 16 h–8 h day–night cycle.

2.2 Chemical Treatments

Table 1 summarizes the endomembrane system disturbing drugs useful in this chapter. For each drug, 4–7-day-old seedlings are transferred to polypropylene 6-well plates (Orange Scientific #5530505) containing 5 mL of chemical at optimal concentration (Table 2) in 0.5× liquid MS media, pH 5.7 containing 2 % sucrose (LMS media). Control treatments are performed in LMS with equal amounts of the respective solvents (Table 2). FM4-64 (Invitrogen #T-3166) dissolved in water is used at a final concentration of 5 μM (*see* **Note 1**). Seedlings are incubated with drugs or dye for the indicated times and procedures (Table 2).

Table 2
Working with bioactive molecules to alter endomembrane system in plant

	Source	Working solution (μM)	Stock solution	Solvent	Incubation time[a]	References
Sortin1	ChemBridge, San Diego, CA, USA	57	25 mg/ml	DMSO	16 h	[35, 36]
Sortin2	ChemBridge, San Diego, CA, USA	108	20 mg/ml	DMSO	6 h	[24, 36]
ES1	ChemBridge, San Diego, CA, USA	33	1 mM	DMSO	2 h	[32, 40]
Tyrphostin A23	Santa Cruz #sc-3554	30	30 mM	DMSO	2 h	[38, 50, 51]
Wortmannin	Sigma-Aldrich #W1628	15	30 mM	DMSO	1 h	[12, 34]
Brefeldin A	Sigma-Aldrich #B6542	50	36 mM	Ethanol	1 h	[16, 28, 29]
Concanamycin A	Sigma-Aldrich #9705	2	2 mM	DMSO	2 h	[52]
TIBA	Duchefa #T0929.0005	25	100 mM	DMSO	2 h	[44]
NAA	Duchefa #N0903.0025	10	10 mM	Water	2 h	[42]
NPA	Duchefa #N0926.0250	200	100 mM	DMSO	2 h	[44]
Latrunculin B	Invitrogen, #L22290	20	20 mM	DMSO	2–24 h	[53, 54]
Oryzalin	Sigma Aldrich #PS410	10	100 mM	Ethanol	2 h	[44, 54]
Cycloheximide	Merck #239763	50	50 mM	Ethanol	1 h	[16, 28, 29]
FM4-64	Invitrogen, #T-3166	4	10 mM	Water	5–60 min	[17]

[a]Incubation times are considering treatments on 4–7-day-old Arabidopsis seedlings

3 Methods

1. Define a particular endomembrane route of interest. Figure 1a shows endomembrane system compartments and trafficking pathways as well as frequently used chemicals to perturb protein routes or organelle morphology (*see* **Note 2**).

2. Protein targeting to different compartments or cell domains can be addressed using GFP-markers by confocal microscopy. Organelle or membrane-localized GFP markers permit compartment morphology to be visualized as well. For example, the tonoplast marker GFP-δTIP can be used to examine the morphology of vacuoles [18–20], or PIN2-GFP can be useful to study defects in endocytosis and recycling [21, 22]. Markers for all or most endomembrane system compartments have been described [listed in Norambuena et al. [23], *see* **Note 3**].
3. The trafficking of the endomembrane system markers can be challenged with drugs altering a determinate endomembrane process in order to characterize a new bioactive chemical. Drug treatments can also be helpful to described trafficking pathway of a new protein.

   Several well-characterized drugs have been used to define certain endomembrane system routes (Table 1). For example, vacuole trafficking has been defined independently with concanamycin A, BFA, or wortmannin (Table 1). Multidrug treatments can also be used to study the effect of a new bioactive chemical (Fig. 1b). This approach allows in some cases the assignment of the cellular target within the endomembrane system by mapping the affected routes with respect to a known drug [24]. As discussed here, this strategy can characterize the trafficking of a specific protein and mutant screens based on multidrug treatment phenotypes.

   The multidrug approach can be combined with different experimental platforms depending on the biological question of interest. For instance, the endosome motility assay [25] will allow the evaluation of cytoskeleton and endosome dynamics. As shown (Fig. 2a), a cytoskeleton impairing drug will also inhibit endosome motility. This assay can be used to identify novel molecules capable of reversing the observed phenotype. These novel molecules could be useful tools to describe the regulatory machinery involved in cytoskeleton and endosome dynamic.

   Furthermore, the multidrug strategy can be used to analyze different tissues or organs to identify the distinctive endomembrane system molecular machinery. Figure 2b shows for example, a known drug (KD) effect can be reversed only in one of two tissues by an unknown drug (UD), suggesting that molecular targets of the UD may be differentially distributed among the tissues. For all these kinds of hypothetical experimental approaches **steps 4–11** can be used.
4. Place five to ten 7-day-old seedlings in each well of a 6-well plate containing 5 mL LMS media supplemented with given drugs (Table 2). Alternatively, it is possible to use plants at different developmental stages and adjust the treatment volumes to the size of plant tissue (*see* **Note 4**).

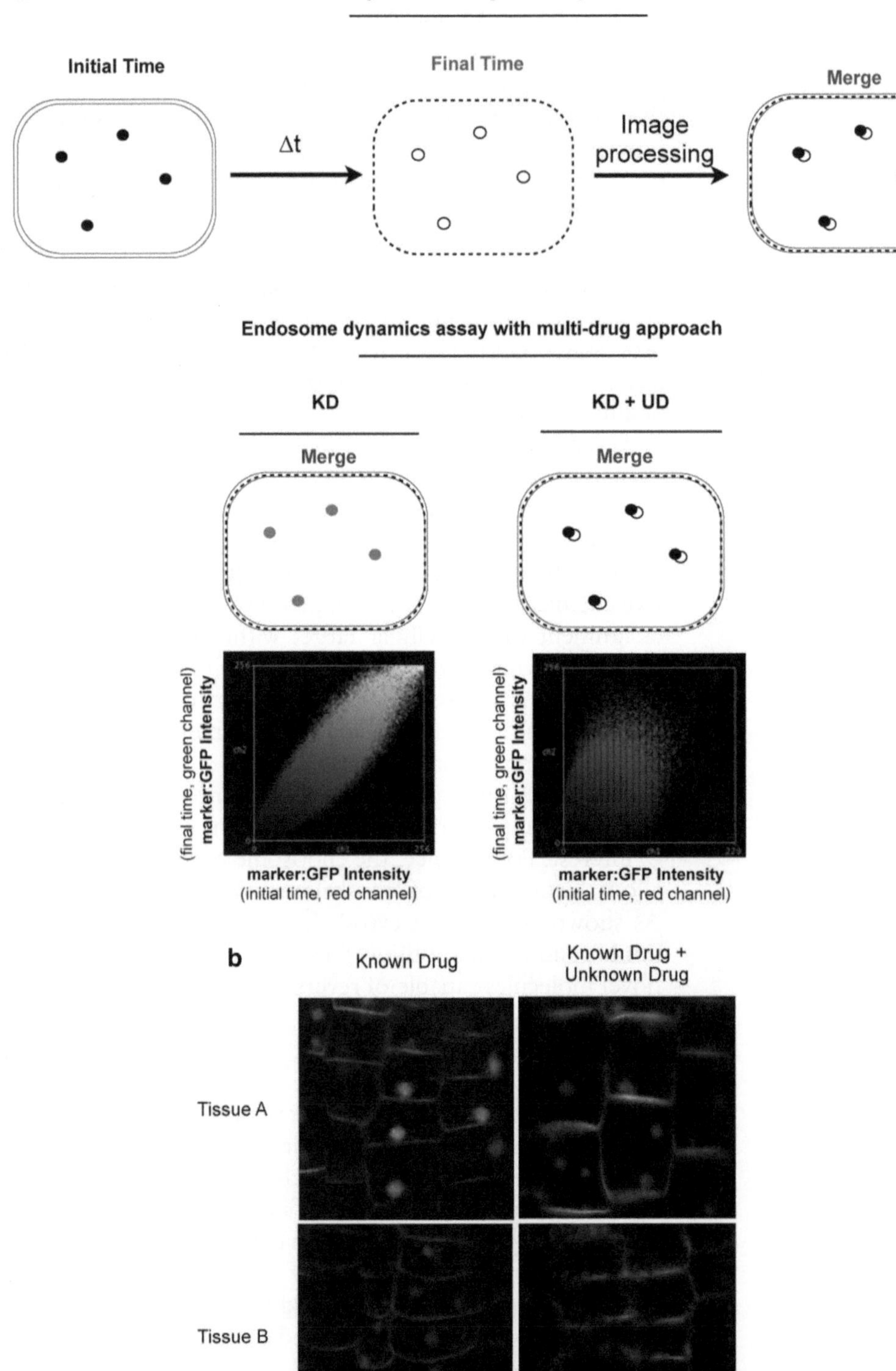

Fig. 2 (**a**) Cytoskeleton dynamic properties assay following PM/endosome protein marker motility with time-lapse analysis. By means of in vivo time-lapse imaging it is possible to assess dynamic properties of any organelle.

5. Incubate plates under standard growth conditions for the period of time suggested in Table 2 for each molecule (*see* **Note 5**). For working with multidrug treatments the trafficking processes and steps of interest must be defined first (Fig. 1) with respect to your own scientific question.
6. For reversibility of drug effects, after single or multiple treatments wash the seedlings for 10 min with 5 mL of fresh LMS. Repeat this three times.
7. Place seedlings on a microscope slide (Fisher Scientific # 12-544-1) with LMS media and glass cover slip (Fisher Scientific, # 12-540C). Image by confocal microscopy. For GFP fluorescence, the laser emission/excitation is 488/505–530 nm (*see* **Note 6**).
8. Select same cell types of interest for all the procedures. This is extremely relevant if you want to compare between several treatments.
9. Take images using 60 or 100× water immersion objective lens and do not overexpose the images (*see* **Note 6**). In further analyses organelle abundance or organelles sizes can be quantified.
10. Image analysis can be done using Macnification 1.5.1 (Orbicule® 2008) or ImageJ 1.40 (Wayne Rasband, NIH, USA).
11. For analyzing endocytosis processes or endosomes and vacuole morphology, FM4-64 dye can be used before or after the drug treatments (*see* **Note 1**).

Fig. 2 (continued) For instance, following a PM/Endosome fluorescent marker is possible to determine endosome position during a period of time for testing its motility. Due to cytoskeleton-assisted endosome dynamism, endosome movement correlates with the positive or negative effects of a given drug upon cytoskeleton. To evaluate endosome movement it is necessary to image at two different times (initial and final time). Pseudo coloring each image will allow comparison between them when merging (merge). Finally, quantifying motility by image correlation analysis (ICA) can be done using the software ImageJ (http://rsbweb.nih.gov/ij/) or any other image processing software. A known drug (KD) treatment affecting dynamic properties will render an exact merged image with colocalized endosome positions. ICA in this scenario will be seen as a diagonal-aligned cluster of points in the correlation graph. Conversely, if the unknown drug is able to revert the phenotype of KD, ICA would be seen as disperse set of points in the correlation graph. (**b**) Different trafficking molecular components between two different types of tissues. Two distinctive tissues can be identified by means of drug effect analysis meaning that molecular components between the tissues are different. For instance, an unknown drug that antagonizes KD-induced endosome aggregation effect can be tested. Tissue A is resistant to the unknown drug since the phenotypes persist even in the presence of the unknown drug. However, tissue B is vulnerable to unknown drug effect since concomitant treatment inhibits KD-phenotype occurrence

4 Notes

1. *Internalization of FM4-64 endocytosis marker dye*. Seedlings are incubated on 5 μM of FM4-64 in MSL media during 10 min at 4 °C in dark conditions. Then, incubate them in the presence or absence of the UD of interest at room temperature during a time period depending on the aim of the experiment [17]. Visualize the fluorescence exciting with 514 nm laser and capturing emission at 560 nm [17].
2. Cycloheximide inhibits de novo protein synthesis (Tables 1 and 2). We recommend use of this drug to isolate the endocytosis process. The hypothetical example (Fig. 1b) shows a study of a BFA compartment reversion phenotype using an UD. The use of cycloheximide would demonstrate that this effect is not dependent on de novo protein synthesis [24]. On the other hand, tyrphostin A23 and the synthetic auxin naphthalene acetic acid (NAA) inhibit endocytosis; therefore this approach is able to isolate the secretory pathway processes to be studied without interference from the endocytic pathway.
3. To visualize endomembrane compartments in a mutant, FM4-64 dye can be used. Alternatively and more informative transient expression of fluorescent markers in the mutants can be done [26] as well as permanent expression by crossing the mutant with endomembrane markers [27–30]. A different approach can be whole plant immunolocalization [2] using antibodies to label different endomembrane compartments [16, 31]. For the whole plant immunolocalization approach, the tissue must be fixed after chemical treatments in order to perform the immunolocalization [2].
4. A standard control treatment must always be defined. Positive control of the drug treatment shows any trouble with drug stability. The solvent controls are the most important controls of your experimental approach. Each drug stock solution is usually prepared in an organic solvent such as DMSO, ethanol, 2-propanol, or others (Table 2). DMSO in high concentration can disturb cell integrity. These kinds of negative controls are relevant to establish the true effects of drugs from the solvent.
5. Each laboratory has different environmental conditions (light and temperature variation). These variations could produce slightly different results temporally and in terms of phenotype severity. This chapter is a guideline, and it may be necessary to standardize drug concentrations, incubation times, or temperature with respect to your own conditions.
6. Long laser or mercury light exposure can produce photobleaching in GFP proteins and fluorescent dyes. Consider this recommendation to obtain good images.

References

1. Geldner N, Anders N, Wolters H, Keicher J, Kornberger W, Muller P, Delbarre A, Ueda T, Nakano A, Jurgens G (2003) The Arabidopsis GNOM ARF-GEF mediates endosomal recycling, auxin transport, and auxin-dependent plant growth. Cell 112:219–230
2. Lauber MH, Waizenegger I, Steinmann T, Schwarz H, Mayer U, Hwang I, Lukowitz W, Jurgens G (1997) The Arabidopsis KNOLLE protein is a cytokinesis-specific syntaxin. J Cell Biol 139:1485–1493
3. Nickle TC, Meinke DW (1998) A cytokinesis-defective mutant of Arabidopsis (cyt1) characterized by embryonic lethality, incomplete cell walls, and excessive callose accumulation. Plant J 15:321–332
4. De Rybel B, Audenaert D, Vert G, Rozhon W, Mayerhofer J, Peelman F, Coutuer S, Denayer T, Jansen L, Nguyen L, Vanhoutte I, Beemster GT, Vleminckx K, Jonak C, Chory J, Inze D, Russinova E, Beeckman T (2009) Chemical inhibition of a subset of *Arabidopsis thaliana* GSK3-like kinases activates brassinosteroid signaling. Chem Biol 16:594–604
5. Dettmer J, Liu TY, Schumacher K (2010) Functional analysis of Arabidopsis V-ATPase subunit VHA-E isoforms. Eur J Cell Biol 89: 152–156
6. Jurgens G (2004) Membrane trafficking in plants. Annu Rev Cell Dev Biol 20:481–504
7. Richter S, Voss U, Jurgens G (2009) Post-Golgi traffic in plants. Traffic 10:819–828
8. Hicks GR, Raikhel NV (2010) Advances in dissecting endomembrane trafficking with small molecules. Curr Opin Plant Biol 13:706–713
9. Nebenfuhr A (2002) Brefeldin A: deciphering an enigmatic inhibitor of secretion. Plant Physiol 130:1102–1108
10. Robinson DG, Langhans M, Saint-Jore-Dupas C, Hawes C (2008) BFA effects are tissue and not just plant specific. Trends Plant Sci 13: 405–408
11. Aniento F, Robinson DG (2005) Testing for endocytosis in plants. Protoplasma 226:3–11
12. Kleine-Vehn J, Leitner J, Zwiewka M, Sauer M, Abas L, Luschnig C, Friml J (2008) Differential degradation of PIN2 auxin efflux carrier by retromer-dependent vacuolar targeting. Proc Natl Acad Sci U S A 105:17812–17817
13. Banbury DN, Oakley JD, Sessions RB, Banting G (2003) Tyrphostin A23 inhibits internalization of the transferrin receptor by perturbing the interaction between tyrosine motifs and the medium chain subunit of the AP-2 adaptor complex. J Biol Chem 278: 12022–12028
14. Samaj J, Baluska F, Voigt B, Schlicht M, Volkmann D, Menzel D (2004) Endocytosis, actin cytoskeleton, and signaling. Plant Physiol 135:1150–1161
15. Kitakura S, Vanneste S, Robert S, Lofke C, Teichmann T, Tanaka H, Friml J (2011) Clathrin mediates endocytosis and polar distribution of PIN auxin transporters in Arabidopsis. Plant Cell 23:1920–1931
16. Reichardt I, Stierhof YD, Mayer U, Richter S, Schwarz H, Schumacher K, Jurgens G (2007) Plant cytokinesis requires de novo secretory trafficking but not endocytosis. Curr Biol 17: 2047–2053
17. Bolte S, Talbot C, Boutte Y, Catrice O, Read ND, Satiat-Jeunemaitre B (2004) FM-dyes as experimental probes for dissecting vesicle trafficking in living plant cells. J Microsc 214: 159–173
18. Hicks GR, Rojo E, Hong S, Carter DG, Raikhel NV (2004) Geminating pollen has tubular vacuoles, displays highly dynamic vacuole biogenesis, and requires VACUOLESS1 for proper function. Plant Physiol 134:1227–1239
19. Cutler SR, Somerville CR (2005) Imaging plant cell death: GFP-Nit1 aggregation marks an early step of wound and herbicide induced cell death. BMC Plant Biol 5:4
20. Avila EL, Zouhar J, Agee AE, Carter DG, Chary SN, Raikhel NV (2003) Tools to study plant organelle biogenesis. Point mutation lines with disrupted vacuoles and high-speed confocal screening of green fluorescent protein-tagged organelles. Plant Physiol 133:1673–1676
21. Laxmi A, Pan J, Morsy M, Chen R (2008) Light plays an essential role in intracellular distribution of auxin efflux carrier PIN2 in *Arabidopsis thaliana*. PLoS One 3:e1510
22. Men S, Boutte Y, Ikeda Y, Li X, Palme K, Stierhof YD, Hartmann MA, Moritz T, Grebe M (2008) Sterol-dependent endocytosis mediates post-cytokinetic acquisition of PIN2 auxin efflux carrier polarity. Nat Cell Biol 10:237–244
23. Norambuena L, Hicks GR, Raikhel NV (2009) The use of chemical genomics to investigate pathways intersecting auxin-dependent responses and endomembrane trafficking in *Arabidopsis thaliana*. Methods Mol Biol 495: 133–143
24. Pérez-Henríquez P, Raikhel N, Norambuena L (2012) Endocytic trafficking towards the vacuole plays a key role in the SCF TIR-independent mechanism of lateral root formation in *Arabidopsis thaliana*. Mol Plant 5(6):1195–1209
25. Voigt B, Timmers AC, Samaj J, Hlavacka A, Ueda T, Preuss M, Nielsen E, Mathur J, Emans N,

Stenmark H, Nakano A, Baluska F, Menzel D (2005) Actin-based motility of endosomes is linked to the polar tip growth of root hairs. Eur J Cell Biol 84:609–621

26. Bandmann V, Kreft M, Homann U (2011) Modes of exocytotic and endocytotic events in tobacco BY-2 protoplasts. Mol Plant 4: 241–251
27. Haseloff J, Siemering KR, Prasher DC, Hodge S (1997) Removal of a cryptic intron and subcellular localization of green fluorescent protein are required to mark transgenic Arabidopsis plants brightly. Proc Natl Acad Sci U S A 94: 2122–2127
28. Feraru E, Paciorek T, Feraru MI, Zwiewka M, De Groodt R, De Rycke R, Kleine-Vehn J, Friml J (2010) The AP-3 beta adaptin mediates the biogenesis and function of lytic vacuoles in Arabidopsis. Plant Cell 22:2812–2824
29. Dhonukshe P, Baluska F, Schlicht M, Hlavacka A, Samaj J, Friml J, Gadella TW Jr (2006) Endocytosis of cell surface material mediates cell plate formation during plant cytokinesis. Dev Cell 10:137–150
30. Ueda T, Yamaguchi M, Uchimiya H, Nakano A (2001) Ara6, a plant-unique novel type Rab GTPase, functions in the endocytic pathway of *Arabidopsis thaliana*. EMBO J 20:4730–4741
31. Grebe M, Xu J, Mobius W, Ueda T, Nakano A, Geuze HJ, Rook MB, Scheres B (2003) Arabidopsis sterol endocytosis involves actin-mediated trafficking via ARA6-positive early endosomes. Curr Biol 13:1378–1387
32. Robert S, Chary SN, Drakakaki G, Li S, Yang Z, Raikhel NV, Hicks GR (2008) Endosidin1 defines a compartment involved in endocytosis of the brassinosteroid receptor BRI1 and the auxin transporters PIN2 and AUX1. Proc Natl Acad Sci U S A 105:8464–8469
33. Stephens L, Cooke FT, Walters R, Jackson T, Volinia S, Gout I, Waterfield MD, Hawkins PT (1994) Characterization of a phosphatidylinositol-specific phosphoinositide 3-kinase from mammalian cells. Curr Biol 4:203–214
34. Wang J, Cai Y, Miao Y, Lam SK, Jiang L (2009) Wortmannin induces homotypic fusion of plant prevacuolar compartments. J Exp Bot 60: 3075–3083
35. Rosado A, Hicks GR, Norambuena L, Rogachev I, Meir S, Pourcel L, Zouhar J, Brown MQ, Boirsdore MP, Puckrin RS, Cutler SR, Rojo E, Aharoni A, Raikhel NV (2011) Sortin1-hypersensitive mutants link vacuolar-trafficking defects and flavonoid metabolism in Arabidopsis vegetative tissues. Chem Biol 18: 187–197
36. Zouhar J, Hicks GR, Raikhel NV (2004) Sorting inhibitors (Sortins): chemical compounds to study vacuolar sorting in Arabidopsis. Proc Natl Acad Sci U S A 101:9497–9501
37. Muller O, Neumann H, Bayer MJ, Mayer A (2003) Role of the Vtc proteins in V-ATPase stability and membrane trafficking. J Cell Sci 116:1107–1115
38. Dhonukshe P, Aniento F, Hwang I, Robinson DG, Mravec J, Stierhof YD, Friml J (2007) Clathrin-mediated constitutive endocytosis of PIN auxin efflux carriers in Arabidopsis. Curr Biol 17:520–527
39. Macia E, Ehrlich M, Massol R, Boucrot E, Brunner C, Kirchhausen T (2006) Dynasore, a cell-permeable inhibitor of dynamin. Dev Cell 10:839–850
40. Sharfman M, Bar M, Ehrlich M, Schuster S, Melech-Bonfil S, Ezer R, Sessa G, Avni A (2011) Endosomal signaling of the tomato leucine-rich repeat receptor-like protein LeEix2. Plant J 68(3):413–423
41. Viotti C, Bubeck J, Stierhof YD, Krebs M, Langhans M, van den Berg W, van Dongen W, Richter S, Geldner N, Takano J, Jurgens G, de Vries SC, Robinson DG, Schumacher K (2010) Endocytic and secretory traffic in Arabidopsis merge in the trans-Golgi network/early endosome, an independent and highly dynamic organelle. Plant Cell 22:1344–1357
42. Paciorek T, Zazimalova E, Ruthardt N, Petrasek J, Stierhof YD, Kleine-Vehn J, Morris DA, Emans N, Jurgens G, Geldner N, Friml J (2005) Auxin inhibits endocytosis and promotes its own efflux from cells. Nature 435: 1251–1256
43. Dhonukshe P, Grigoriev I, Fischer R, Tominaga M, Robinson DG, Hasek J, Paciorek T, Petrasek J, Seifertova D, Tejos R, Meisel LA, Zazimalova E, Gadella TW Jr, Stierhof YD, Ueda T, Oiwa K, Akhmanova A, Brock R, Spang A, Friml J (2008) Auxin transport inhibitors impair vesicle motility and actin cytoskeleton dynamics in diverse eukaryotes. Proc Natl Acad Sci U S A 105:4489–4494
44. Geldner N, Friml J, Stierhof YD, Jurgens G, Palme K (2001) Auxin transport inhibitors block PIN1 cycling and vesicle trafficking. Nature 413:425–428
45. Noh B, Murphy AS, Spalding EP (2001) Multidrug resistance-like genes of Arabidopsis required for auxin transport and auxin-mediated development. Plant Cell 13:2441–2454
46. Baluska F, Hlavacka A, Samaj J, Palme K, Robinson DG, Matoh T, McCurdy DW, Menzel D, Volkmann D (2002) F-actin-dependent endocytosis of cell wall pectins in meristematic root cells. Insights from brefeldin A-induced compartments. Plant Physiol 130: 422–431

47. Spector I, Shochet NR, Blasberger D, Kashman Y (1989) Latrunculins—novel marine macrolides that disrupt microfilament organization and affect cell growth: I. Comparison with cytochalasin D. Cell Motil Cytoskeleton 13: 127–144
48. Zhang Y, He J, Lee D, McCormick S (2010) Interdependence of endomembrane trafficking and actin dynamics during polarized growth of Arabidopsis pollen tubes. Plant Physiol 152: 2200–2210
49. Baskin TI, Wilson JE, Cork A, Williamson RE (1994) Morphology and microtubule organization in Arabidopsis roots exposed to oryzalin or taxol. Plant Cell Physiol 35:935–942
50. Ortiz-Zapater E, Soriano-Ortega E, Marcote MJ, Ortiz-Masia D, Aniento F (2006) Trafficking of the human transferrin receptor in plant cells: effects of tyrphostin A23 and brefeldin A. Plant J 48:757–770
51. Fujimoto M, Arimura S, Ueda T, Takanashi H, Hayashi Y, Nakano A, Tsutsumi N (2010) Arabidopsis dynamin-related proteins DRP2B and DRP1A participate together in clathrin-coated vesicle formation during endocytosis. Proc Natl Acad Sci U S A 107:6094–6099
52. Dettmer J, Hong-Hermesdorf A, Stierhof YD, Schumacher K (2006) Vacuolar H+-ATPase activity is required for endocytic and secretory trafficking in Arabidopsis. Plant Cell 18: 715–730
53. Kim H, Park M, Kim SJ, Hwang I (2005) Actin filaments play a critical role in vacuolar trafficking at the Golgi complex in plant cells. Plant Cell 17:888–902
54. Wright KM, Wood NT, Roberts AG, Chapman S, Boevink P, Mackenzie KM, Oparka KJ (2007) Targeting of TMV movement protein to plasmodesmata requires the actin/ER network: evidence from FRAP. Traffic 8:21–31

Chapter 15

Cheminformatic Analysis of High-Throughput Compound Screens

Tyler W.H. Backman and Thomas Girke

Abstract

This article gives an overview of basic computational methods that are commonly used for analyzing small molecule screening data in the chemical genomics field. First, we introduce cheminformatic concepts for analyzing drug-like small molecule structures and their properties. Second, we introduce compound selection approaches for assembling screening libraries using compound property and diversity analyses. Finally, we discuss methods for interpreting screening hits by analyzing compound structures and induced phenotypes using similarity search and clustering approaches. These are critical steps for optimizing screening hits, and relating structure to bioactivity and phenotype.

Key words Small molecule, Cheminformatics, Clustering, Phenotype, Chemical genomics screen

1 Introduction

Software tools for analyzing small molecules play an important role in many bioscience and biomedical areas, including drug discovery, chemical biology, chemical genomics, and medicinal chemistry [1–5]. These tools are essential for analyzing the structural similarities, physicochemical properties, and bioactivity profiles of natural and synthetic compounds to gain insight into their modes of action in biological systems. These insights are important for developing effective small molecule probes used for studying the function of protein and cellular networks in drug discovery and chemical genomics research [6]. Additionally, software tools can assist in the identification of structural and physicochemical relationships among compounds in metabolic or signaling pathways [7–9]. Moreover, the growing relevance of chemical genomics approaches in biomedical research has increased the demand for small molecule analysis software in academia [10, 11].

The number of small molecule structures in the publicly available chemical space is increasing rapidly. At the time of writing, the structures of over 30 million distinct small molecules are available

Glenn R. Hicks and Stéphanie Robert (eds.), *Plant Chemical Genomics: Methods and Protocols*, Methods in Molecular Biology, vol. 1056, DOI 10.1007/978-1-62703-592-7_15, © Springer Science+Business Media New York 2014

in open-access databases, like PubChem [12–14], ChemBank [15], NCI [16], ChemDB [17], and ZINC [18]. Additionally, preliminary bioactivity data from hundreds of high-throughput screening (HTS) experiments against a wide spectrum of target sites have recently become available for over one million compounds in bioassay databases curated by PubChem, ChemBank, BindingDB [19], ChemMine [20], and others. Additional reference databases for bioactivity information of small molecules and their target sites are KEGG LIGAND [9], BindingDB [19], PDBind [21], ChEBI [22], AffinDB [23], STITCH [24], DrugBank [25], and SuperTarget [26]. Other noncommercial databases provide access to 3D structures, physicochemical property predictions or commercial availability data on compounds. These include online services such as ChemDB [17], ZINC [18], and SuperDrug [27]. To mine these data resources efficiently, a broad knowledge of existing bioinformatics and cheminformatics resources is necessary. The most essential software utilities for analyzing these data types are summarized here (Table 1).

Structural similarity search tools are important for retrieving and organizing chemical and bioactivity information from databases [39, 40], as well as predicting bioactivity properties of small molecules [41]. The most commonly used structural search approaches in this area are substructure, superstructure, and fragment-based similarity searches [39]. Physicochemical property predictions are important for assessing the drug-likeness and lead-likeness of compounds in silico [42–46]. These make it possible to predict bioactivity and other properties of small molecules using machine learning approaches. Additionally, they are fundamental to the development of quantitative structure–activity relationship (QSAR) models [35, 47, 48].

Clustering and machine learning are useful techniques in the discovery of bioactive compounds [49, 50]. Commonly, they utilize the information generated by the above compound search and property prediction methods as multidimensional input for the actual clustering and classification steps. Important applications of these methods include diversity analyses of compound collections, design of custom screening libraries, prioritization of lead compounds, and predictions of compounds with bioactive or other properties of interest. Some examples of freely available software for analyzing small molecule screening data are given in Table 1. Despite the importance of these analysis utilities for many bioscience areas, only very few public domain resources are available, as the majority of cheminformatics tools are currently commercial software.

The following gives a brief overview of the basic concepts used to analyze small molecule structures and screening hits from chemical genomics experiments. Among the various software tools available for the individual analysis steps, we have chosen to focus

Table 1
Examples of software tools commonly used in small molecule discovery

Software	Key functionality	Compound input
Web tools		
ChemMine Tools [28]	format interconversion, structure visualization, similarity comparison, similarity clustering, property prediction, similarity searching	SDF, SMILES, Mouse clicks
PubChem [12]	compound database, bioactivity database, structure visualization, similarity clustering, similarity searching	SDF, SMILES, CID, Mouse clicks, InChI
WENDI—Web engine for Non-obvious drug information [29]	similarity searching, compound–gene relationship searching	SMILES, Mouse clicks
Chembench [30]	QSAR	SDF, ACT (activity data)
Downloadable applications		
Bioclipse [31]	library management, property prediction, structure visualization, QSAR	SDF, SMILES, Mouse clicks
KNIME—the Konstanz information miner desktop [32]	graphical cheminformatics pipeline development, format interconversion, property prediction, structure visualization	SDF, SMILES, Mol2
Low level cheminformatics		
ChemmineR [33]	format interconversion, structure visualization, similarity comparison, similarity clustering, similarity searching	SDF, SMILES
CDK—Chemistry development kit [34, 35]	Java library: format interconversion, structure visualization, property prediction	SDF, SMILES, CML, InChI, and others
OpenBabel [36]	Perl, Python, C++, and graphical interfaces: format interconversion, property prediction	SDF, SMILES, Mol2, PDB, XYZ, CML, and others
CDL—Chemical Descriptors Library [37]	C++ library for property prediction	SDF, SMILES, InChI
JOELib [38]	Java library: format interconversion, property prediction	SDF, SMILES, GAUSSIAN, CML, MOPAC

Due to space considerations, this list is not complete

here on three relatively popular environments that are freely available in the public domain (Table 1). For more advanced users, the command-line driven statistical environment R combined with its many add-on packages is by far the most flexible solution [51]. In addition, we have chosen ChemMine Tools and PubChem as examples of graphical online environments that are very easy to learn, especially for occasional, non-expert users [52].

2 Materials

2.1 Important Data Formats

The most common data format to represent the structures and annotations of small molecules is the Structure Data Format (SDF; Fig. 1a). This format stores many compounds in a single text file. Each compound is represented by a header, atom, bond, and data block, while the delimiter "$$$$" serves as separator between compounds. The atom and bond blocks define the two- or three-dimensional structure of each compound. The header block contains various technical specifications and the data block stores annotation and meta-information associated with each compound. A more condensed format to store compound structures is the SMILES (Simplified Molecular Input Line Entry Specification) format (Fig. 1b). It specifies the two-dimensional structure of a compound using a simple line notation where each compound is stored on a single line. Most cheminformatic tools support these and many other input formats. The compound structures from public databases or small molecule vendors can usually be downloaded in large batches in either of these formats.

a

```
702
  -OEChem-09131121112D

  9  8  0     0  0  0  0  0  0999 V2000
    2.5369   -0.2500    0.0000 O   0  0  0  0  0  0  0  0  0  0  0  0
    3.4030    0.2500    0.0000 C   0  0  0  0  0  0  0  0  0  0  0  0
    4.2690   -0.2500    0.0000 C   0  0  0  0  0  0  0  0  0  0  0  0
  1  2  1  0  0  0  0
  2  3  1  0  0  0  0
M  END
> <PUBCHEM_COMPOUND_CID>
702

$$$$
```

b

```
CCO
```

c

```
21680720     44.50
13942129     18.22
Caffeine      0.00
```

Fig. 1 Examples of common cheminformatics text file formats. (**a**) Ethanol in SD format (SDF) representing a chemical structure with atom and bond tables. (**b**) Ethanol in simplified molecular-input line-entry system (SMILES) format representing a chemical structure as a single text string. (**c**) Example ACT file showing three compounds and associated scores from a compound activity screening experiment

2.2 Activity Data

Most of the cheminformatics tools listed (Table 1) will support physiochemical properties and bioactivity data represented as a tab delimited file with one row per compound, and one column per property. The first column stores the compound names or their identifiers. The subsequent columns contain numeric values which can represent bioactivity data from high-throughput screens or their physiochemical properties. An "ACT" file with the extension "act" is a special case of this format containing only an identifier column and a single numeric column representing the activity of each compound from a given compound screen (see example in Fig. 1c).

3 Methods

3.1 Selecting Compounds for Screening

Apart from the usual bioassay development steps, many screening projects start with the selection of a small molecule library. In academic environments, the compounds are often acquired from commercial small molecule vendors and then managed by a central screening facility. Depending on the specific needs of a project, the user can cherry-pick compounds from larger collections or choose to screen preformatted libraries. The latter is often a more cost effective approach. Based on various economic and practical considerations, the main objective in the compound selection process is usually to enrich a screening collection for structural diversity, drug-likeness, and lead-likeness, and to minimize the number of candidates with undesirable properties, such as compounds with problematic features, promiscuous binding properties, or instability characteristics [53, 54]. For instance, the famous Lipinski Rule of Five [42] is often applied to enrich compound collections with drug-like candidates. This rule filters for compounds with ≤ 5 hydrogen bond donors, ≤ 10 hydrogen acceptors, a molecular weight ≤ 500 Da, and an octanol–water partition coefficient log $P \leq 5$. To obtain from a primary screen a representative number of active compounds, e.g., 50–100 candidates, it is usually necessary to screen at least 100–200 times as many compounds. This estimate assumes an average hit frequency of 0.5–1 %. The latter frequency can greatly vary depending on the properties represented in a compound library and the target site(s) used for screening. The following gives a brief overview of basic library assembly and assessment steps.

1. Drug- and Lead-Likeness Filters
   Filtering compounds for drug-likeness and lead-likeness requires the prediction of various physicochemical property descriptors for each candidate. Often these descriptors are readily available in the data block of their SD files. If this is not the case then they can be generated in silico with different software tools [44, 45]. For instance, compounds downloaded from

PubChem contain in their SD files' 20 numeric descriptors, and ChemMine Tools provides an online interface to the property prediction module of the JOELib package [38]. This service can calculate 38 physicochemical property values for custom compound sets including the Lipinski descriptors. In the R environment the descriptors can be calculated using various molecular property functions provided by the ChemmineR and rcdk libraries [55]. Once the descriptors are available, one can perform the actual filtering on their values using standard spreadsheet programs or data subsetting routines in R.

2. Removal of Compounds with Problematic Properties
   The descriptor sets generated in the previous step, such as enumerations of functional groups and other relevant structural patterns, can be used to eliminate many compounds with undesirable properties [53, 54]. Usually, most of these problematic compounds have been removed from commercial screening libraries already. Confirming that this is the case can be important to avoid promiscuous binders in a collection (*see* **Note 1**).
3. Diversity Selection
   Many diversity selection approaches exist to maximize the structural and property distances in compound collections [53, 56, 57]. Compound descriptors (structure, properties, or both) are the most relevant parameter for diversity selection, as they are used by a suitable distance metric to calculate the distance between compounds within a set. A selection algorithm is then used to select a user definable number of compounds in which the inter-compound distances are maximized. For the latter selection step, many non-supervised clustering algorithms, principal component analyses (PCA) and binning approached can be used. After the clustering, one can choose from each cluster of similar compounds only one or a small number of their most distant members. To improve the confidence in screening hits, it is also common to assemble "twin" libraries (where every compound is represented by two or more structurally highly similar members), while the pair-to-pair distances are maximized. This can be useful for prioritizing screening hits. If both partners in a pair of related compounds score as active in an HTS experiment then they are less likely to be false positives than single compound hits within a pair. More details on distance measures and clustering methods are provided in the search and clustering sections of this article, where very similar analysis routines are commonly used.

3.2 Scoring Screening Hits

The raw data generated by small molecule screens can be very diverse, but certain analysis routines are common in most cases. These include normalization, scoring, and replicate summarization steps.

For instance, the data from single target screens (e.g., binding and enzyme assays) are usually numeric values from plate readers, while the data types from high-content screens can reach from plate reader intensity values to complex phenotype results, including growth inhibition rates or image data from macroscopic assays or high-throughput microscope screens. To consolidate these diverse data sets and to allow comparisons among compound activities across different screens, simple scaling approaches are often used in public screening databases like PubChem. Here the plate reader results or image measurements are scaled to relative activity values from 0 to 100. With some limitations, a relative scale can also handle binary screening outcomes, such as simple active or non-active calls. A more detailed description of the methods and statistical analysis steps used for HTS raw data analysis goes beyond the scope of this article.

1. Analysis Tools for Screening Data
   Excellent resources for performing these raw data level analyses are the R libraries HTSanalyzeR, cellHTS2, and imageHTS, which are all available for free download from the Bioconductor depository [58, 59]. In addition, the cellHTS2 project provides a user-friendly Web interface for its raw data analysis pipeline (http://web-cellhts2.dkfz.de).

3.3 Structural Similarity Searching

Structural similarity searches are an efficient approach to identify similar compounds in reference or screening databases of compound structures. The obtained search results often provide critical information for many downstream analysis steps, such as prioritizing screening hits and assembling analog sets for secondary screens of related compounds for QSAR studies. They can also be used for identifying target site candidates in high-content screens. For instance, if a screening hit of unknown activity is structurally related to a compound with well-characterized bioactivity then it is probable that it binds to the same or a related target site. A common approach to compare and search small molecules with respect to their structural similarities is to enumerate their structural features. These numerical counts of structural features are called "structural descriptors." The numbers of common and unique features are then used to calculate a similarity measure of a query compound against the database entries. The results are then sorted for the user from the most to the least similar search hits. The most commonly used descriptor types are atom pairs and fingerprints.

1. Atom Pair Searches
   Atom pairs are a structural descriptor that is defined by the shortest paths among the non-hydrogen atoms in a molecule. Each path is described by the types of atoms in a pair, the length of their shortest bond path, the number of their π electrons, and the non-hydrogen atoms bonded to them. The number of atom pairs describing a molecule grows with its number of atoms. To use atom pairs for similarity comparisons,

one can simply enumerate their common and unique atom pairs, and then use these numbers to compute a similarity coefficient, such as the Tanimoto coefficient. The latter coefficient is defined as $c/(a+b+c)$, which is the proportion of the features shared among two compounds divided by their union. The variable c is the number of features (or on-bits in binary fingerprints) common in both compounds, while a and b are the number of features that are unique in one or the other compound, respectively. The Tanimoto coefficient has a range from 0 to 1 with higher values indicating greater similarity than lower ones. It is important to emphasize that a Tanimoto coefficient of 1 does not necessarily mean that two compounds are identical. It only means that they have identical structural descriptors or identical on-bits in a binary fingerprint. Related similarity coefficients are the Tversky and Dice indices. The Tversky index is defined as $c/(\alpha \times a+\beta \times b+c)$. It extends the Tanimoto index by two weighting variables α and β. If α and β are set to 1 then the index returns the same result as the Tanimoto coefficient. Setting α and β in the Tversky index to 0.5 returns the Dice index. The atom pair search algorithm is implemented in the ChemmineR library, which allows the user to search any custom compound collection. Its time performance is usually sufficient to search compound sets in size ranges of tens to several hundred thousand molecules. For larger data sets it is recommended to use the computationally less demanding fingerprint approach.

2. Fingerprint Searches
   Fingerprints and structural keys are constant size representations of structural features of compounds. Among the many different types of fingerprints available, only the widely used PubChem fingerprints will be covered here. These fingerprints are a binary representation of the presence and absence of a library of 881 substructure features (ftp://ftp.ncbi.nih.gov/pubchem/specifications/pubchem_fingerprints.txt). In this system every molecular structure is described by 881 bits where 1 indicates the presence and 0 the absence of a feature. Compared to atom pairs, the PubChem fingerprints are a knowledge-based system that stores less information than the much more complex and unbiased atom pair concept. For database searching fingerprints are often much more time and memory efficient, but they are less sensitive than atom pair descriptors [40, 60]. PubChem fingerprint searches can be performed online against the PubChem database using the online search services provided by PubChem and ChemMine Tools. Alternatively, one can access it from R with the ChemmineR library. The same search method can be used in rcdk, which is useful for searching custom databases.

3. Maximum Common Substructure for Compound Similarity Comparisons
   The maximum common substructure (MCS) method is a graph-based similarity concept that is defined as the largest substructure (sub-graph) shared among two compounds [61]. It is a pair-wise concept that is not directly related to the above structural descriptors, but its results (e.g., size of MCS relative to source structures) can be used for the computation of the same similarity coefficients. Compared to descriptor-based similarity concepts, the MCS method provides the most accurate and sensitive similarity measure, especially for compounds with large size differences. Similarity scores between compound pairs can be computed with the Similarity Toolbox provided by ChemMine Tools. The interface calculates atom pair and maximum common substructure (MCS) similarities with the Tanimoto coefficient, Dice's coefficient, and Tversky index as similarity measures.
4. Database Selection
   The choice of a suitable reference database is an important decision in structural similarity searching (*see* Introduction for database overview). To retrieve information about the bioactivity of related compounds, it is often useful to perform comprehensive searches against the bioactivity space available in public databases, including PubChem, ChemBank, BindingDB, etc. Searching databases of known drugs and metabolic compounds, including DrugBank and KEGG, can provide additional useful information about screening hits. In order to assemble analog sets for downstream QSAR analyses, one can search databases containing commercial availability and supplier information, such as eMolecules, ChemSpider, Zink, and many others. Most public compound databases provide online similarity search interfaces. In many cases it is also possible to download the complete compound set in SDF format to build custom databases for local search strategies. This can also be useful for batch searches of large numbers of query compounds.

3.4 Clustering Compounds

Clustering of compounds by the similarity of their structural, physicochemical, or bioactivity profiles is a powerful approach for correlating structural features of compounds with their activities. Many clustering approaches require as input a distance matrix that can be easily generated by all-against-all compound similarity comparisons and then subtracting the obtained similarity coefficients (e.g., Tanimoto coefficient Tc) from one (1—Tc). The resulting distance matrix is then passed on to the actual clustering program. In some cases the clustering software calculates the distance matrix internally. In addition to structural similarity, one can cluster compounds by the similarity of their physiochemical property or bioactivity profiles. There are many choices for computing similarity/dissimilarity measures between these profiles. For instance, with binary (active/non-active) phenotype data one can use the Hamming distance, and for quantitative results one can use the Euclidean distance or correlation-based approaches, such as the Pearson correlation coefficient. Prior to clustering, correlation-based values need to be transformed into distances by subtracting them from one. Frequently used clustering algorithms in this area are hierarchical clustering, multidimensional scaling (MDS), binning, and Jarvis–Patrick clustering [60].

1. Hierarchical Clustering
   Most hierarchical clustering algorithms join the most to least similar items (e.g., compound structures) in an agglomerative manner using as cluster joining rule either single or average or complete or other linkage methods. The output of hierarchical clustering is a similarity tree (dendrogram) that can be used to sort the rows of the underlying distance matrix, property, or bioactivity table according to the order defined in the dendrogram. For visualization purposes, one can present the dendrogram next to a heat map where the tabular data are represented by a color scheme. When clustering a second data type (i.e., activities), one can also include a column dendrogram and sort the table columns accordingly. Hierarchical clustering by structural compound similarities is available in PubChem, but only for compounds represented in its database. The ChemMine Tools and ChemmineR environments are more flexible by supporting clustering of custom compound collections by structural similarity, physicochemical properties and bioactivity profiles.

2. MDS Clustering
   Similar to hierarchical clustering, MDS starts with a matrix of item–item distances and then assigns coordinates for each item in a two- to three-dimensional space to represent the distances graphically in a scatter plot which resembles a PCA plot. MDS clustering can be performed in ChemMine Tools and ChemmineR using structural similarities, physicochemical properties or bioactivity profiles as inputs.

3. Binning Clustering
   Binning clustering assigns compounds to similarity groups based on a user-definable similarity cutoff. For instance, if a Tanimoto coefficient of 0.6 is chosen then compounds will be joined into groups that share a similarity of this value or greater using a single linkage rule for cluster joining. Because an optimum similarity threshold is often not known, most implementations can return the clustering results for multiple cutoffs. Binning clustering is currently supported by ChemMine Tools and ChemmineR.

4 Notes

1. One method to check a screening library for potentially problematic compounds is to compute physicochemical properties and molecular descriptors, and filter for compounds which contain frequently problematic functional groups, or have other properties rarely associated with drug-likeness and lead-likeness [53, 54]. A quick first step is to compute a spreadsheet of physicochemical properties using a tool such as the Chemmine Tools property toolbox (*see* Table 1). One can then import the results into spreadsheet software, or a programming environment such as R where one can filter and sort compounds by properties commonly associated with drug-likeness and lead-likeness. The Lipinski Rule of Five [42] is a common predictor of drug-likeness which can be applied by identifying compounds with ≤5 hydrogen bond donors, ≤10 hydrogen acceptors, a molecular weight ≤500 Da, and an octanol–water partition coefficient log $P \leq 5$.

Acknowledgments

We thank the community software development projects listed in Table 1. We also acknowledge support from the core facilities at the Institute for Integrative Genome Biology (IIGB) at UC Riverside. The authors cheminformatics tools (ChemMine Tools and ChemmineR) were developed with support from the National Science Foundation [grant numbers: ABI-0957099, 2010-0520325 and IGERT-0504249].

References

1. Oprea TI (2002) Chemical space navigation in lead discovery. Curr Opin Chem Biol 6(3): 384–389
2. Strausberg RL, Schreiber SL (2003) From knowing to controlling: a path from genomics to drugs using small molecule probes. Science 300(5617):294–295
3. Savchuk NP, Balakin KV, Tkachenko SE (2004) Exploring the chemogenomic knowledge space with annotated chemical libraries. Curr Opin Chem Biol 8(4):412–417
4. Haggarty SJ (2005) The principle of complementarity: chemical versus biological space. Curr Opin Chem Biol 9(3):296–303
5. Oprea TI, Tropsha A, Faulon JL, Rintoul MD (2007) Systems chemical biology. Nat Chem Biol 3(8):447–450
6. Dobson CM (2004) Chemical space and biology. Nature 432(7019):824–828
7. Hattori M, Okuno YY, Goto S, Kanehisa M (2003) Heuristics for chemical compound matching. Genome Inform 14:144–153
8. Zhang P, Foerster H, Tissier CP, Mueller L, Paley S, Karp PD, Rhee SY (2005) MetaCyc and AraCyc. Metabolic pathway databases for plant research. Plant Physiol 138(1):27–37
9. Kanehisa M, Goto S, Hattori M, Aoki-Kinoshita KF, Itoh M, Kawashima S, Katayama T, Araki M, Hirakawa M (2006) From genomics to chemical genomics: new developments in KEGG. Nucleic Acids Res 34(Database issue): 354–357
10. Schreiber SL (1998) Chemical genetics resulting from a passion for synthetic organic chemistry. Bioorg Med Chem 6(8):1127–1152
11. Olah MM, Bologa CG, Oprea TI (2004) Strategies for compound selection. Curr Drug Discov Technol 1(3):211–220

12. Li Q, Cheng T, Wang Y, Bryant SH (2010) PubChem as a public resource for drug discovery. Drug Discov Today 15(23–24):1052–1057
13. Austin CP, Brady LS, Insel TR, Collins FS (2004) NIH molecular libraries initiative. Science 306(5699):1138–1139
14. PubChem Team (2008) PubChem is a NCBI database that provides information on the biological activities of small molecules. http://pubchem.ncbi.nlm.nih.gov
15. Seiler KP, George GA, Happ MP, Bodycombe NE, Carrinski HA, Norton S, Brudz S, Sullivan JP, Muhlich J, Serrano M, Ferraiolo P, Tolliday NJ, Schreiber SL, Clemons PA (2008) ChemBank: a small-molecule screening and cheminformatics resource database. Nucleic Acids Res 36(Database issue):351–359
16. Ihlenfeldt WD, Voigt JH, Bienfait B, Oellien F, Nicklaus MC (2002) Enhanced CACTVS browser of the open NCI database. J Chem Inf Comput Sci 42(1):46–57
17. Chen JH, Linstead E, Swamidass SJ, Wang D, Baldi P (2007) ChemDB update-full-text search and virtual chemical space. Bioinformatics 23(17):2348–2351
18. Irwin JJ, Shoichet BK (2005) ZINC-a free database of commercially available compounds for virtual screening. J Chem Inf Model 45(1): 177–182
19. Liu T, Lin Y, Wen X, Jorissen RN, Gilson MK (2007) BindingDB: a web-accessible database of experimentally determined protein-ligand binding affinities. Nucleic Acids Res 35(4): 198–201
20. Girke T, Cheng LC, Raikhel N (2005) ChemMine. A compound mining database for chemical genomics. Plant Physiol 138(2): 573–577
21. Wang R, Fang X, Lu Y, Wang S (2004) The PDBbind database: collection of binding affinities for protein-ligand complexes with known three-dimensional structures. J Med Chem 47(12):2977–2980
22. Degtyarenko K, de Matos P, Ennis M, Hastings J, Zbinden M, McNaught A, Alcántara R, Darsow M, Guedj M, Ashburner M (2008) ChEBI: a database and ontology for chemical entities of biological interest. Nucleic Acids Res 36(Database issue):344–350
23. Block P, Sotriffer CA, Dramburg I, Klebe G (2006) AffinDB: a freely accessible database of affinities for protein-ligand complexes from the PDB. Nucleic Acids Res 34(Database issue): 522–526
24. Kuhn M, von Mering C, Campillos M, Jensen LJ, Bork P (2008) STITCH: interaction networks of chemicals and proteins. Nucleic Acids Res 36(Database issue):684–688
25. Wishart DS, Knox C, Guo AC, Cheng D, Shrivastava S, Tzur D, Gautam B, Hassanali M (2008) DrugBank: a knowledgebase for drugs, drug actions and drug targets. Nucleic Acids Res 36(Database issue):901–906
26. Günther S, Kuhn M, Dunkel M, Campillos M, Senger C, Petsalaki E, Ahmed J, Urdiales EG, Gewiess A, Jensen LJ, Schneider R, Skoblo R, Russell RB, Bourne PE, Bork P, Preissner R (2008) SuperTarget and Matador: resources for exploring drug-target relationships. Nucleic Acids Res 36(Database issue):919–922
27. Goede A, Dunkel M, Mester N, Frommel C, Preissner R (2005) SuperDrug: a conformational drug database. Bioinformatics 21(9): 1751–1753
28. Backman TW, Cao Y, Girke T (2011) Chemmine tools: an online service for analyzing and clustering small molecules. Nucleic Acids Res 39(Web Server issue):486–491
29. Zhu Q, Lajiness MS, Ding Y, Wild DJ (2010) WENDI: a tool for finding non-obvious relationships between compounds and biological properties, genes, diseases and scholarly publications. J Cheminform 2:6
30. Walker T, Grulke CM, Pozefsky D, Tropsha A (2010) Chembench: a cheminformatics workbench. Bioinformatics 26(23):3000–3001
31. Spjuth O, Helmus T, Willighagen EL, Kuhn S, Eklund M, Wagener J, Murray-Rust P, Steinbeck C, Wikberg JE (2007) Bioclipse: an open source workbench for chemo- and bioinformatics. BMC Bioinforma 8:59
32. Berthold MR, Cebron N, Dill F, Gabriel TR, Kotter T, Meinl T, Ohl P, Sieb C, Thiel K, Wiswedel B (2007) KNIME: the Konstanz information miner. Springer, New York
33. Cao Y, Charisi A, Cheng LC, Jiang T, Girke T (2008) ChemmineR: a compound mining framework for R. Bioinformatics 24(15):1733–1734
34. Steinbeck C, Han Y, Kuhn S, Horlacher O, Luttmann E, Willighagen E (2003) The chemistry development kit (cdk): an open-source java library for chemo- and bioinformatics. J Chem Inf Comput Sci 43(2):493–500
35. Steinbeck C, Hoppe C, Kuhn S, Floris M, Guha R, Willighagen EL (2006) Recent developments of the chemistry development kit (CDK)—an open-source java library for chemo- and bioinformatics. Curr Pharm Des 12(17):2111–2120
36. Guha R, Howard MT, Hutchison GR, Murray-Rust P, Rzepa H, Steinbeck C, Wegner J, Willighagen EL (2006) The blue obelisk-interoperability in chemical informatics. J Chem Inf Model 46(3):991–998
37. Sykora VJ, Leahy DE (2008) Chemical Descriptors Library (CDL): a generic, open

source software library for chemical informatics. J Chem Inf Model 48:1931–1942

38. Wegner JK, Fröhlich H, Zell A (2004) Feature selection for descriptor based classification models. 2. Human intestinal absorption (HIA). J Chem Inf Comput Sci 44(3):931–939
39. Sheridan RP, Kearsley SK (2002) Why do we need so many chemical similarity search methods? Drug Discov Today 7(17):903–911
40. Chen X, Reynolds CH (2002) Performance of similarity measures in 2D fragment-based similarity searching: comparison of structural descriptors and similarity coefficients. J Chem Inf Comput Sci 42(6):1407–1414
41. Cheng AC, Coleman RG, Smyth KT, Cao Q, Soulard P, Caffrey DR, Salzberg AC, Huang ES (2007) Structure-based maximal affinity model predicts small-molecule druggability. Nat Biotechnol 25(1):71–75
42. Lipinski CA, Lombardo F, Dominy BW, Feeney PJ (1997) Experimental and computational approaches to estimate solubility and permeability in drug discovery and development settings. Adv Drug Deliv Rev 23(1–3):3–25
43. Baurin N, Baker R, Richardson C, Chen I, Foloppe N, Potter A, Jordan A, Roughley S, Parratt M, Greaney P, Morley D, Hubbard RE (2004) Drug-like annotation and duplicate analysis of a 23-supplier chemical database totalling 2.7 million compounds. J Chem Inf Comput Sci 44(2):643–651
44. Tetko IV, Gasteiger J, Todeschini R, Mauri A, Livingstone D, Ertl P, Palyulin VA, Radchenko EV, Zefirov NS, Makarenko AS, Tanchuk VY, Prokopenko VV (2005) Virtual computational chemistry laboratory-design and description. J Comput Aided Mol Des 19(6):453–463
45. Monge A, Arrault A, Marot C, Morin-Allory L (2006) Managing, profiling and analyzing a library of 2.6 million compounds gathered from 32 chemical providers. Mol Divers 10(3):389–403
46. Hajduk PJ, Sauer DR (2008) Statistical analysis of the effects of common chemical substituents on ligand potency. J Med Chem 51(3):553–564
47. Gedeck P, Rohde B, Bartels C (2006) QSAR-how good is it in practice? Comparison of descriptor sets on an unbiased cross section of corporate data sets. J Chem Inf Model 46(5):1924–1936
48. Sutherland JJ, O'Brien LA, Weaver DF (2004) A comparison of methods for modeling quantitative structure-activity relationships. J Med Chem 47(22):5541–5554
49. van der Walt C, Barnard E (2006) Data characteristics that determine classifier performance. Proceedings of 16th annual symposium of the pattern recognition association of South Africa, pp 160–165
50. Ivanciuc O (2007) Applications of support vector machines in chemistry. Rev Comput Chem 23:291
51. Gentleman R, Carey V, Dudoit S, Irizarry R, Huber W (2005) Bioinformatics and computational biology solutions using R and bioconductor. Springer, New York
52. Backman TW, Cao Y, Girke T (2011) Chemmine tools: an online service for analyzing and clustering small molecules. Nucleic Acids Res 39:W4386–W491
53. Verheij HJ (2006) Leadlikeness and structural diversity of synthetic screening libraries. Mol Divers 10(3):377–388
54. Baell JB, Holloway GA (2010) New substructure filters for removal of pan assay interference compounds (PAINS) from screening libraries and for their exclusion in bioassays. J Med Chem 53(7):2719–2740
55. Guha R (2007) Chemical Informatics functionality in R. J Stat Softw 18(8):1–16
56. Landon MR, Schaus SE (2006) JEDA: Joint entropy diversity analysis. An information-theoretic method for choosing diverse and representative subsets from combinatorial libraries. Mol Divers 10(3):333–339
57. Perez JJ (2005) Managing molecular diversity. Chem Soc Rev 34(2):143–152
58. Pau G, Fuchs F, Sklyar O, Boutros M, Huber W (2010) EBImage-an R package for image processing with applications to cellular phenotypes. Bioinformatics 26(7):979–981
59. Wang X, Terfve C, Rose JC, Markowetz F (2011) HTSanalyzeR: an R/Bioconductor package for integrated network analysis of high-throughput screens. Bioinformatics 27(6):879–880
60. Cao Y, Jiang T, Girke T (2010) Accelerated similarity searching and clustering of large compound sets by geometric embedding and locality sensitive hashing. Bioinformatics 26(7):953–959
61. Cao Y, Jiang T, Girke T (2008) A maximum common substructure-based algorithm for searching and predicting drug-like compounds. Bioinformatics 24(13):366–374

Chapter 16

Endomembrane Dissection Using Chemically Induced Bioactive Clusters

Natasha Worden, Thomas Girke, and Georgia Drakakaki

Abstract

Chemical genomics is a novel approach that allows for the rapid functional analysis of plant proteins, complexes, pathways, and networks. Systematic screens for bioactive small molecules causing specific subcellular phenotypes have been successfully performed in mammalian cells, but thus far, are limited in plants. This protocol describes a systematic chemical screen of plasma membrane recycling markers in plants, using confocal microscopy and the subsequent clustering of subcellular phenotypes, to identify chemicals with desired effects. The method provides an approach to identify novel chemicals for pathway dissection, making chemical genomics more accessible to the scientific community. The matrix of novel chemicals described in this protocol can be expanded and analyzed continuously as more data is collected, increasing our knowledge of the endomembrane system, and accumulating compartment-specific markers and chemical probes that perturb specific aspects of endomembrane trafficking.

Key words Chemical genomics, Bioactive clusters, Endomembrane trafficking

1 Introduction

The endomembrane system is a complex network of interconnected compartments, including the endoplasmic reticulum, Golgi apparatus, vesicles, tonoplast, and plasma membrane [1]. It plays an integral role in plant growth and development by facilitating the uptake and intracellular transport of proteins and small molecules [2]. A key characteristic of the endomembrane system is the ability of the plasma membrane to internalize external molecules in vesicles and reorganize its protein content [3], which is required for cell signaling [4], polarity [5, 6], transport [6], and the cell cycle [7]. By determining the role of individual endomembrane compartments, we can learn more about specific cellular functions and their role in plant growth and development.

Chemical genomics is a novel approach based on the use of small molecules (rather than mutations, as in classical genetics) to disrupt the activities of targeted proteins and identify their function.

Glenn R. Hicks and Stéphanie Robert (eds.), *Plant Chemical Genomics: Methods and Protocols*, Methods in Molecular Biology, vol. 1056, DOI 10.1007/978-1-62703-592-7_16, © Springer Science+Business Media New York 2014

This approach is valuable because it is rapid, reversible and can be achieved in a spatiotemporally restricted manner, overcoming the complexities caused by genetic lethality and redundancy [8]. Chemical genomics allows the functional characterization of proteins by dissecting vesicle networks at the subcellular level [9]. Small molecules can interrupt pathways at different points, providing functional details that would remain undetermined using classical genetics. An example of such inhibitors include brefeldin A (BFA), which is used to define Golgi-dependent trafficking and the internalization of proteins into BFA-induced bodies [10]. Although BFA predominantly targets the SEC7 family of exchange factors for ADP ribosylation factor GTPases (ARF-GEFs) [11], it also interferes with other vesicle trafficking processes [10]. Recently a pollen germination screen identified endosidin 1 (ES1), which arrests PIN-FORMED 2 (PIN2) and the brassinosteroid receptor (BRI1) in endosomal aggregates (ES1 bodies), defined by SYNTAXIN OF PLANTS 61 (SYP61). ES1 displays more compartment specificity than BFA, affecting a subpopulation of the endomembrane vesicles aggregated by BFA [8, 12]. Both of these chemicals have been extensively utilized to characterize endomembrane pathways, the corresponding endosomal populations, and to investigate the role of receptor endocytosis in plant signaling [12–14]. The value of endosidins to dissect the endomembrane system was further explored in a high-content screen to identify additional endosidins affecting the endomembrane system [9].

A systematic chemical genomic screen carried out at the subcellular level can identify novel small molecules that can be used to probe different cellular components, and help characterize their roles in plant growth and development. Individual subcellular images obtained by confocal microscopy are difficult to quantify, but the systematic clustering of image data from bioactive screens offers a basic quantification. The protocol in this chapter provides a basic framework for subcellular phenotype characterization that can be expanded to suit additional needs and accommodate new quantification methods and software as they are developed [9]. Researchers can evaluate combinations of subcellular phenotypes by clustering data from high-throughput subcellular screens, leading to an improved understanding of the overall endomembrane system. The ability of small molecules to act quickly and probe different subcellular markers makes them ideal for high-throughput subcellular screening. Systematic screens will benefit from automated microscopy and digital image analysis platforms that are currently under development [15].

In the screen described in this chapter, we examine the subcellular effects (specifically plasma membrane recycling markers in *Arabidopsis thaliana*) of chemicals on the endomembrane system, observed by confocal microscopy. Using this approach, many bioactive small molecules can be discovered in a single screen, while phenotypic clustering efficiently identifies chemicals with particular effects. Plasma membrane recycling proteins such as PIN2::PIN1:GFP [16], PIN2::PIN2:GFP [17], and BRI1::BRI1:GFP [18] expressed in *Arabidopsis* root tips are excellent candidates for this screen because they are trafficked through the secretory pathway and are recycled in endosomes. The disruptions of such markers, by small molecules, are categorized as distinct phenotypes, and can be used to characterize their trafficking routes. Clustering of small molecules based on the similarity of induced phenotypes reveals specific bioactive clusters that may indicate pathway intersection between subcellular markers, providing systems-level insights into endomembrane trafficking.

Overall, this protocol describes a high-throughput screen to cluster phenotypes of plant cells expressing different *Arabidopsis* endomembrane markers to categorize the bioactive chemicals and improve knowledge of the cellular endosomal processes. The phenotype scores and matrix provide a very basic way to quantify large subcellular image datasets. The clusters serve as a starting point for studying specific phenotypes or systems, and provide valuable information about endomembrane recycling pathways. The roles of bioactive chemicals and our knowledge of vesicle networks will continue to evolve as more markers are screened and compared to our initial matrix. Systemization will allow researchers to find molecules inducing specific phenotypes, based on the matrix, and predict the effects of untested molecules (and perhaps even the functions of any natural analogs in plants) based on their chemical structures.

2 Materials

Sterile materials and double-distilled sterile water should be used during the germination and growth stages to prevent the contamination of growth plates.

2.1 Germination and Growth Materials

1. Sterilization solution: 30 % sodium chlorate (bleach) in 100 % ethanol with 30 μL Triton X-100 (SIGMA) per 50 mL of solution.
2. 1.5-mL microcentrifuge tubes.
3. 0.5 *Arabidopsis* growth medium (AGM): 2.3 g/L Murashige and Skoog Basal Salts with minimal organics (Sigma), 10 g/L sucrose (Fisher), and 8 g/L Phytoagar (Invitrogen). Adjust to pH to 5.7 using KOH. Autoclave for 30 min at 120 °C and store at RT.
4. Sterile, disposable petri dishes (Fisher).
5. Growth chamber set at 22 °C with 18 h light cycle.
6. Aluminum foil.
7. Seed stock containing a fluorescent marker attached to gene of interest.

2.2 Microscopy Materials

1. Chemicals to be screened in a 5 mg/mL stock solution dimethylsulfoxide (DMSO).
2. DMSO for control.
3. Leica TCS SP2 confocal microscope or other high resolution confocal microscope.
4. 96-well plate (Fisher).
5. 0.5 *Arabidopsis* growth medium (AGM) (described in Subheading 2.1).
6. Forceps.
7. Slides and coverslips.

3 Methods

3.1 Sterilization

Perform work with sterile seeds and plates in the laminar flow hood. All procedures may be carried out at room temperature, unless otherwise specified.

1. Add 500 μL of sterilization solution to 50 mg of *Arabidopsis* seeds in a 1.5-mL microcentrifuge tube.
2. Shake on speed 3 for 10 min in a Vortex Genie 2 (Fisher).
3. In the hood, remove sterilization solution and rinse three times with 100 % ethanol, allow to air-dry.
4. Store sterilized seeds at 4 °C.

3.2 Growth of Arabidopsis Seedlings

1. Prepare 0.5 AGM plates by dispensing a 0.5-cm layer into a sterile petri dish in a sterile hood and allow medium to solidify for 30 min.
2. Place sterile seeds on the solidified AGM plates for germination.
3. Cover plates in foil and cold-vernalize at 4 °C for 48 h to ensure the same germination efficiency (*see* **Note 1**).
4. Place plates of vernalized seeds in the 22 °C growth chamber for germination.

3.3 Chemical Treatment

1. Dilute 1 μL of the chemical stock solution (typically 1–10 mM in 100 % DMSO) in 200 μL of 0.5 AGM and dispense into the wells of a 96-well plate, using one well for the 1 μL of DMSO for control. Repeated pipetting is required to mix the chemical thoroughly. Allow the medium to solidify.
2. Transfer 3-day-old seedlings to the 96-well plate for chemical treatment, five seedlings per well.
3. Place the 96-well plate in the 22 °C growth chamber and incubate for 3 h under light.
4. When ready to observe, use forceps to transfer seedlings to a slide with 100 μL of water and cover with a coverslip.

Table 1
List of potential subcellular phenotypes observed after chemical treatment and their associated codes

Code	Subcellular phenotype
0	No change (DMSO control)
P1	Increased abundance of endosome-like localization with partial aggregation (*see* **Notes 2, 3**)
P2	ES1-like bodies (*see* **Notes 2, 3**)
P3	BFA-like bodies (*see* **Note 3**)
P4	Diffused aggregations (*see* **Note 4**)
P5	Dispersed endosome-like vesicles combined with cytoplasmic localization
P6	Altered cell shape as defined by PM markers (*see* **Note 4**)
P7	P6 plus appearance of endomembrane structures (*see* **Note 4**)
P8	Cytoplasmic localization (*see* **Notes 2, 5**)
P9	Tonoplast localization and aggregates (*see* **Note 6**)
P10	Defective cell plate (*see* **Note 7**)
P11	Vacuolar localization (*see* **Note 8**)
P12	Vacuolar bodies (*see* **Notes 2, 9**)
P13	Increased fluorescence at plasma membrane (*see* **Note 10**)
P14	Reduced fluorescence at plasma membrane (*see* **Note 10**)
P15	Altered cell polarity (*see* **Note 11**)
16	Compound autofluorescence
17	No data

5. Subcellular phenotypes are observed in cortical or epidermal cells of the seedling root tips using the 63× water immersion objective.
6. Images are captured using the 2× and 4× digital zoom.

3.4 Phenotype Characterization and Clustering

1. To evaluate and characterize the intracellular effects induced by small molecules, a scoring system for chemically induced subcellular phenotypes has been developed. The scoring system assigns a number and a corresponding color code to each subcellular phenotype. The scoring system includes 15 distinct phenotypes after chemical treatment (Table 1, Fig. 1). Code 0 represents no chemical effect and code 17 represents the absence of data. Compound autofluorescence is represented by code 16.
2. Using phenotype data, construct a matrix where rows represent compounds and columns represent marker types (Fig. 2).

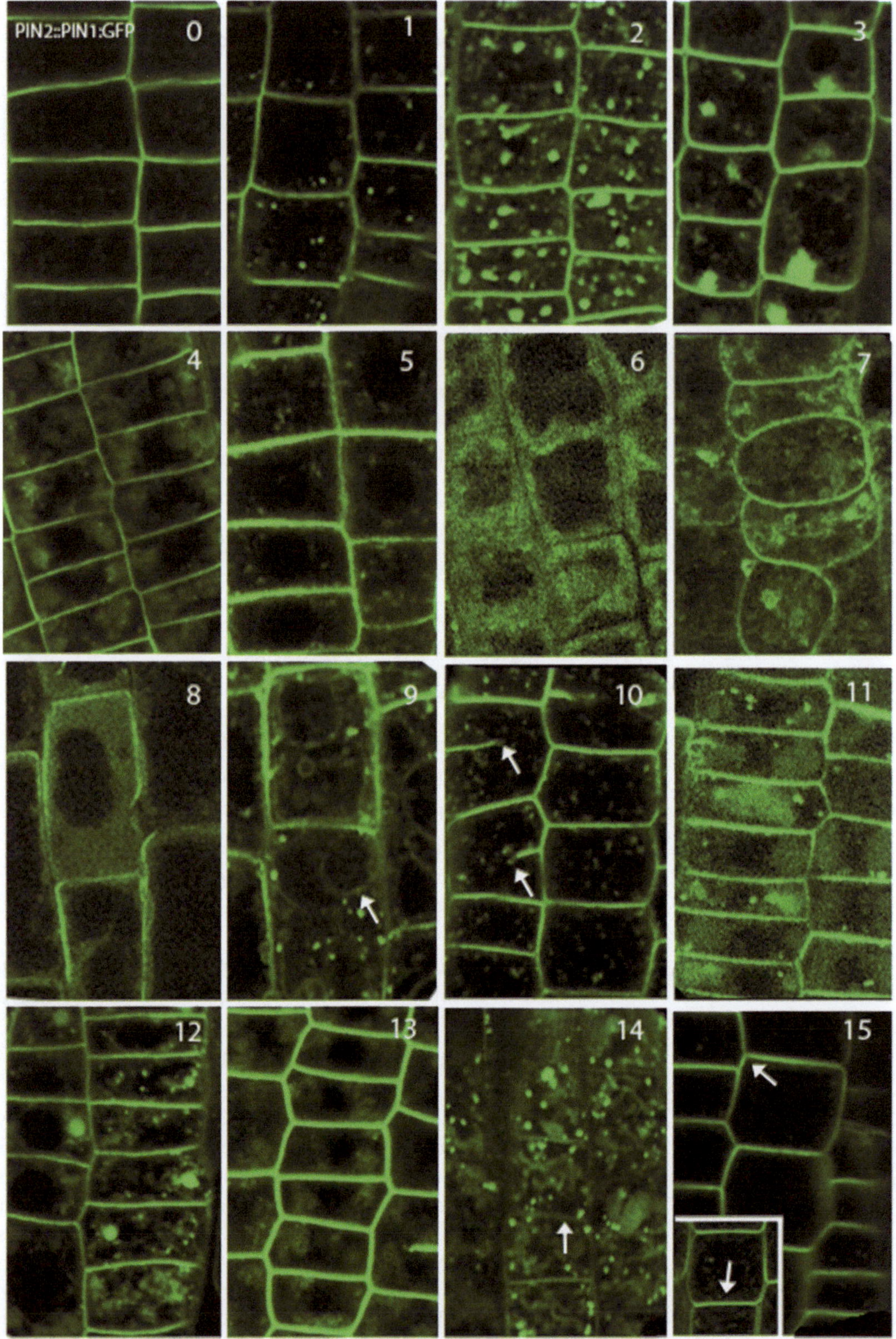

Fig. 1 Examples of subcellular phenotypes corresponding to codes in Table 1. Confocal images taken on cortical or epidermal cells of PIN2::PIN1:GFP. The *inset* of 15 shows the normal basal polarity seen in PIN2::PIN1:GFP, while the figure shows apical polarity

Chemical name	BRI1::BRI1 2hr	PIN2::PIN1 2hr	PIN2::PIN2 2hr	PIN2::PIN1 24hr	PIN2::PIN2 24hr
LATCA_S_09398	1	0	1	5	0
LATCA_SPB_06981	1	0	1	0	0
CBDS20K_5885031	1	0	0	5	0
LATCA_SPB_04446	1	1	1	5	0
Timtek10K_ST026563	1	2	1	5	0
Timtek10K_ST027827	2, 8	0	1	6	0
CBDS20K_5809280	1	2	1	0	0
CBDS20K_5673258	1,8	2	1	0	0
Timtek10K_R412910	1,8	2,11	1	0	0
CBDS20K_5616952	1	2	1	0,1,2,3,4,5,6,7,8,9,10, 11,12,13,14,15,16,17	0,1,2,3,4,5,6,7,8,9,10, 11,12,13,14,15,16,17
LATCA_S_10775	1	2	1	0,1,2,3,4,5,6,7,8,9,10, 11,12,13,14,15,16,17	0,1,2,3,4,5,6,7,8,9,10, 11,12,13,14,15,16,17
CBDS20K_5959766	2	2,9	1	0	0
CBDS20K_6054675	1,8	2,6	2,6	0,1,2,3,4,5,6,7,8,9,10, 11,12,13,14,15,16,17	0,1,2,3,4,5,6,7,8,9,10, 11,12,13,14,15,16,17
CBDS20K_5521757	11,8,16	2,16	2	0	0
CBDS20K_6288782	2,8,9	2	2	0	0
CBDS20K_5302717	2	2	2	0	0
CBDS20K_5326476	2	2	2	0,1,2,3,4,5,6,7,8,9,10, 11,12,13,14,15,16,17	0
CBDS20K_5474792	2	2	2	0,1,2,3,4,5,6,7,8,9,10, 11,12,13,14,15,16,17	0
LATCA_5528968	2,8	2	0	0	0
CBDS20K_5473301	2,8	2	5	0,1,2,3,4,5,6,7,8,9,10, 11,12,13,14,15,16,17	0

Fig. 2 Example of matrix representing small molecules and their corresponding phenotypic scores. Hierarchical complete linkage clustering with the hclust function from R is used to form clusters. Similar phenotypic scores are color-coded

Populate cells in the matrix with the corresponding numerical phenotype classes (1–17). If multiple phenotypes occur, include all phenotypes in the corresponding fields of the matrix. Use the score of 17 to indicate missing data points. In the downstream distance function, the value 17 is allowed to match all other scores 0–17 including itself (*see* **Note 12**).

3. The Hamming distance [19] is used to convert the phenotype data to a distance matrix. Each distance value in the matrix is the number of phenotype differences observed for each compound pair divided by the total number of phenotype values (Table 1).
4. Use the distance matrix to perform agglomerative hierarchical complete linkage clustering with the hclust function from R [6] (*see* **Note 13**). A dendrogram is formed as a result of clustering, with rows representing the chemicals and columns representing the phenotypic effects of small molecule treatments. Create a map by color coding specific phenotypes differently for visualization of data (*see* **Note 14**), an example of which is shown in Fig. 3.

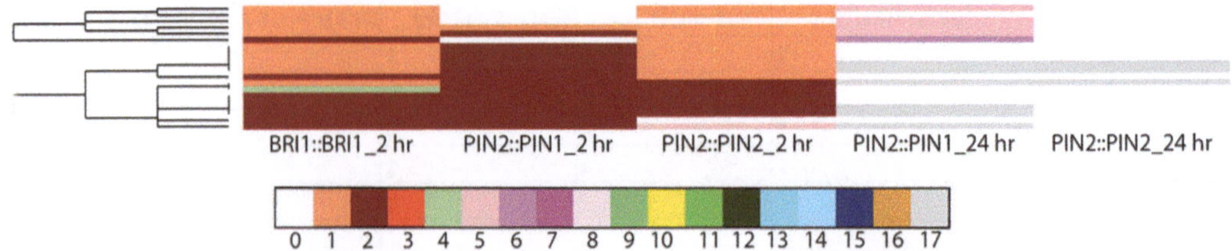

Fig. 3 Example of a close up section of a heat map created from phenotype clusters. Colors represent phenotypes, and similar phenotypes clustered together. This heat map shows clusters of phenotype P2, ES1-like bodies, and phenotype P1, increased fluorescence in endosome-like vesicles with partial aggregation. The dendrogram on the left shows the similarities of the chemicals according to their phenotypes and not their structures

4 Notes

1. Use longer vernalization periods for seeds that are more difficult to germinate.
2. Different phenotypes, for example, endomembrane bodies (P1 and P2), apparent cytoplasmic localization (P8) or vacuolar body formation (P12) are possible for the examined markers in different combinations.
3. Phenotypes P1, P2, and P3 represent endosomal aggregations, which in plasma membrane markers are small, medium, and large bodies, respectively.
4. P6 and P7 describe "unhealthy looking cells" including absence of initial PM localization or the presence of fluorescence in unidentified membrane structures. Further dilution of these chemical might reveal specific phenotypes.
5. Cytoplasmic localization (P8) is the appearance of fluorescence marker in the cytoplasm.
6. Fluorescence outlining vacuoles and vacuolar structures indicates tonoplast localization (P9).
7. Defective cell plate (P10) may include an incomplete or misshapen cell plate.
8. Vacuolar localization (P11) describes increased fluorescence in the vacuole as indicated by large fluorescent bodies.
9. P12 describes distinct vacuolar body fluorescence that could represent lytic bodies [20, 21].
10. P13 describes an increase of plasma membrane localization, while P14 corresponds to reduction of PM fluorescence, both of which may occur in combination with aberrant endosomal localization.
11. Polarity for specific markers is a change in fluorescence on a polar membrane. Measure polarity after 24 h; polarity changes

are not rapid enough to be observed at 3 h. For example, in PIN2::PIN1:GFP it describes the absence of basal polarity.

12. Screening with multiple subcellular markers will create a stronger cluster.
13. For details on algorithm used *see* ref. [22]. R is a language and environment for statistical computing and graphics. It provides a wide variety of statistical and graphical techniques and is freely available from www.r-project.org.
14. When interpreting the map it is important to put emphasis on identical color patterns and not a gradient of related traits, because meaningful quantitative relationships do not exist between the different phenotypic classes.

Acknowledgements

We acknowledge the National Science Foundation Grant IOS-1258135 to GD. NW was supported by a Plant Sciences GSR and the CREATE-IGERT NSF DGE-0653984 grant.

References

1. Bonifacino JS, Glick BS (2004) The mechanisms of vesicle budding and fusion. Cell 116(2):153–166
2. Hicks GR, Raikhel NV (2012) Small molecules present large opportunities in plant biology. Annu Rev Plant Biol 63:261–282
3. Sorkin A (2000) The endocytosis machinery. J Cell Sci 113(Pt 24):4375–4376
4. Geldner N, Robatzek S (2008) Plant receptors go endosomal: a moving view on signal transduction. Plant Physiol 147(4):1565–1574
5. Kleine-Vehn J, Friml J (2008) Polar targeting and endocytic recycling in auxin-dependent plant development. Annu Rev Cell Dev Biol 24:447–473
6. Takano J et al (2005) Endocytosis and degradation of BOR1, a boron transporter of Arabidopsis thaliana, regulated by boron availability. Proc Natl Acad Sci U S A 102(34):12276–12281
7. Van Damme D, Inzé D, Russinova E (2008) Vesicle trafficking during somatic cytokinesis. Plant Physiol 47(4):1544–1552
8. Drakakaki G et al (2009) Chemical dissection of endosomal pathways. Plant Signal Behav 4(1):57–62
9. Drakakaki G et al (2011) Clusters of bioactive compounds target dynamic endomembrane networks in vivo. Proc Natl Acad Sci USA 108(43):17850–17855
10. Nebenfuhr A, Ritzenthaler C, Robinson DG (2002) Brefeldin A: deciphering an enigmatic inhibitor of secretion. Plant Physiol 130(3): 1102–1108
11. Geldner N et al (2003) The Arabidopsis GNOM ARF-GEF mediates endosomal recycling, auxin transport, and auxin-dependent plant growth. Cell 112(2):219–230
12. Robert S et al (2008) Endosidin1 defines a compartment involved in endocytosis of the brassinosteroid receptor BRI1 and the auxin transporters PIN2 and AUX1. Proc Natl Acad Sci U S A 105(24):8464–8469
13. Geldner N et al (2007) Endosomal signaling of plant steroid receptor kinase BRI1. Genes Dev 21(13):1598–1602
14. Irani NG et al (2012) Fluorescent castasterone reveals BRI1 signaling from the plasma membrane. Nat Chem Biol 8(6):583–589
15. Salomon S et al (2010) High-throughput confocal imaging of intact live tissue enables quantification of membrane trafficking in Arabidopsis. Plant Physiol 154(3):1096–1104
16. Wisniewska J et al (2006) Polar PIN localization directs auxin flow in plants. Science 312(5775):883
17. Muller A et al (1998) AtPIN2 defines a locus of Arabidopsis for root gravitropism control. Embo J 17(23):6903–6911

18. Friedrichsen DM et al (2000) Brassinosteroid-insensitive-1 is a ubiquitously expressed leucine-rich repeat receptor serine/threonine kinase. Plant Physiol 123(4):1247–1256
19. Hamming RW (1950) Error detecting and error correcting codes. Bell Syst Tech J 29(2): 147–160
20. Zheng H, Staehelin LA (2011) Protein storage vacuoles are transformed into lytic vacuoles in root meristematic cells of germinating seedlings by multiple, cell type-specific mechanisms. Plant Physiol 155(4):2023–2035
21. Kleine-Vehn J et al (2008) Differential degradation of PIN2 auxin efflux carrier by retromer-dependent vacuolar targeting. Proc Natl Acad Sci U S A 105(46):17812–17817
22. Murtagh F (1985) Multidimensional clustering algorithms. COMPSTAT Lectures 4. Physica-Verlag, Wuerzburg

Chapter 17

Statistical Molecular Design: A Tool to Follow Up Hits from Small-Molecule Screening

Anders E.G. Lindgren, Andreas Larsson, Anna Linusson, and Mikael Elofsson

Abstract

In high-throughput screening (HTS) a robust assay is used to interrogate a large collection of small organic molecules in order to find compounds, hits, with a desired biological activity. The hits are then further explored by an iterative process where new compounds are designed, purchased, or synthesized, followed by an evaluation in one or more assays. Statistical molecular design (SMD) is a useful method to select a balanced, varied, and information-rich compound collection based on hits from HTS in order to create a foundation for development of optimized compounds with improved properties. In this chapter, we describe the use of SMD to explore a hit obtained from small-molecule screening.

Key words Statistical molecular design, Library design, Principal component analysis, D-optimal design

1 Introduction

In basic and applied research small organic molecules play an important role as chemical probes to study complex biological processes. In addition, such molecules serve as starting points for development of commercial products, for instance pharmaceuticals, herbicides, and pesticides. The process towards a probe or a commercial product requires a chemical starting point. High-throughput screening (HTS) is today a well-established and in principle unbiased technique to obtain such starting points. Typically a robust assay is used to screen a large collection of small organic molecules to identify active compounds, hits. The hits from primary screening are further subjected to dose–response analysis and often tested in additional assays to confirm the desired activity profile. Validated hits rarely meet the criteria to be used as a chemical probe and are very far from being ready for commercial applications. The hits need to be optimized using an iterative process based on the design of analogs that can be synthesized or purchased. Biological evaluation will then provide structure–activity

Glenn R. Hicks and Stéphanie Robert (eds.), *Plant Chemical Genomics: Methods and Protocols*, Methods in Molecular Biology, vol. 1056, DOI 10.1007/978-1-62703-592-7_17, © Springer Science+Business Media New York 2014

relationships (SARs) that link chemical properties with biological activity and thereby facilitate the continued optimization process.

The number of compounds available either by direct purchase or synthesis from commercially available building blocks is however daunting, and it is clear that efficient strategies are needed to select representative subsets. Traditionally, new compounds are designed one by one instead of in predetermined sets. This often leads to the development being skewed in one direction already from the start, and large volumes of the chemical space are excluded or never even considered. If statistical molecular design (SMD) is instead used to decide which compounds to synthesize, a more balanced and diverse set will be obtained and more information can be extracted from future analysis of the biological results [1]. This is of particular importance if the intention is to calculate quantitative structure–activity relationship (QSAR) models [2] that computationally link chemistry and biology. The use of QSAR models has often been met with incredulity amongst medicinal chemists as the models often fail to predict the activity of new compounds [3]. However, one of the main reasons (*see* **Note 1**) why the QSAR strategy often fails are the poorly designed compound sets on which the models are built. Hence, it is very important that the SMD concept is being used throughout the whole process, starting already with the first set of compounds designed around the original hit (or an equivalent starting point) [4].

This chapter describes the procedure of using SMD to create a balanced and varied library of analogues to a known hit from HTS (or a comparable method of finding a chemical starting point). The goal is that this library should be representative and contain enough chemical diversity in order to establish SARs and facilitate the computation of QSAR models.

When a suitable chemical starting point has been identified preliminary SARs can sometimes be established based on commercial compounds (*see* **Note 2**). In the current case study we chose to directly focus on the design of compounds from commercially available building blocks. The first step is to decide on a synthetic strategy and thus also which classes of building blocks to investigate. This step, called retrosynthetic analysis, is critical since the synthetic strategy needs to tolerate a wide range of functional groups and a large quantity of building blocks should be commercially available [4].

Once this initial step is completed, substructure searches of databases from suitable vendors of commercially available compounds are performed. Depending on the nature of the building blocks some resellers might be more pertinent than others (*see* **Note 3**). For this project it was found that an appropriate amount of building blocks were sold by Sigma-Aldrich, and thus no additional resellers were considered.

Given enough time and financial support it is nowadays possible to synthesize almost any compound found to be of interest to the project. However, resources are always limited and in reality only a small subset can be synthesized (*see* **Note 4**). Filters are therefore used to reduce the number of building blocks further (*see* **Note 5**). In the pharmaceutical industry filters known as Lipinski's rules [5] use molecular weight and properties relating to polarity to remove compounds and building blocks that are unlikely to become orally administrable. Filters can be designed to eliminate the majority of these compounds, but many will still remain. It is therefore beneficial to manually inspect all building blocks and remove the ones containing undesirable structural features. Although filters and manual inspection can reduce the numbers dramatically the number of theoretical target compounds is still unmanageable. The reality is that a small subset needs to be selected and the subset needs to cover as much as possible of the relevant chemical space.

Once an appropriate list of building blocks is obtained, molecular descriptors can be calculated (*see* **Note 6**) and a principal component analysis (PCA) [6] can be made in order to get an overview of the chemical diversity within the set (*see* **Note 7**). Suitable building blocks are then chosen (*see* **Note 8**) and combined so that a balanced library of analogues to the original hit is created.

2 Materials

1. Building block searches were performed using the substructure search feature available on www.sigma-aldrich.com [7].
2. OpenEye FILTER 2.0.2 was used to remove compounds containing unwanted functional groups from the list of building blocks (*see* **Note 9**) [8].
3. Chemical Computing Group (CCG) MOE 2010.10 was used for manual inspection of compounds and calculation of molecular descriptors for all approved building blocks [9].
4. Umetrics SIMCA-P+12.0.1 was used to calculate PCAs [10].
5. Umetrics MODDE 9.0 was used for the D-optimal onion designs (DOOD) [11].

3 Methods

This chapter uses a case study in order to illustrate the methodology of designing a library of analogues to a new hit compound, **ES3** (Fig. 1).

Even if the chapter tries to be as thorough as possible, it is impossible to cover every situation that might arise. If the reader is unsure how to carry out a particular procedure (e.g., if a different

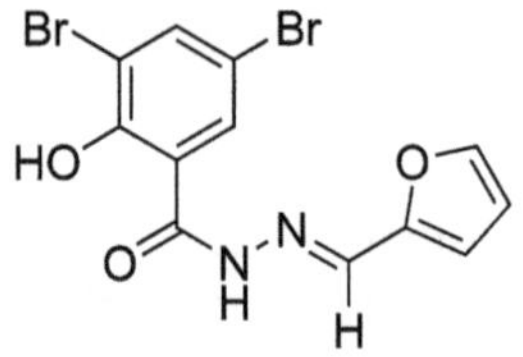

Fig. 1 The chemical structure of **ES3**

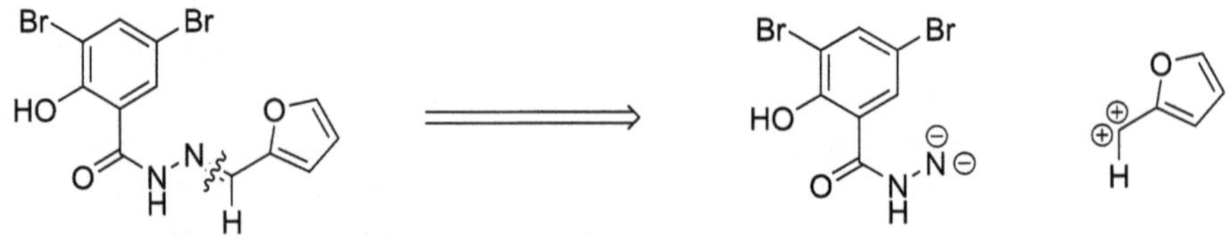

Scheme 1 Retrosynthetic analysis of **ES3**

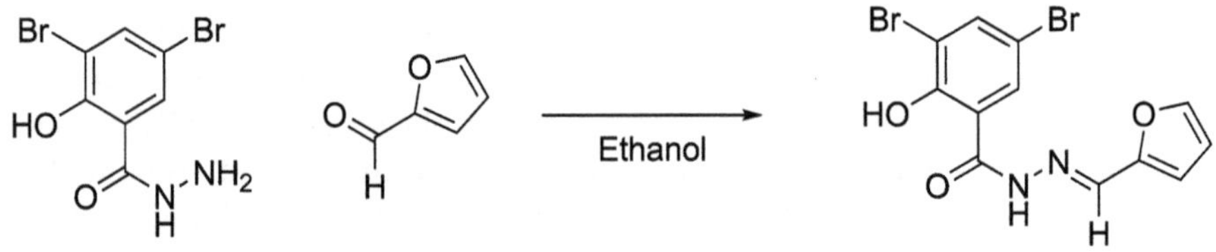

Ethanol

Scheme 2 Synthesis of **ES3** starting from commercially available building blocks

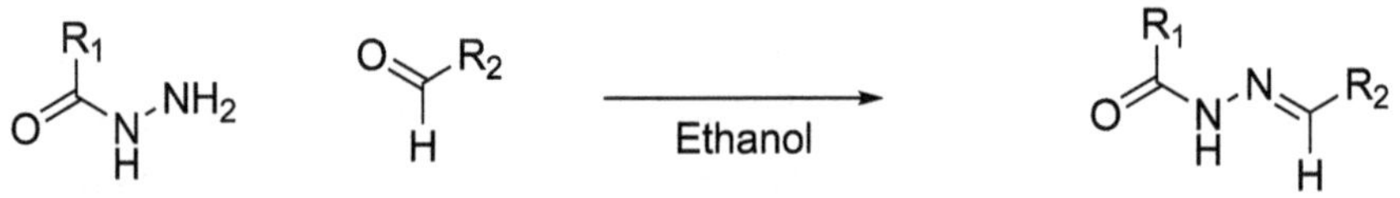

Ethanol

Scheme 3 Generalized synthesis of acylhydrazones starting from commercially available hydrazides and aldehydes

version of the program is being used or if problems occur when attempting to modify the method for a different set of building blocks), please refer to the documentation contained within the programs themselves.

3.1 Synthetic Strategy

1. Perform a retrosynthesis of **ES3** in order to realize which bonds can easily be disconnected (Scheme 1) so that appropriate synthons can be chosen (*see* **Note 10**).
2. Replace these synthons with actual reagents (i.e., hydrazides and aldehydes) in order to find a preliminary synthetic strategy for **ES3** (Scheme 2).
3. Investigate whether or not this is a suitable strategy for library synthesis of analogues to **ES3** by analyzing the scope and limitations of the method (*see* **Note 11)** (Scheme 3).

3.2 Building Blocks

3.2.1 Aldehydes

1. Use the substructure search feature available on www.sigma-aldrich.com in order to find as many suitable aldehydes as possible. Set the limit for molecular weight to 100–300 Da as compounds outside this interval are undesirable for our purposes (*see* **Note 12**).
2. Export the list containing all 1,572 aldehydes as an .sdf file.
3. The resulting list of compounds contains a large amount of undesirable building blocks (e.g., dialdehydes or compounds containing toxic functional groups). OpenEye FILTER will be used to remove the majority of these. Start the program, and navigate to the folder containing your .sdf files. Enter the command "filter -in AllAldehydes.sdf -out FilteredAldehydes.sdf -filter AldehydeFilter.filter -fail FailedAldehydes.sdf" (*see* **Note 13**).
4. Examine the remaining 1,330 aldehydes manually using MOE in order to remove unwanted building blocks. In our case, this resulted in 958 remaining aldehydes that were deemed suitable for further handling within MOE (*see* **Note 14**).
5. Calculate partial charges for all compounds using MMFF94 (*see* **Note 15**).
6. Use the "wash" function (with standard settings) in order to obtain the correct protonation state and remove counterions.
7. Calculate all 186 2D descriptors (*see* **Notes 16** and **17**). Finally, save the database as a .txt file (remember to include "Field Titles") in preparation for the next step. The calculated descriptors will be used as variables in the PCA described in Subheading 3.3.

3.2.2 Hydrazides

1. A single general search for hydrazides will result in a large number of uninteresting compounds. This number can be heavily reduced if instead multiple searches with more stringent criteria are carried out. Perform substructure searches for all compounds shown (Fig. 2) (*see* **Note 18**). The molecular weight should be limited to below 300 Da for **1**–**8** and below 125 Da for **9**. Export the results as .sdf files
2. Use MOE to combine all the .sdf files into a single database containing all 232 hydrazides.
3. As before, OpenEye FILTER will be used to remove some of the remaining undesirable building blocks (e.g., dihydrazides or compounds containing toxic functional groups). Start the program, and navigate to the folder containing your .sdf files. Enter the command "filter -in AllHydrazides.sdf -out FilteredHydrazides.sdf -filter HydrazideFilter.filter –fail FailedHydrazides.sdf" (*see* **Note 13**).
4. Examine the remaining 204 hydrazides manually using MOE in order to remove unwanted building blocks. In our case, this

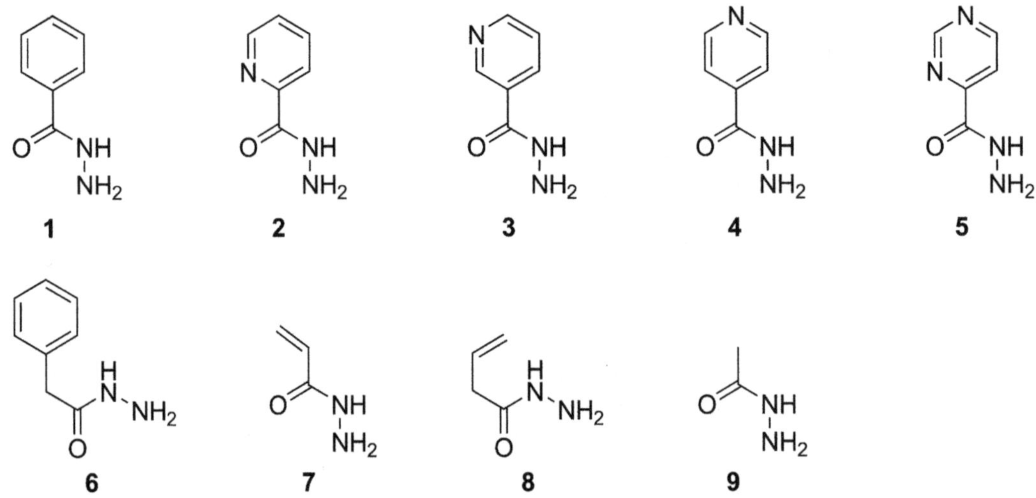

Fig. 2 Overview of suitable substructure searches for finding appropriate hydrazides

resulted in 192 remaining building blocks that were deemed suitable for further handling within MOE (*see* **Note 14**).

5. Calculate partial charges for all compounds using MMFF94 (*see* **Note 15**).
6. Use the "wash" function (with standard settings) in order to obtain the correct protonation state and remove counterions.
7. Calculate all 186 2D descriptors (*see* **Notes 16** and **17**). Finally, save the database as a .txt file (remember to include "Field Titles") in preparation for the next step. The calculated descriptors will be used as variables in the PCA described in Subheading 3.3.

3.3 Principal Component Analysis

3.3.1 Aldehydes

1. Start SIMCA-P+ 12.0.1, and create a new project. Choose the .txt file containing your database as input source. Do not exclude the columns containing only text.
2. Choose MDL_NUMBER as Primary Observation ID and mol, COMMON_NAME, IUPAC_NAME, and CAS_NUMBER as Secondary Observation IDs (*see* **Note 19**). Exclude Similarity, MOLECULAR_WEIGHT, BOILING_POINT, DENSITY, and PRODUCTS. When this is done, click Next.
3. The next screen should show 186 "Variable ID Names" and 958 "Observation ID Names." The "Missing Value Map" should be empty. If this is not the case, go back and redo **step 2**. Choose a name and folder for your project, and then click Finish.
4. You will be asked to exclude some variables that contain no values different from the median; answer "yes to all" (these variables are not permanently excluded and can be easily included again should there be a reason to).

5. Edit your model and exclude the following variables (*see* **Note 16**): BCUT_PEOE_0, BCUT_PEOE_1, BCUT_PEOE_2, BCUT_PEOE_3, BCUT_SLOGP_0, BCUT_SLOGP_1, BCUT_SLOGP_2, BCUT_SLOGP_3, BCUT_SMR_0, BCUT_SMR_1, BCUT_SMR_2, BCUT_SMR_3, GCUT_PEOE_0, GCUT_PEOE_1, GCUT_PEOE_2, GCUT_PEOE_3, GCUT_SLOGP_0, GCUT_SLOGP_1, GCUT_SLOGP_2, GCUT_SLOGP_3, GCUT_SMR_0, GCUT_SMR_1, GCUT_SMR_2, GCUT_SMR_3, opr_violation, PC+, PC-, PEOE_PC+, PEOE_PC-, PEOE_RPC+, PEOE_RPC-, PEOE_VSA+0, PEOE_VSA+1, PEOE_VSA+2, PEOE_VSA+3, PEOE_VSA+4, PEOE_VSA+5, PEOE_VSA+6, PEOE_VSA-0, PEOE_VSA-1, PEOE_VSA-2, PEOE_VSA-3, PEOE_VSA-4, PEOE_VSA-5, PEOE_VSA-6, PEOE_VSA_FHYD, PEOE_VSA_FNEG, PEOE_VSA_FPNEG, PEOE_VSA_FPOL, PEOE_VSA_FPOS, PEOE_VSA_FPPOS, PEOE_VSA_HYD, PEOE_VSA_NEG, PEOE_VSA_PNEG, PEOE_VSA_POL, PEOE_VSA_POS, PEOE_VSA_PPOS, RPC+, RPC-, rsynth, SlogP_VSA0, SlogP_VSA1, SlogP_VSA2, SlogP_VSA3, SlogP_VSA4, SlogP_VSA5, SlogP_VSA6, SlogP_VSA7, SlogP_VSA8, SlogP_VSA9, SMR_VSA0, SMR_VSA1, SMR_VSA2, SMR_VSA3, SMR_VSA4, SMR_VSA5, SMR_VSA6, SMR_VSA7.
6. Autofit a PCA model to your work set, which should now contain 101 variables.
7. Take a look at eigenvalues of the different principal components (PCs). Start removing PCs until only the first eight are left (*see* **Note 20**).
8. Investigate the Summary *X/Y* Overview plot, and remove variables with negative Q^2 values (i.e., a_nB, b_triple, chiral_u, and petitjeanS).
9. Make sure that your model has an appropriate number of PCs (i.e., eight), and start looking for outliers using the Score Scatter Plot and the Distance to Model X Block plot (*see* **Note 21**).
10. Study the chemical structure of any outliers before removing these, if the building block seems reasonable and interesting from a chemical point of view it can be kept. Exclude MFCD00006961, MFCD00036214, MFCD00063509, and MFCD00450438.
11. Repeat **steps 6–10** until no more outliers or variables with negative Q^2 values remain. This should result in a five-component model after the removal of four additional variables (i.e., a_nS, a_base, a_acid, and vsa_acid) and six more outliers (i.e., MFCD11502254, MFCD01859948, MFCD00010579, MFCD01859948, MFCD01095819, and MFCD00091834).

12. Finally, fit a model with five PCs (*see* **Note 22**), and try to interpret what each component represents using the "Loading Scatter Plot" (*see* **Note 23**). The first component represents molecular "size," and the second describes polarity. The third and fourth components describe flexibility, density, and degree of halogenation. The fifth component does not seem to describe a specific property of the building blocks.
13. Create a list of the compound database having the five principal component score vectors as descriptors for each entry, and save the first identifier column (Obs ID (Primary)) together with the principal components (M4.t[1], M4.t[2], M4.t[3], M4.t[4] and M4.t[5]) in a *.xl* or a *.txt format.

3.3.2 Hydrazides

1. Start SIMCA-P+ 12.0.1, and create a new project. Choose the .txt file containing your database as input source. Do not exclude the columns containing only text.
2. Choose MDL_NUMBER as Primary Observation ID and mol, COMMON_NAME, IUPAC_NAME, and CAS_NUMBER as Secondary Observation IDs (*see* **Note 19**). Exclude Similarity, MOLECULAR_WEIGHT, and PRODUCTS. When this is done, click Next.
3. The next screen should show 186 "ID Names" and 192 "Observation ID Names." The "Missing Value Map" should be empty. If this is not the case, go back and redo **step 2**. Choose a name and folder for your project, and then click Finish.
4. You will be asked to exclude some variables that contain no values different from the median; answer "yes to all" (these variables are not permanently excluded and can be easily included again should there be a reason to).
5. Edit your model and exclude the following variables (*see* **Note 16**): BCUT_PEOE_0, BCUT_PEOE_1, BCUT_PEOE_2, BCUT_PEOE_3, BCUT_SLOGP_0, BCUT_SLOGP_1, BCUT_SLOGP_2, BCUT_SLOGP_3, BCUT_SMR_0, BCUT_SMR_1, BCUT_SMR_2, BCUT_SMR_3, GCUT_PEOE_0, GCUT_PEOE_1, GCUT_PEOE_2, GCUT_PEOE_3, GCUT_SLOGP_0, GCUT_SLOGP_1, GCUT_SLOGP_2, GCUT_SLOGP_3, GCUT_SMR_0, GCUT_SMR_1, GCUT_SMR_2, GCUT_SMR_3, PC+, PC-, PEOE_PC+, PEOE_PC-, PEOE_RPC+, PEOE_RPC-, PEOE_VSA+0, PEOE_VSA+1, PEOE_VSA+2, PEOE_VSA+3, PEOE_VSA+4, PEOE_VSA+5, PEOE_VSA+6, PEOE_VSA-0, PEOE_VSA-1, PEOE_VSA-2, PEOE_VSA-3, PEOE_VSA-4, PEOE_VSA-5, PEOE_VSA-6, PEOE_VSA_FHYD, PEOE_VSA_FNEG, PEOE_VSA_FPNEG, PEOE_VSA_FPOL, PEOE_VSA_FPOS, PEOE_VSA_FPPOS, PEOE_VSA_HYD, PEOE_VSA_NEG, PEOE_VSA_PNEG, PEOE_VSA_POL, PEOE_VSA_POS, PEOE_VSA_PPOS,

RPC+, RPC-, rsynth, SlogP_VSA0, SlogP_VSA1, SlogP_VSA2, SlogP_VSA3, SlogP_VSA4, SlogP_VSA5, SlogP_VSA6, SlogP_VSA7, SlogP_VSA8, SlogP_VSA9, SMR_VSA0, SMR_VSA1, SMR_VSA2, SMR_VSA3, SMR_VSA4, SMR_VSA5, SMR_VSA6, SMR_VSA7.

6. Autofit a PCA model to your work set, which should now contain 96 variables.
7. Take a look at eigenvalues of the different PCs. Start removing PCs until only the first five are left (*see* **Note 20**).
8. Investigate the Summary *X/Y* Overview plot, and remove variables with negative Q^2 values (i.e., a_acid, a_base, a_nF, b_double, chiral, petitjeanS, and vsa_acid).
9. Make sure that your model has an appropriate number of PCs (i.e., five), and start looking for outliers using the Score Scatter Plot and Distance to Model X Block plot (*see* **Note 21**).
10. Study the chemical structure of any outliers before removing these, if the building block seems reasonable and interesting from a chemical point of view it can be kept. Exclude MFCD011065 and MFCD00257.
11. Repeat **steps 6–10** until no more outliers or variables with negative Q^2 values remain. This should result in the removal of one more variable, FCharge.
12. Finally, fit a model with five PCs and try to interpret what each component represents using the "Loading Scatter Plot" (*see* **Note 23**). This is done in order to verify and form an opinion of the chemical relevance of the PCA. The first component represents molecular "size," and the second describes polarity. The third and fourth components describe flexibility, density, and degree of halogenation. The fifth component does not seem to describe a specific property of the building blocks.
13. Create a list of the compound database having the five principal component score vectors as descriptors for each entry, and save the first identifier column (Obs ID (Primary)) together with the principal components (M4.t[1], M4.t[2], M4.t[3], M4.t[4] and M4.t[5]) in a *.xl* or a *.txt format.

3.4 D-Optimal Onion Design

1. Start Modde 9.0, and create a new investigation to open up the design wizard dialog. Name the design, e.g., DOOD hydrazide, and then choose Advanced Design and Next. On the next page choose "Onion or D-optimal design from imported candidate set," locate your previously saved *.xl* or *.txt file, and then choose Next. In the import preview, make sure that the identifier column is colored in yellow, the PC columns in white, and the headers row in blue before pressing OK and Next. Continue without defining any response, and then choose "screening" and Next.

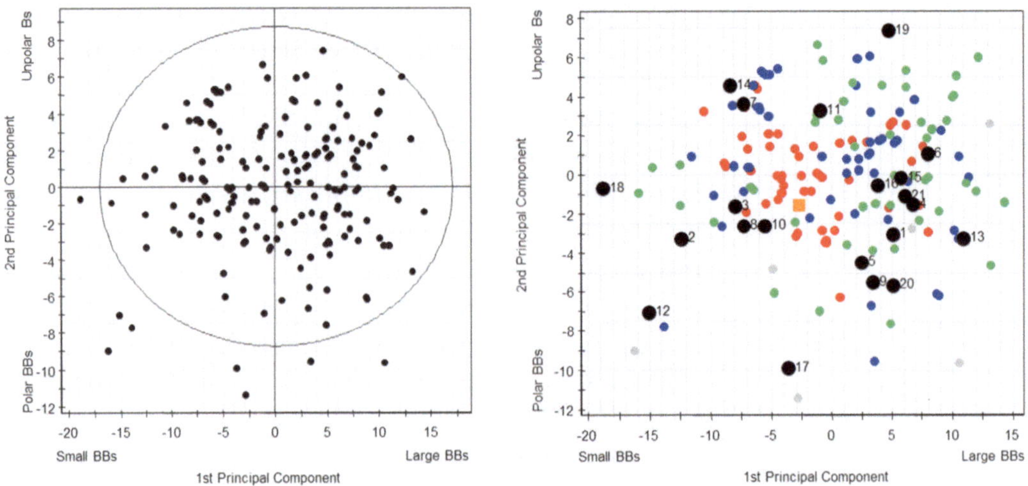

Fig. 3 Results of the DOOD on the hydrazide building blocks (BBs). Selected building blocks are *black*, and the different layers are identified by color. Layer 1 = *red*, layer 2 = *blue*, and layer 3 = *green*

2. Select a linear onion D-optimal design (*see* **Note 24**); adjust the number of desired design runs, i.e., the number of building blocks to select, center points, replicates, and layers to 21, 0, 0, and 3, respectively; and then press Next.
3. An overview of the layers is now shown. Use the default settings that are 1.05–33.68 % for the first layer (from the center point to the object furthest away) and 34.21–66.84 % for the second. Change the third layer to 67.37–97 % (*see* **Note 25**). The number of design runs for each layer is set to 7, 7, and 7, respectively (*see* **Note 26**). Change the number of repetition from 1 to 20 for each layer, and run the D-optimal calculations by clicking Next.
4. The onion D-optimal results page will now show 80 generated designs for each layer. Select a design with 7 runs for layer 1 (G-efficiency 68.8 and condition number 3.57), 8 runs for layer 2 (G-efficiency 71.7 and condition number 2.26), and 6 runs for layer 3 (G-efficiency 64.3 and condition number 2.62) so as to keep the condition number as low as possible, the G-efficiency high, and still maintain the number of objects as 21 (*see* **Note 27**). Click finish to generate your worksheet containing 21 hydrazides. Figure 3 shows the score plot based on the first and second principal components for the remaining 190 hydrazides (Fig. 3, left) and the 21 hydrazides selected by DOOD (Fig. 3, right).
5. The worksheet now consists of 21 building blocks. Sort in a randomized order, e.g., by ascending run order, and copy them into a *.xl* file. Add the starting building block derived from **ES3** at the end of the column (*see* Fig. 4).

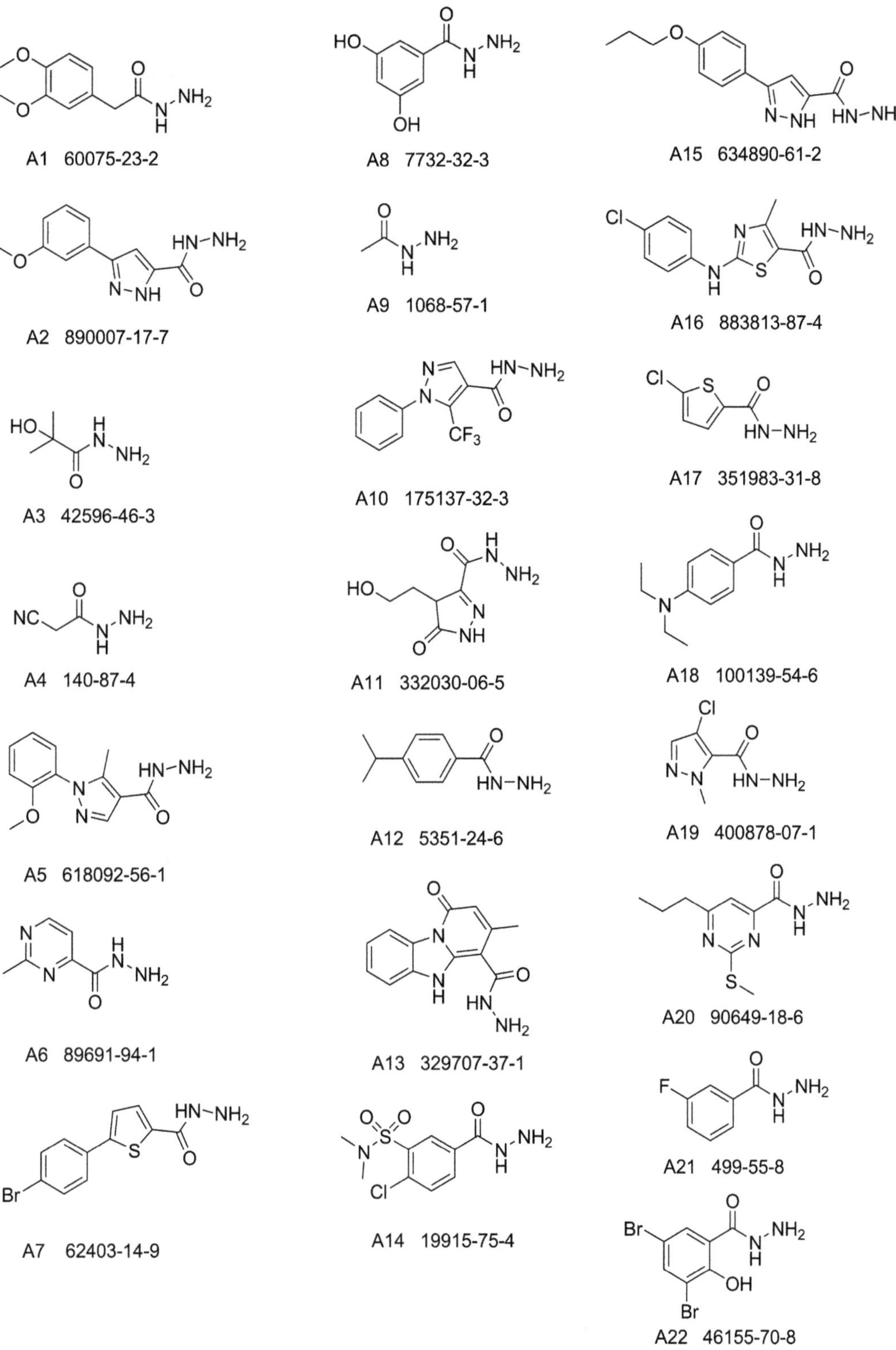

Fig. 4 Overview of the selected hydrazide building blocks (A1–A22) with CAS numbers

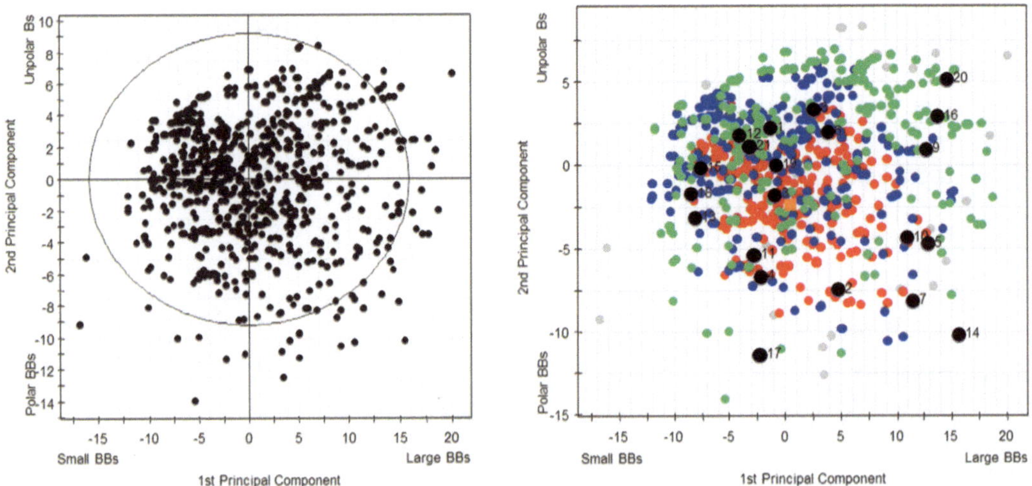

Fig. 5 Results of the DOOD on the aldehyde building blocks (BBs). Selected building blocks are *black*, and the different layers are identified by color. Layer 1 = *red*, layer 2 = *blue*, and layer 3 = *green*

6. Repeat the above procedure for the aldehyde building blocks. The three generated layers will be set at 0.21–33.40, 33.51–66.70, and 66.81–97 %, with 7 design runs in each layer.
7. Choose a design with 6 runs (G-efficiency 69.5 and condition number 2.40) for layer 1, 7 runs for layer 2 (G-efficiency 83.6 and condition number 2.53), and 8 runs (G-efficiency 76.4 and condition number 2.15) for the last layer (*see* **Note 28**). Figure 5 shows the score plot based on the first and second principal components for the remaining 948 aldehydes (Fig. 5, left) and the 21 aldehydes selected by DOOD (Fig. 5, right). One of the selected aldehydes, (5-formyl-2-furyl)methyl thiocyanate (MFCD00098948, 633300-34-2), containing a reactive thiocyanate, was removed based on anticipated incompatibility with the synthetic protocol and was therefore replaced by its closest neighbor, (2-formylphenoxy)-acetic acid (MFCD00003315, 6280-80-4), using the Euclidian distance in all five principal components.
8. Sort the 21 aldehyde building blocks (Fig. 6) in the worksheet in a randomized order, and copy them into the same *.xl* file as the hydrazide building blocks above. Add the aldehyde building block derived from **ES3** at the end of the column.

3.5 Systematic Combination of Building Blocks (See Note 29)

1. Combine the building block lists for the aldehydes and hydrazides to form the first 22 acylhydrazone products.
2. Shift the aldehyde building block column one step downward so that the last entry in the column becomes the first, and then combine it again with the hydrazide column to form a second set of 22 acylhydrazones.

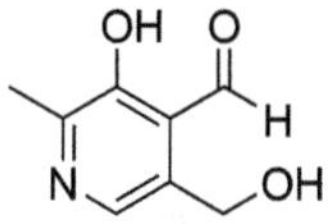

B1 65-22-5

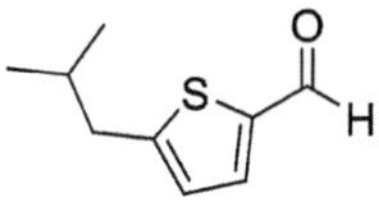

B2 104804-16-2

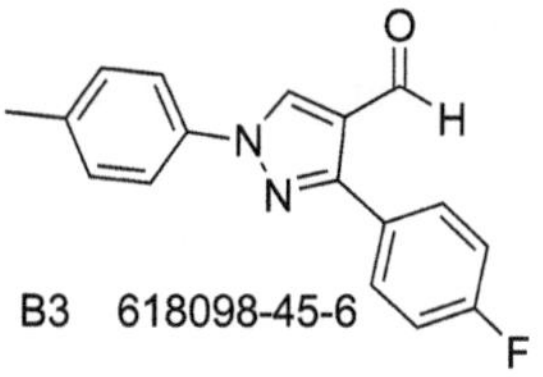

B3 618098-45-6

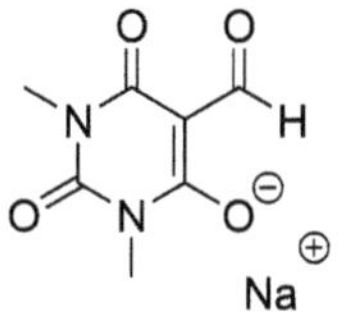

B4 118632-20-5

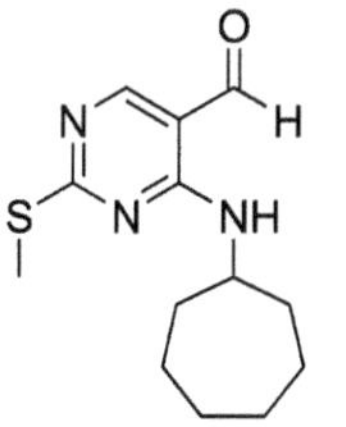

B5 1026362-86-6

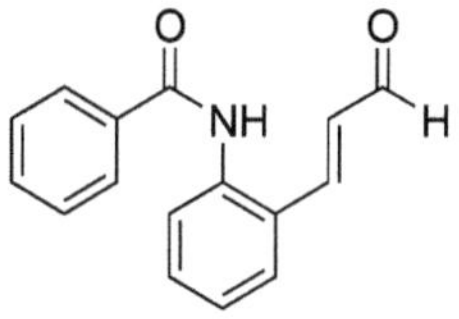

B6 94298-53-0

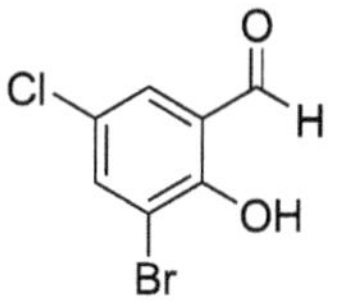

B7 19652-32-5

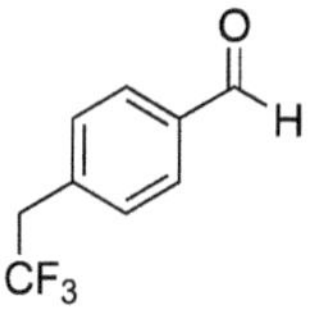

B8 4021-50-5

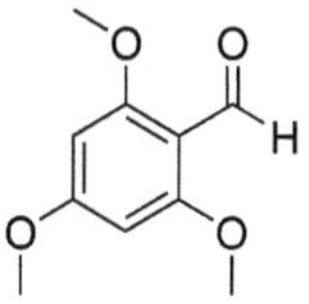

B9 59652-88-9

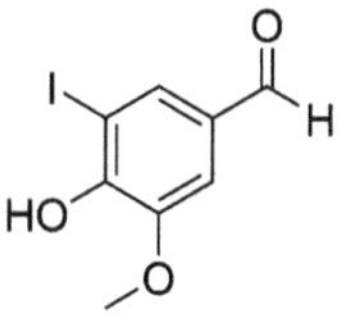

B10 5438-36-8

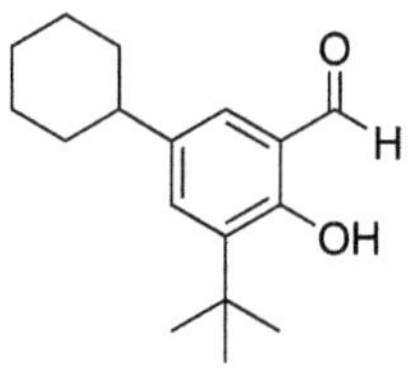

B11 245066-96-0

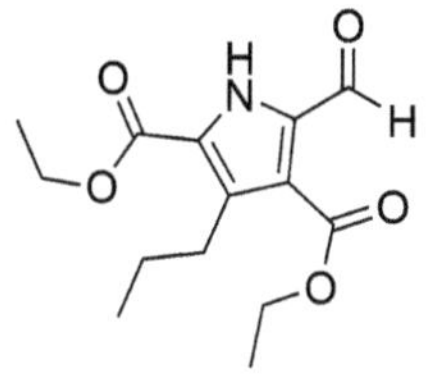

B12 142038-33-3

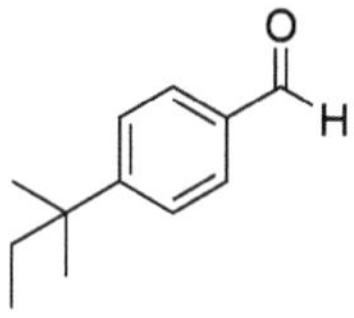

B13 67468-54-6

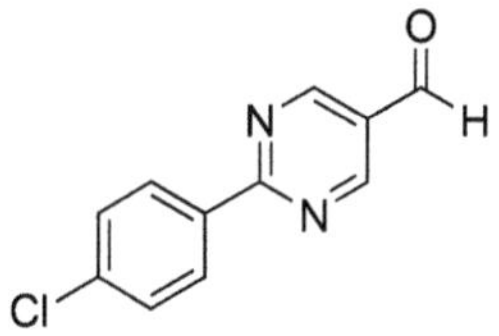

B14 928713-84-2

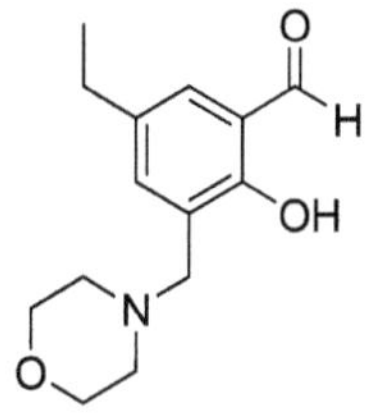

B15 834884-98-9

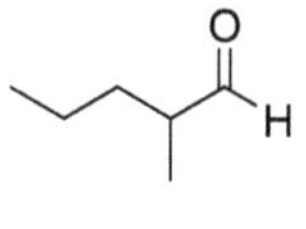

B16 123-15-9

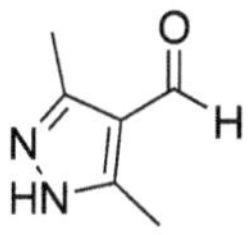

B17 201008-71-1

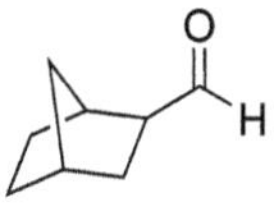

B18 5453-80-5

B19 50910-55-9

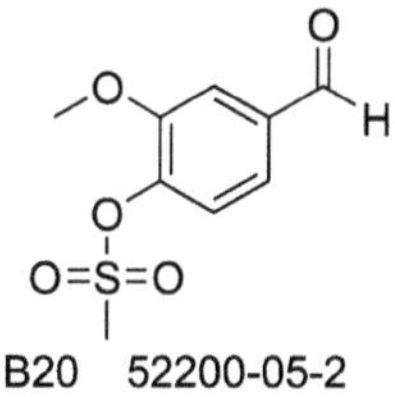

B20 52200-05-2

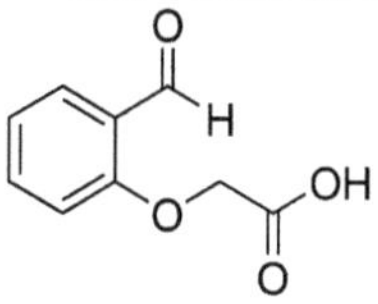

B21 633300-34-2

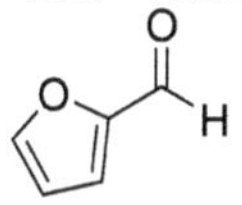

B22 98-01-1

Fig. 6 Overview of the selected aldehyde building blocks. BBs are identified by CAS number

Table 1
The 66 acylhydrazones obtained by SMD

Acylhydrazone	BB combination	Acylhydrazone	BB combination	Acylhydrazone	BB combination
1	A1B1	23	A1B22	45	A1B21
2	A2B2	24	A2B1	46	A2B22
3	A3B3	25	A3B2	47	A3B1
4	A4B4	26	A4B3	48	A4B2
5	A5B5	27	A5B4	49	A5B3
6	A6B6	28	A6B5	50	A6B4
7	A7B7	29	A7B6	51	A7B5
8	A8B8	30	A8B7	52	A8B6
9	A9B9	31	A9B8	53	A9B7
10	A10B10	32	A10B9	54	A10B8
11	A11B11	33	A11B10	55	A11B9
12	A12B12	34	A12B11	56	A12B10
13	A13B13	35	A13B12	57	A13B11
14	A14B14	36	A14B13	58	A14B12
15	A15B15	37	A15B14	59	A15B13
16	A16B16	38	A16B15	60	A16B14
17	A17B17	39	A17B16	61	A17B15
18	A18B18	40	A18B17	62	A18B16
19	A19B19	41	A19B18	63	A19B17
20	A20B20	42	A20B19	64	A20B18
21	A21B21	43	A21B20	65	A21B19
22	A22B22	44	A22B21	66	A22B20

3. Shift the aldehyde building block another step downward, and combine it again with the hydrazides to form the last 22 acylhydrazones.
4. You now have a final set of 66 acylhydrazones (Table 1), where each building block is represented three times (*see* **Note 30**).

3.6 Chemistry and Biology

The resulting 66 acylhydrazones constitute a representation of the chemical space composed by all theoretical compounds that can be formed by the original 958 aldehydes and 192 hydrazides. The number of target compounds is manageable and they can be synthesized, isolated, and characterized as described previously [12].

Since the number of compounds is limited, a focused effort can be made to ensure that each compound is synthesized in sufficient amount and high purity. If a building block fails to produce the target compound due to unanticipated or even unknown reasons it can be easily exchanged by manual selection of an equivalent or a very similar building block in the plots in Figs. 3 and 5, respectively. Preferably all acylhydrazones in the set are then tested simultaneously at multiple concentrations in sufficient replicates. This strategy results in robust data with respect to both chemistry and biology and allows reliable conclusions to be drawn regarding chemical properties affecting biological activity.

4 Concluding Remarks

In this chapter we illustrated SMD as a powerful tool to explore screening hits with the goal of obtaining SARs. Such information is important for further optimization and development of potent and selective chemical probes. Importantly, the strategy is independent of the biological system the small molecules act on. We have successfully employed SMD within several areas including thrombin inhibitors, inhibitors of bacterial toxin secretion, antigenic peptides, and peptides involved in protein–protein interactions [4].

5 Notes

1. Other reasons include inadequate biological assays resulting in inconsistent biological data and the lack of thorough validation of the QSAR model itself using for instance test sets, i.e., prediction and evaluation of compounds that were not used to calculate the model.
2. The number of available analogues can range from a handful to even thousands depending on the compound class. This clearly affects the ability to select representative subsets that allow a relevant exploration of SARs.
3. Some vendors specialize in a certain class of compounds (e.g., Bachem or PolyPeptide Group with focus on peptide chemistry), while others carry a wide range of more general reagents and building blocks (e.g., Sigma-Aldrich, Acros, or Alfa-Aesar).
4. Other factors, such as a time-consuming or an expensive biological assay can also set an upper limit to the amount of compounds that can be meaningfully produced and tested.
5. Unwanted building blocks can be removed based on several criteria. Filters based on, e.g., molecular weight and polarity, can be applied. In addition more stringent filters that,

e.g., prevent substituents in certain positions based on structural information on the target protein and the structure of the binding site can be applied. Building blocks containing functional groups that are known to be toxic, reactive, or incompatible with the synthetic strategy are also removed. Additional factors like amounts that can be purchased, price, and timely delivery are generally also considered.

6. Molecular descriptors are numerical values used to describe the physical and chemical properties of a compound. Descriptors are often described as being 1D, 2D, or 3D. 1D descriptors capture intrinsic properties such as molecular weight and number of hydrogen bond acceptors and donors. 2D descriptors capture features relating to atom connectivity, e.g., number of aromatic rings and number of rotatable bonds. 3D descriptors depend on the conformation of the compound and include, e.g., solvent-accessible surface area and molecular volume. Today hundreds of descriptors can be computed, but not all are relevant.
7. When hundreds of building blocks each are characterized with more than a hundred descriptors the information content and complexity of the resulting matrix cannot be handled without powerful computational tools. PCA ([6]) is a method that analyzes multidimensional data such as the matrix formed by many molecules characterized with many descriptors and compresses the data into fewer dimensions, e.g., three dimensions defined by three new orthogonal principal components that capture the variation in the original data set. Each molecule will then be positioned in the three-dimensional space, and compounds close to each other have similar properties while molecules far apart differ substantially. Based on this three-dimensional presentation a representative subset can be selected.
8. A biased selection can be done "by hand," but there are also methods available that automate this process and attempt to choose a subset capturing as much as possible of the available chemical variation (e.g., DOOD or factorial design).
9. OpenEye FILTER contains an excellent documentation explaining how the program operates and how the *.filter files are structured.
10. A synthon is an idealized reagent representing a fragment of a molecule with an associated polarity (+ or −). For instance, both anhydrides and acyl chlorides would be represented by the same synthon.
11. It is well known within the field that acylhydrazones can be synthesized from hydrazides and aldehydes. The method is compatible with a wide variety of functional groups, and a large amount of building blocks are commercially available. Thus, the method is well suited for library synthesis.

12. The general trend when optimizing compounds for a biological target is that over time they will increase in size and lipophilicity. Therefore, we want to set an upper limit on the molecular weight in order to somewhat deal with this issue. Also, if the compounds are too small they are unlikely to show good affinity towards the target, and thus we also need a lower limit to the molecular weight.
13. Explanation of the arguments used for OpenEye FILTER:
    - Filter: used to start the program.
    - -in: defines the input file containing all compounds.
    - -out: defines the output file containing the filtered list of building blocks.
    - -filter: defines the .filter file containing the rules used to filter compounds.
    - -fail: defines the output file where the rejected compounds are saved. This argument is optional but useful if you want to investigate which compounds were removed by the program.

    The "Lead-like" filter enclosed within the documentation was used with some modifications.
    - "RULE 1 aldehyde" limits the number of aldehydes per molecule to one.
    - "NEWRULE hydrazine 0 NN" removes all compounds containing hydrazine groups.
14. The manual inspection of building blocks is based mostly on experience and is not strictly necessary. However, it will most likely improve the subsequent PCA model as uninteresting compounds do not have to be taken into account when the model is built. Building blocks that were removed include, for instance, those containing long flexible aliphatic chains or large lipophilic groups.
15. A force field is a set of parameters used to model molecular systems using classical physics. The parameters are always optimized against a certain class of compounds, for instance proteins. Hence, a force field is only relevant for the class it was optimized against. MMFF94 is optimized for small molecules and is thus suitable for our needs.
16. If a variable is excluded before the work set is created it cannot easily be included again. It is therefore good practice to wait until the work set is created before removing any variables in case they might be needed when investigating the data more carefully. Variables can be excluded for several reasons. Some (e.g., PC+, PC−, RPC+, and RPC−) are duplicates and are still available only for the sake of compatibility; others might not be considered relevant for our purposes (e.g., rsynth which

describes how easily a compound can be synthesized). Some also describe the same molecular property, but slightly different methods are used for the calculations (e.g., PEOE_PC+ vs. Q_PC+). Finally, some are removed due to the fact that they are not readily interpretable.

17. Knowing the 3D structure is not necessary when calculating 2D descriptors. Using 3D descriptors lies outside the scope of this chapter. However, if 3D descriptors were to be calculated it would be necessary to first find the active conformation of the relevant compounds. This can be done by performing a conformational search followed by an energy minimization of the entire database.
18. Compounds **1–5** will find all six-membered aromatic rings and **6** adds compounds with one extra methylene group. Compounds **7** and **8** will find all other aromatic compounds, and compound **9** includes all low-molecular-weight, nonaromatic compounds. Additional substructure searches can be performed, but in our case they did not give any additional hits.
19. Ideally, CAS_NUMBER should be used as primary observation ID. However, a large number of the compounds do not have a CAS number, and thus this is not possible.
20. No additional components are needed when the eigenvalues of the PCs start to level out. An eigenvalue above 2 can also be used as a rule of thumb. Where this cutoff should be placed can also be decided from the Q^2 values. The same principle applies when the increase from one component to the next starts to level out: no additional components are needed.
21. Observations above 3*D-Crit(0.05) or drastically outside Hotelling T2 (0.95) can be considered as outliers.
22. After removing outliers the number of necessary PCs is usually reduced since the amount of variation is diminished.
23. Variables furthest from the center have the largest impact on a principal component, while variables close to the center affect a component only to a small degree. For instance, volume, diameter, weight, and number of atoms are different ways of describing molecular "size." If many variables describing the same molecular property are found at the fringe of a principal component, it is likely that the component describes mostly this property. Variables found close together are positively correlated while those found at the opposite sides are negatively correlated.
24. A D-optimal design is based on the premise that a subset of samples from a given candidate set is covering the maximum space when the determinant of the candidate set matrix is maximized. Hence, one common problem is that the D-OPTIMAL design only

picks out the extreme points in the candidate set, especially if the number of objects to pick out is much lower than the candidate set. A way to get a more thorough sampling of the molecular space is to use a DOOD, where D-optimal designs are made in separate layers from the center and out.

25. Setting the outer layer to end at 97 % is a good way to avoid the most extreme objects to be picked up by the D-optimal design.
26. The number of desired design runs in each layer must be at least equal to the number of terms in the model. This means that in most cases a DOOD with 2–3 layers is sufficient.
27. G-efficiency compares the prediction performance of the D-optimal design to that of a fractional factorial design, where the fractional factorial design would have a value of 100 %. The condition number represents a measure of the sphericity of the design (orthogonality). Linear factorial designs, without center points, have a condition number of 1 where the design points are situated on the surface of a sphere. Although there is the possibility in the program to auto-select designs in the layers according to any of the above criteria, a manual selection balancing the two criteria while maintaining a proper amount of objects in each layer is often preferable.
28. Due to the large candidate set and the relatively few numbers of objects to select, there are numerous solutions for the D-optimal algorithm. Hence, the exact list which is displayed herein might not be retrieved. If not, try to increase the number of design runs from 20.
29. The generation of the final set by systematic combination of the randomized ordered building blocks will in most cases result in a sufficiently well distribution of the combination of the different building blocks. It is also possible to perform a second D-optimal design to determine the combinations that should be made. In such a case the two sets of building blocks should be treated as two qualitative factors (representing aldehydes and hydrazides), and each selected building block should be assigned to a qualitative setting.
30. By assuring that each building block is represented three times, the risk of making incorrect or false conclusions about building blocks in the following SAR analysis is minimized.

Acknowledgment

This work was supported by the Swedish Research Council, the Swedish Foundation for Strategic Research, and the Knut and Alice Wallenberg Foundation.

References

1. Wold S, Josefson M, Gottfries J, Linusson A (2003) The utility of multivariate design in PLS modeling. J Chemom 18:156–165
2. Eriksson L, Johansson E (1996) Multivariate design and modeling in QSAR. Chemom Intell Lab Syst 34:1–19
3. Johnson SR (2006) The trouble with QSAR (or how I learned to stop worrying and embrace fallacy). J Chem Inf Model 48: 25–26
4. Linusson A, Elofsson M, Andersson IE, Dahlgren MK (2010) Statistical molecular design of balanced compound libraries for QSAR modeling. Curr Med Chem 17: 2001–2016
5. Lipinski CA, Lombardo F, Dominy BW, Feeney PJ (1979) Experimental and computational approaches to estimate solubility and permeability in drug discovery and development settings. Adv Drug Deliv Rev 23:3–25
6. Jackson JE (1991) A user's guide to principal components. Wiley, New York
7. http://www.sigmaaldrich.com/catalog/search/substructure/SubstructureSearchPage. Accessed 14 Sep 2011
8. FILTER 2.0.2., OpenEye Scientific Software, 9 Bisbee Court, Suite D, Santa Fe, NM 87508, USA. http://www.eyesopen.com. Accessed 14 Sep 2011
9. MOE 2010.10., Chemical Computing Group, 1010 Sherbrooke St. W, Suite 910, Montreal, Quebec, Canada H3A 2R7. http://www.chemcomp.com. Accessed 14 Sep 2011
10. SIMCA-P+ 12.0.1., Umetrics AB, Box 7960, SE-907 19, Umeå, Sweden. http://www.umetrics.com. Accessed 14 Sep2011
11. MODDE 9.0., Umetrics AB, Box 7960, SE-907 19, Umeå, Sweden. http://www.umetrics.com. Accessed 14 Sep 2011
12. Dahlgren MD, Zetterström CE, Gylfe Å, Linusson A, Elofsson M (2010) Design, synthesis and multivariate quantitative structure-activity relationship of salicylidenacylhydrazides—inhibitors of type III Secretion in *Yersinia*. Bioorg Med Chem 18:2686–2703

Part IV

Target Identification

Chapter 18

Early Stage Hit Triage for Plant Chemical Genetic Screens and Target Site Identification

Terence A. Walsh

Abstract

The increasing use of plant biological screens of large compound libraries to discover informative chemical probes for plant chemical genetics requires efficient methods for hit selection and advancement. Downstream target identification and validation studies with selected chemistries can also be resource-intensive and have a significant failure rate. Several steps and considerations for early stage hit triage are outlined to increase the probability of success that downstream studies with the chemical probe will be robust and productive, especially for target site discovery. Conversely, problematic compounds can be shelved or avoided entirely, saving time and resources. These steps include assessment of compound availability, purity, stability and solubility; determination of the biological dose–response; early and iterative evaluation of analogs; avoidance of promiscuous "frequent-hitters"; consideration of physicochemical parameters affecting compound bioavailability and mobility, use of "low-barrier" biological testing systems; and assessing the potential for compound metabolism or bioconversion.

Key words Plant chemical genetics, Hit selection, Hit triage, Target identification

1 Introduction

The use of chemical probes in plant biology has a long history with the development and study of herbicides, phytotoxins, and plant growth regulators, and in the discovery and study of a wide range of chemical effectors of plants [1, 2]. These studies have expanded into the increasing use of small molecules as a convenient surrogate for and complement to genetic mutant analysis. This has led to the discovery and investigation of many unique chemically induced phenotypes and new avenues of research in plant biology [3, 4] with the result that well-characterized chemical probes [5] are becoming increasingly relevant and useful in plant biology.

The most insightful information can be derived from the use of a chemical probe when the target protein(s) and sites of chemical interaction within the plant are known. However, target site discovery still remains a rate-limiting step in plant chemical

Glenn R. Hicks and Stéphanie Robert (eds.), *Plant Chemical Genomics: Methods and Protocols*, Methods in Molecular Biology, vol. 1056, DOI 10.1007/978-1-62703-592-7_18, © Springer Science+Business Media New York 2014

biology and chemical genetics [6]. There are few widely applicable rapid methods for target site elucidation and most require a significant investment of time and resources and can have a significant failure rate. Our lab has investigated many plant-active chemistries that induce a variety of phenotypic effects for which we have identified the target sites and other sites of chemical interaction [7, 8]. We have used mutant screening in genetically tractable organisms such as *Arabidopsis thaliana* as a well-validated method for target site elucidation [9]. Recovery of resistant (or hypersensitive) mutants using probe chemistry enables subsequent identification of the site of mutation and candidate sites of chemical interaction. This is a relatively unbiased target site discovery method with a reasonable success rate.

Although methods for identification of the site of resistance mutations are becoming more rapid with advances in sequencing and rapid genomic characterization, there is still a significant effort required to perform these analyses and for the studies required to validate that the site of mutation is indeed the site of chemical interaction. Thus, special attention needs to be paid to ensuring that compounds of interest (COIs) that are promoted from an initial screen and early-stage characterization into target site discovery have the best possible chance of yielding informative results. In the pharmaceutical area there is a wealth of knowledge on the behavior of exogenously applied compounds in intact animal and cell systems and in in vitro biochemical assays, and an awareness of the many pitfalls in analysis that can occur [10, 11]. This knowledge is not as mature in the plant arena and principally resides in agricultural chemical companies that have experience in the screening of a wide variety and large number of novel small molecules from internal and externally sourced compound libraries. These institutions also have accumulated knowledge about the delivery and behavior of xenobiotic compounds and natural products within plants. In recent years, evolution in compound screening and testing has extended to the need for triage of early stage hits from miniaturized high-throughput plant screens to select and promote compounds for conventional analysis on larger plants toward the development of herbicides, other pesticides, and various plant growth regulators [12]. However, plant chemical genetic screens and studies using exogenous xenobiotic compounds are increasingly being deployed in academic labs with a background in plant molecular or cell biology where this type of in-house knowledge is not as developed or readily accessible.

A key element for any plant biology laboratory undertaking plant chemical genetic studies and the selection of interesting compounds for further evaluation is to build an effective collaboration with key chemistry-skilled partners. The insights and capabilities provided by an astute organic and/or analytical chemist can save a lot of time and potential frustration in structure evaluation and confirmation, judicious analog selection, and analytical assessment of the plant fate of new COIs. Although many successful plant chemical genetic and target identification studies have been published with novel or previously uncharacterized chemical effectors, many more have failed because of poor compound behavior, inability to recover informative mutants, or even the lack of available compound for further study.

This article aims to provide some simple triage steps and considerations that can assist in making effective COI selections for target site determination in plant chemical genetic studies (**Note 1**). Many of these processes can be iterative and will provide increasing evidence that a COI is worthy of further investigation. There are increasingly sophisticated computational and analytical tools to assist in many of these steps but these are generally specific to the types of libraries, chemistries, and biological systems being investigated and so are not detailed.

2 Methods

2.1 Establish Compound Availability

Early stage micronized (small volume) plant chemical genetic screens typically require very little compound (sub-microgram amounts) to identify an initial hit [12, 13]. However, the amount of compound required will escalate after a hit has been identified and confirmed (preferably with a new lot of the compound) as more follow-up biological studies are performed. If the compound is of sufficient interest to be promoted to target site determination, then additional compound is consumed that may eventually require tens of mg or more. Thus, it is important to pre-estimate the amount of compound likely to be needed for target determination studies (which will be dependent on the potency of the COI) and to confirm with the source that there are sufficient additional quantities of the COI available for the anticipated studies. It is an advantage if the molecule can be readily resynthesized (and many libraries can be biased in this way) but many commercially available compounds can be in limited supply. This is particularly the case for natural products that can be extremely difficult to re-source once an initial supply has been depleted.

2.2 Ensure Purity and Stability

Some misleading or serendipitous effects in chemical screens in plant biology can sometimes be traced to the potent effect of a small amount of a contaminant. Many commercial sources supply some guarantee of chemical purity for the compounds they provide, but this may not always be the case. The age and provenance of chemical samples can vary considerably. Simple confirmation tests of purity and chemical integrity, for example by HPLC, NMR, and MS analysis, are essential to confirm the purity and structure of a COI prior to investment in extensive biological and target site studies [14].

For natural products this can be a much more challenging proposition due to their structural complexity and difficulty in isolation and purification.

Another source of early confounding results in chemical screens in plant biological systems can be the instability of a compound in the system of interest. The potential biotransformation of COIs will be discussed later, but an early worthwhile assessment is to establish that the COI does not contain chemical elements that may be intrinsically unstable in mild aqueous conditions (consult your local organic chemist) and that it remains unchanged in the biological media used over the time scale of a typical experiment or mutant screen. This can be followed using many standard analytical techniques such as HPLC or LC-MS analyses.

2.3 Evaluate Solubility

Many sources of assay variability and lack of repeatability especially in micronized plant systems can be attributed to variation in compound solubility over time or in different media [15]. Simple observations of precipitation on addition of a compound dissolved in an organic solvent such as methanol or DMSO into the aqueous media of the assay or precipitation over the time scale of the experiments are warning signs that the activity of the compound may be being compromised by insolubility. Experimentation with different solvents or solvent/media ratios may be required to maximize solubility, or another active compound that avoids these problems should be selected.

2.4 Evaluate Dose Response

The dose response of a plant biological system to COIs should be established early, especially as early-stage testing is often conducted at a single high dose. This will serve to distinguish the most potent chemical hits from the weaker ones and also establish effective doses for subsequent mutant selection screens. The shape of the dose response curves can also give indications that various hit compounds are behaving similarly. Conversely, anomalously steep dose response curves could be an indication of issues with compound solubility or reactivity that may lead to difficulties in reproducibility or artifactual phenomena in the assay.

2.5 Test Analogs

Once the chemical authenticity and response behavior of a COI has been established, it is worthwhile to establish a preliminary structure–activity relationship (SAR) that relates the principal chemical features of COIs with their bioactivity. This provides several early indications of interest: Is the compound a "one-compound wonder" that is structurally unique? Is it part of a broader class of bioactive chemistry? Are there particular chemical features that are required for activity, or other features that appear to be more flexible? Many chemical screening libraries are assembled and offered based on "diversity" or "novelty," so there may be an explicit bias for the members of the library to be fairly unique and differentiated from each other in order to cover a wide area of chemical space. In such cases, it is unlikely that there will be other molecules within the initial library that explore adjacent chemical space effectively. Therefore it is worthwhile to search for, acquire, and test available analogs of COIs "early and often" to establish if the first compound that showed interesting activity is the most potent, most effective, and/or the most specific one available. It is also possible to uncover an analog that has similar bioactivity but is more readily available or stable than the initial hit. It can be frustrating to quickly undertake extensive studies with the first-hit compound, only to find later that a related compound exerts a more potent or specific effect and is a better chemical probe.

The ability to build an SAR is a compelling and indicative piece of evidence that the biological effect of a COI is well defined and specific. If the SAR is broad or ill-defined, this may suggest that there are several sites of interaction that may be more difficult to deconvolute in subsequent target site studies. Alternatively, it may be found that the SAR eventually overlaps with other compounds that have a known or previously characterized mode of action, so that additional work can be discontinued. The development of a novel biological response SAR can be of great utility in subsequent validation that a candidate target protein is indeed the site of chemical interaction. If the response of the isolated protein to the compounds in the SAR recapitulates that identified in the biological screen, this is a strong indication that the candidate protein is indeed the target.

2.6 Be Aware of "Frequent Hitters"

Some of the features that may emerge from a preliminary SAR may not be necessarily unique to your target or bioassay. As the science of high throughput chemical screening has matured, it has become apparent that some chemical classes of compounds (beyond intrinsically reactive compounds) have been found to hit in many screens, especially for in vitro biochemical screens such as enzyme assays. Such promiscuous compounds have been termed "frequent-hitters." Several publications have now outlined many of the prime offending attributes of these types of compounds and some intriguing explanations for their broad activity have been put forward [16, 17]. Many commercially available and in-house libraries have now been purged of these types of compounds. Nevertheless, it is worthwhile becoming acquainted with frequent hitter structures and to be circumspect about pursuing compounds that have frequent hitter attributes. Their effects may be nonspecific and lead to fruitless target identification studies.

It is possible (but time-consuming) to run additional counter-screen assays on differing targets or biological processes and therefore provide accumulated evidence of specificity for COIs for their effect on a given assay. As a plant biology lab runs more assays and screens and accumulates more data across a range of chemistry inputs and screens, trends for promiscuous or frequent hitter compounds may emerge across assays and screens. The attributes of frequent hitters have been primarily deduced from pharmaceutical screens so as the field matures, it will be of interest to see if frequent hitters are identified that are more prevalent or unique to plant-based screens. In Arabidopsis phenotype screens, there are a wide range of compounds that appear to exert mild phenotypic effects and seem to occur with quite high frequency in compound libraries. For example, many types of compound appear to give phenotypes that can be attributed to weak auxinic/gravitropic effects [18] or effects on cytoskeletal processes and cell wall biosynthesis, or inhibition of Photosystem II [2]. These plant-specific physiological processes appear to have a wide range of xenobiotic chemistries that can interfere with them.

2.7 Consider Compound Bioavailability and Mobility

The potency of compound in a given cellular or intact plant assay is highly dependent on its uptake and bioavailability to the ultimate site of interaction within the plant. In plant chemical genetic studies, a basic understanding of the physicochemical features of compounds that can significantly affect their bioavailability in plant systems is useful to guide the selection and testing of lead COIs. It is also important to be aware of the intrinsic differences in exposure to a compound between various typical assay systems in intact plant and plant cell-based screen systems. Compounds applied to an expanded leaf must be phloem-mobile to effectively move out of the leaf to growing points such as the root and meristem. The physicochemical parameters for xenobiotic phloem mobility have been well characterized [19, 20]. Weakly acidic compounds of moderate lipophilicity are especially conducive for mobility in the phloem as they accumulate and are trapped in the phloem stream. In contrast, the xylem stream (with a considerably higher volume of solute movement) will move leaf-applied compounds toward the leaf margins. There will be no significant basipetal relocation of an uncharged lipophilic compound out of a leaf to the roots [21].

2.8 Use Low Barrier Systems

When target site elucidation is of primary interest in a model system such as Arabidopsis, it is most effective to have as few barriers as possible between the applied compound and the site of interaction. These barriers (UV degradation, cuticular penetration, long-distance translocation, tissue dilution, sequestration, etc.) can lead to variable results due to the behavior or exclusion of a compound on its way to the site of action, for example by poor leaf penetration or lack of mobility from roots into leaves. Removal of these barriers increases sensitivity to the introduced compound, decreases potential variability in the phenotypic response being monitored, and enhances the ability to select COIs with the best chance to yield mutants at or near the target site in low-barrier mutant screens.

In many micronized Arabidopsis screens, plants are totally immersed in media that is typically mildly acidic (pH 5–6). At this pH, many compounds with acidic moieties such as carboxyl groups are predominantly charged and so cellular uptake can be relatively poor, or primarily mediated by certain transporters (e.g., 2,4-D acid is recognized by the auxin permease [22]). In contrast, neutral slightly lipophilic molecules can readily permeate into cells. Xenobiotic compounds with carboxylic acids that are esterified can be more active than the corresponding free acids as they are more readily taken up and subsequently undergo intracellular cleavage to the free acid by the action of carboxylesterases [23]. Thus, the compound can be effectively ion-trapped within the cell. For example this effect was noted in chemical genetic studies of a series of novel phenyltriazoloacetic acid inhibitors targeting Arabidopsis purine biosynthesis [8].

COIs should be inspected for potentially ionizable groups with pKa values between pH 5 and 9 that could influence the uptake and redistribution of the compound. Judicious derivatization, for example esterification of a carboxylic acid, may increase the potency of a lead compound in a bioassay sufficient to make it a preferred probe for target identification. Lipophilic uncharged compounds can be accumulated in plant tissue very efficiently by partitioning out of the aqueous media phase into the lipophilic portions of cells and plant tissue. However, highly lipophilic compounds will be trapped within these matrices and be relatively sequestered and unavailable to many biological targets or sites of action. Thus, bioavailability may be improved by choosing compounds in an active series with lower lipophilic character. In contrast, highly charged compounds such as dicarboxylates and phosphates may be excluded from cellular uptake unless recognized by specific plasma membrane pumps.

2.9 Be Aware of Potential Compound Metabolism and Bioconversion

Intact plants and plant tissues can be remarkably effective at metabolizing exogenously added compounds via a variety of mechanisms [24–26]. They can also be effective at preventing intracellular accumulation of xenobiotic compounds by either pumping them back out of the cell or into a vacuole to sequester them away from sensitive sites of interaction [27–29]. Therefore preliminary assessment of the stability and longevity of a compound in the plant system under investigation is worthwhile prior to extensive target discovery work. This at least requires an analytical method to monitor the continued presence of the applied compound. This can show that the unaltered compound can be recovered by extraction from the plant tissue. Analytical LC/MS methods are now readily developed to analyze the recovery specific compounds. If recoveries are less than anticipated (and the compound has been shown to be stable in the biological media in the absence of plant tissues), this is an indication of metabolism of the compound by the plant tissue.

If the COI is rapidly metabolized over the timescale of the experiments or screening procedure but bioactivity is retained or increases, this can be a preliminary indication that the compound is being converted to an active form and the applied compound may not be the active moiety. This is of particular significance for downstream target site discovery by mutational analysis as loss-of-function mutations in the metabolic conversion processes may readily confer resistance to the COI. These type of mutations may not be particularly informative about the actual site of action, although can add useful knowledge about all the sites of chemical interaction of a particular COI within the organism. Many bioactive molecules can be presented in a "pro-"form ("prodrugs" in the pharma field) that is bioactivated to reveal a final active form. Some examples in plant systems are 6-methylanthranilate that is metabolized through the Trp biosynthetic pathway to 3-methyl-trytophan, a strong inhibitor of tryptophan synthase [30]; a pro-herbicide that is activated by a Cyp450 [31]; 2,4-D butyrate that is β-oxidized to the more active auxin surrogate 2,4-D [32]; and other pro-auxins [33]. In many cases, the pro-compound can be more bioactive than the final "target-active" compound when applied exogenously as the "pro-" elements of the compound enable effective cellular uptake and subsequent delivery and activation at or close to the site of action.

3 Notes

1. Application of these steps to the evaluation of novel hits in plant chemical genetic screens can aid in selection of the best candidate chemistries for advancement into more in-depth studies. It will also increase the probability of success that downstream studies with the chemical probe will be robust and productive, especially for target site discovery. Conversely, problematic compounds can be shelved or avoided entirely, saving time and resources.

References

1. Kaschani F, van der Hoorn R (2007) Small molecule approaches in plants. Curr Opin Chem Biol 11:88–98
2. Dayan FE, Duke SO, Grossmann K (2010) Herbicides as probes in plant biology. Weed Sci 58:340–350
3. Tóth R, van der Hoorn RAL (2010) Emerging principles in plant chemical genetics. Trends Plant Sci 15:81–88
4. Robert S, Raikhel NV, Hicks GR (2009) Powerful partners: Arabidopsis and chemical genomics. Arabidopsis Book 7:e0109
5. Frye SV (2010) The art of the chemical probe. Nat Chem Biol 6:159–161
6. Burdine L, Kodadek T (2004) Target identification in chemical genetics: the (often) missing link. Chem Biol 11:593–597
7. Walsh TA, Neal R, Merlo AO, Honma M, Hicks GR, Wolff K, Matsumura W, Davies JP (2006) Mutations in an auxin receptor homolog AFB5 and in SGT1b confer resistance to synthetic picolinate auxins and not to 2,4-dichlorophenoxyacetic acid or indole-3-acetic acid in Arabidopsis. Plant Physiol 142:542–552

8. Walsh TA, Bauer T, Neal R, Merlo AO, Schmitzer PR, Hicks GR, Honma M, Matsumura W, Wolff K, Davies JP (2007) Chemical genetic identification of glutamine phosphoribosylpyrophosphate amidotransferase as the target for a novel bleaching herbicide in Arabidopsis. Plant Physiol 144: 1292–1304
9. Walsh TA (2007) The emerging field of chemical genetics: potential applications for pesticide discovery. Pest Manag Sci 63:1165–1171
10. Kerns EH, Di L (2003) Pharmaceutical profiling in drug discovery. Drug Discov Today 8: 316–323
11. Eddershaw PJ, Beresford AP, Bayliss MK (2000) ADME/PK as part of a rational approach to drug discovery. Drug Discov Today 5:409–414
12. Tietjen K, Drewes M, Stenzel K (2005) High throughput screening in agrochemical research. Comb Chem High Throughput Screen 8: 589–594
13. Kolukisaoglu U, Thurow K (2010) Future and frontiers of automated screening in plant sciences. Plant Sci 178:476–484
14. Di L, Kerns EH (2009) Stability challenges in drug discovery. Chem Biodivers 6:1875–1886
15. Di L, Kerns EH (2006) Biological assay challenges from compound solubility: strategies for bioassay optimization. Drug Discov Today 11: 446–451
16. Baell JB (2011) Redox-active nuisance screening compounds and their classification. Drug Discov Today 16:840–841
17. Feng BY, Shelat A, Doman TN, Guy RK, Shoichet BK (2005) High-throughput assays for promiscuous inhibitors. Nat Chem Biol 1: 146–148
18. De Rybel B, Audenaert D, Beeckman T, Kepinski S (2009) The past, present, and future of chemical biology in Auxin research. ACS Chem Biol 4:987–998
19. Hsu FC, Kleier DA (1996) Phloem mobility of xenobiotics. 8. A short review. J Exp Bot 47: 1265–1271
20. Lichtner F (2000) Phloem mobility of crop protection products. Aust J Plant Physiol 27: 609–614
21. Liu ZQ (2006) Leaf epidermal cells: a trap for lipophilic xenobiotics. J Integr Plant Biol 48: 1063–1068
22. Delbarre A, Muller P, Imhoff V, Guern J (1996) Comparison of mechanisms controlling uptake and accumulation of 2,4-dichlorophenoxy acetic acid, naphthalene-1-acetic acid, and indole-3-acetic acid in suspension-cultured tobacco cells. Planta 198:532–541
23. Gershater M, Edwards R (2007) Regulating biological activity in plants with carboxylesterases. Plant Sci 173:579–588
24. Marrs KA (1996) The functions and regulation of glutathione S-transferases in plants. Annu Rev Plant Physiol Plant Mol Biol 47: 127–158
25. Gachon CMM, Langlois-Meurinne M, Saindrenan P (2005) Plant secondary metabolism glycosyltransferases: the emerging functional analysis. Trends Plant Sci 10:542–549
26. Mizutani M, Ohta D (2010) Diversification of P450 genes during land plant evolution. In: Merchant S, Briggs WR, Ort D (eds) Annual review of plant biology. Annual Reviews, Palo Alto, CA, USA. vol 61, pp 291–315
27. Crouzet J, Trombik T, Fraysse AS, Boutry M (2006) Organization and function of the plant pleiotropic drug resistance ABC transporter family. FEBS Lett 580:1123–1130
28. Conte SS, Lloyd AM (2011) Exploring multiple drug and herbicide resistance in plants—spotlight on transporter proteins. Plant Sci 180:196–203
29. Buss DS, Callaghan A (2008) Interaction of pesticides with p-glycoprotein and other ABC proteins: a survey of the possible importance to insecticide, herbicide and fungicide resistance. Pest Biochem Physiol 90:141–153
30. Siehl DL, Subramanian MV, Walters EW, Blanding JH, Niderman T, Weinmann C (1997) Evaluating anthranilate synthase as a herbicide target. Weed Sci 45:628–633
31. Thies F, Backhaus T, Bossmann B, Grimme LH (1996) Xenobiotic biotransformation in unicellular green algae—involvement of cytochrome P450 in the activation and selectivity of the pyridazinone pro-herbicide metflurazon. Plant Physiol 112:361–370
32. Hayashi M, Toriyama K, Kondo M, Nishimura M (1998) 2,4-dichlorophenoxybutyric acid-resistant mutants of Arabidopsis have defects in glyoxysomal fatty acid beta-oxidation. Plant Cell 10:183–195
33. Savaldi-Goldstein S, Baiga TJ, Pojer F, Dabi T, Butterfield C, Parry G, Santner A, Dharmasiri N, Tao Y, Estelle M, Noel JP, Chory J (2008) New auxin analogs with growth-promoting effects in intact plants reveal a chemical strategy to improve hormone delivery. Proc Natl Acad Sci U S A 105:15190–15195

Chapter 19

Screening for Gene Function Using the FOX (*F*ull-Length cDNA *O*vere*X*pressor Gene) Hunting System

Mieko Higuchi-Takeuchi and Minami Matsui

Abstract

Mutant resources are indispensable for the characterization of the functions of genes. There are two types of mutants, loss-of-function and gain-of-function mutants. Recently, we have developed a novel system in plants that uses a gain-of-function approach and is named as the FOX (*f*ull-length cDNA *o*vere*x*pressor gene) hunting system. In this system, *Arabidopsis* full-length cDNAs (fl-cDNAs) are randomly overexpressed under the control of the *cauliflower mosaic virus* (*CaMV*) *35S* promoter in *Arabidopsis* plants. These transgenic plants, or *Arabidopsis* FOX lines, possess ectopically expressed fl-cDNAs in their genome. Chemical genomics is a newly emerging field that connects chemical biology with genomes. Since each FOX line expresses an excess amount of the protein from the transgene it can be resistant or hypersensitive to bioactive chemicals when the protein is the target for the chemical. In this protocol, we describe the procedure for identification of the fl-cDNAs responsible for the target of the chemical or for the signal transduction pathway involving the chemical.

Key words Full-length cDNA, *Arabidopsis*, Overexpression, FOX hunting system

1 Introduction

Mutant resources are indispensable for the characterization of the functions of genes. There are two types of mutant resources, loss-of-function and gain-of-function mutants [1, 2]. The generation of loss-of-function mutants has been carried out by T-DNA or transposon insertion, chemical and physical mutagenesis. On the other hand, activation tagging is a very valuable tool for generating gain-of-function mutants [3–5]. However, the over-expression depends on the site of the T-DNA or transposable element insertion. In addition, the *cauliflower mosaic virus* (*CaMV*) *35S* enhancer on the T-DNA or transposon can influence the expression of genes several kb from the insertion site, thereby causing difficulties in identifying the genes responsible for the mutants.

Glenn R. Hicks and Stéphanie Robert (eds.), *Plant Chemical Genomics: Methods and Protocols*, Methods in Molecular Biology, vol. 1056, DOI 10.1007/978-1-62703-592-7_19, © Springer Science+Business Media New York 2014

We therefore have developed a different approach to systematically generate gain-of-function mutants using full-length cDNAs (fl-cDNAs). We generated gain-of-function mutant lines caused by ectopic expression of fl-cDNAs under the control of the *CaMV 35S* promoter in *Arabidopsis thaliana* and designated the method as the FOX hunting system (Full-length cDNA Over-eXpressor gene hunting system) [6]. We used around 10,000 fl-cDNA clones originating from the RIKEN *Arabidopsis* full-length cDNA collection [7]. To construct an *Agrobacterium* expression library each fl-cDNA was mixed at approximately equal molar ratio. This *Agrobacterium* library was used to transform *Arabidopsis* plants by in planta transformation resulting in *Arabidopsis* FOX lines. Several genes have been identified using these FOX lines [8, 9]. The seed pool consisting of the *Arabidopsis* FOX lines is available to researchers from the RIKEN BioResource Center (http://www.brc.riken.jp/lab/epd/Eng/). We have also generated rice FOX *Arabidopsis* lines in which rice fl-cDNAs are expressed in *Arabidopsis.*

Gain-of-function technology can be applied to the chemical genomics approach that has been developed recently to facilitate the understanding of gene function [10]. Multicopy gene suppressor screening based on ectopic expression of transgenes has been successfully carried out to isolate mutants resistant to growth inhibitors in budding yeast [11]. They screened for growth inhibitors from about 1,000 compounds and isolated genes that conferred resistance to these chemicals. A novel plant chemical compound, cobtorin, was identified as an inhibitor of parallel alignment of the cortical microtubules and cellulose microfibrils [12]. Using the *Arabidopsis* FOX lines they found that overexpression of two enzymes involved in pectin degradation (pectin methylesterase and polygalacturonase) caused resistance to cobtorin treatment [13]. They proposed that cobtorin affects the modification and distribution of pectin, and that pectin has an important role in the deposition or the maintenance of cellulose microfibrils. Thus, the combination of chemical genomics and gain-of-function mutants provides a novel tool with which to dissect various signaling pathways.

In this protocol, we describe the procedures to isolate mutants from the seed pool of the *Arabidopsis* FOX lines and to identify fl-cDNAs. The overall process is shown in Fig. 2. The introduced fl-cDNA is isolated from the candidate *Arabidopsis* FOX line and then reintroduced into *Arabidopsis* to confirm the observed phenotype. Thus, the FOX hunting system provides a novel approach for chemical genetic screening.

2 Materials

2.1 *Arabidopsis* FOX Lines Pool Set

1. The *Arabidopsis* FOX lines are distributed from the RIKEN BioResource Center (http://www.brc.riken.jp/lab/epd/Eng/).
2. They are provided as a seed pool set. A seed pool contains about 400 seeds from 50 lines (equivalent to 8 seeds per line). One seed pool set contains 20 seed pools (equivalent to 1,000 lines).
3. The FOX lines are generated in the Columbia-0 ecotype.

2.2 Cloning of Isolated fl-cDNA into pBIGS2113SF

1. pBIGS2113SF (Fig. 1) is used as the expression vector for transformation of *Arabidopsis* [6]. Detailed vector information is described in our website (http://pfgweb.psc.riken.jp/fh-line.html).
2. *Sfi*I (20,000 U/mL) is purchased from New England Biolabs Japan, Inc. (Tokyo, Japan). 10× H buffer and 0.2 % BSA solution are supplied with the restriction enzyme.
3. T4 DNA ligase (400 U/μL) is purchased from New England Biolabs. 10× ligation buffer is supplied with the ligase.

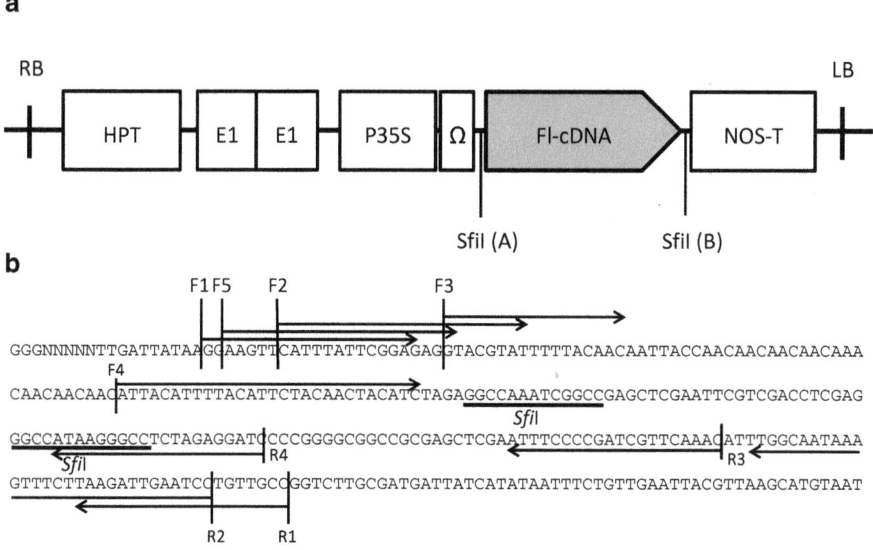

Fig. 1 The structure of the expression vector used for the FOX hunting system (pBIG2113SF). (**a**) Full-length cDNAs are cloned into the *SfiI* site of pBIG2113SF. *HPT* hygromycin resistance gene, *E1* 5′-upstream sequence of the *CaMV 35S* promoter (−419 to −90), *35S CaMV 35S* promoter (−90 to −1), *Ω* 5′-upstream sequence of TMV (Tobacco mosaic virus), *NOS-T* polyadenylation signal of the gene for nopaline synthase in the Ti plasmid, *RB* T-DNA right border, *LB* T-DNA left border. Primer sets described in Table 1 are shown with *arrows*. (**b**) Location of primer sets in pBIG2113SF used for fl-cDNA amplification and sequencing. Primer sequences are shown in Table 1

4. 3 M sodium acetate: Dissolve 40.8 g of $NaOAc \cdot 3H_2O$ in 100 mL of double distilled water (DDW) and adjust the pH to 4.8 with glacial acetic acid. The solution is used after autoclaving.

2.3 Culture of E. coli and Agrobacterium Cells

1. *E. coli* DH10B and *Agrobacterium* GV3101pMP90 are used for transformation.
2. LB medium: Dissolve 10 g of tryptone peptone, 5 g of yeast extract, and 5 g of NaCl in 1,000 mL of DDW and autoclave. To make LB agar plates, add 10 g of agarose before autoclaving.
3. A stock solution of kanamycin (Wako Pure Chemical Industries, Ltd. Tokyo, Japan) is prepared at 100 mg/mL in DDW and sterilized by filtration through a 0.22 μm Millipore membrane filter.

2.4 Transformation of Arabidopsis Plants

1. Infiltration medium: Dissolve half-strength Murashige and Skoog (MS) and 50 g of sucrose in 1,000 mL of DDW. Add 112 μL of Gamborg's 1,000× vitamin solution (Sigma-Aldrich Japan, Tokyo, Japan), 10 μL of benzylaminopurine stock solution, and 200 μL of Silwet L-77 (Agri-Turf Supplies, Inc., Santa Barbara, CA).
2. Benzylaminopurine stock solution: Dissolve 1 mg of benzylaminopurine (Wako Pure Chemical Industries, Ltd. Tokyo, Japan) in 1 mL of dimethyl sulfoxide (DMSO).

2.5 Selection of Arabidopsis Plants Expressing fl-cDNAs

1. BAM plates [14]: Dissolve 101 mg of KNO_3 in 1,000 mL of DDW and add 8 g of Bacto Agar (Becton, Dickinson and Company, Tokyo, Japan) and then autoclave (*see* **Note 1**).
2. 0.2 % agar: Add 2 g of Bacto Agar to 1,000 mL of DDW and autoclave.
3. Bleach solution: Mix 10 mL of sodium hypochlorite solution and 100 μL of Triton X-100 in 90.9 mL DDW.
4. Antibiotics: Hygromycin B (Wako Pure Chemical Industries, Ltd. Tokyo, Japan) and cefotaxime (Sigma-Aldrich Japan, Tokyo, Japan) are dissolved in DDW to concentrations of 20 mg/mL and 100 mg/mL, respectively. Sterilize the solutions by filtering through 0.22 μm Millipore membrane filters.

3 Methods

3.1 Screening of Arabidopsis FOX Lines

1. Grow each seed pool of the *Arabidopsis* FOX lines under screening conditions to isolate mutants. In this protocol, the isolation of a chemical-resistant *Arabidopsis* FOX line [13] is described as an example and is shown in Fig. 2.

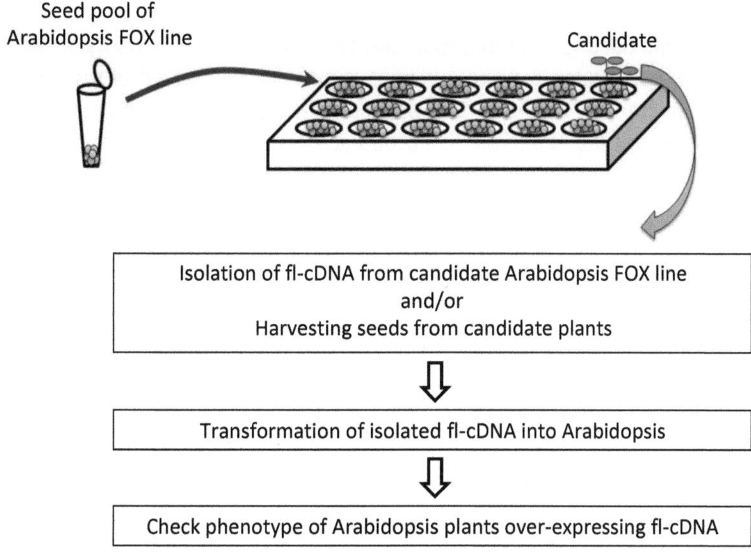

Fig. 2 Identification of fl-cDNAs responsible for the phenotypes. Seed pools of the *Arabidopsis* FOX lines are screened under various conditions. The fl-cDNA is isolated from the candidate *Arabidopsis* FOX line. The introduced fl-cDNA is transformed back into *Arabidopsis* to confirm reproduction of the phenotype. Seed pools of the *Arabidopsis* FOX lines are available from the RIKEN BioResource Center

2. Put seeds of *Arabidopsis* FOX line in a tube and add 70 % ethanol and leave for 1 min.
3. Discard the 70 % ethanol and then add the bleach solution and leave for 10 min.
4. Remove the bleach solution completely and wash the seeds with sterile water three times.
5. Sow the seed pools of the *Arabidopsis* FOX lines on plates containing the chemical compound (*see* **Note 2**). Grow under suitable conditions.
6. Select the *Arabidopsis* seedlings that show a phenotype (*see* **Note 3**).
7. After selection of candidate lines, transfer them to soil and then grow for extraction of genomic DNA and/or to harvest seeds. Confirm the phenotype observed in the progeny.
8. At the second screening, check the candidate lines to determine whether the phenotype changes with the concentration of the chemical. Analyze the phenotype in both the absence and presence of the chemical to exclude the possibility that the phenotype observed is not caused by the chemical compound.

Table 1
Primer sets used for amplification and sequencing of fl-cDNAs using the FOX lines. The primers given are arranged internally or nested with respect to the first set of primers. (Locations of primers are shown in Fig. 1b). The F1 and R1 primer set is used for amplification of fl-cDNAs and F5 is used for sequencing of the amplified PCR product. All primer sequences are indicated 5′ to 3′

Forward primer		Reverse primer	
Name	Sequence	Name	Sequence
F1	GGAAGTTCATTTATTCGGAGAG	R1	GGCAACAGGATTCAATCTTAAG
F2	CATTTATTCGGAGAGGTACGTAT	R2	GGATTCAATCTTAAGAAA CTTTATTGCCAA
F3	GTACGTATTTTTACAACAATT ACCAACAAC	R3	CAAATGTTTGAACGATCGGGGAAAT
F4	ATTACATTTTACATTCTACAAC TACATCT	R4	GATCCTCTAGAGGCCCTTAT
F5	AAGTTCATTTATTCGGAGAG		

3.2 Isolation of Full-Length cDNA from Arabidopsis FOX Lines

1. Extract genomic DNA from the candidate *Arabidopsis* FOX lines. Pure genomic DNA of high molecular weight is required to amplify the introduced fl-cDNA.
2. Make the PCR solution according to the instructions for the PCR enzyme (*see* **Note 4**). The F1 and R1 primer set in Table 1 is usually used for amplification of the introduced fl-cDNAs in *Arabidopsis* FOX plants. Other primer sets in Table 1 can be used when PCR with F1 and R1 is unsuccessful.
3. PCR program: After denaturation at 94 °C for 3 min, 30 cycles of 94 °C for 30 s, 52 °C for 30 s, and 72 °C for 3 min are carried out. At the end the PCR mixture is incubated at 72 °C for 10 min to complete the extension.
4. Sequence the PCR fragments directly after purification on an agarose gel using F5 primer (*see* **Notes 5–7**).

3.3 Cloning of Isolated fl-cDNA into pBIGS2113SF

1. Add 0.9 μL of pBIGS2113SF (5–20 ng/μL), 1 μL of 10× H buffer, 0.1 μL of 0.2 % BSA solution, and 0.5 μL of *Sfi*I to 7 μL of the purified PCR solutions and incubate at 50 °C overnight.
2. After incubation at 50 °C overnight, add 0.5 μL of *Sfi*I to the reaction solution and incubate the tubes at 50 °C for 3 h.
3. Add 1 μL of 3 M sodium acetate and 10 μL of isopropanol to the reaction tubes and mix well. After centrifugation of the tubes at 4 °C for 20 min, the pellets are washed twice with 70 % ethanol and dried.

4. Add 1 μL of DDW and 0.5 μL of 10× ligation buffer to the tubes and then add 0.5 μL of T4 DNA ligase.
5. Incubate the tubes at 16 °C overnight.
6. Mix 20 μL of DH10B *E. coli* competent cells and 1 μL of ligation products and mix well (*see* **Note 8**). Transfer the cells into an ice-cold cuvette. Electroporate the cells at 1.5 kV, 200 Ω and 25 μF for 4 ms. Add 500 μL of SOC medium and transfer the *E. coli* cells into tubes. Chemically prepared competent cells can also be used for transformation.
7. Incubate the *E. coli* cells at 37 °C for 1 h. Spread all the *E. coli* cells on LB agar plates with 50 μg/mL of kanamycin and incubate at 37 °C overnight.
8. Pick up kanamycin-resistant colonies with a toothpick and transfer the *E. coli* cells to tubes containing PCR solution.
9. Carry out the PCR under the same conditions as for the amplification of the fl-cDNA to confirm the size of the PCR fragments.
10. Pick up the *E. coli* cells from the colonies and put into 2 mL of LB medium with 50 μg/mL kanamycin.
11. After incubation at 37 °C overnight, extract the plasmid harboring the fl-cDNA.
12. Sequence the fl-cDNA in pBIGS2113SF.

3.4 Transformation of Expression Constructs into *Agrobacterium*

1. Add 1 μL of the plasmid DNA harboring the fl-cDNA to 20 μL of competent cells of the *Agrobacterium* strain GV3101pMP90. After mixing, transfer the *Agrobacterium* cells to a cuvette on ice.
2. Electroporate the cells at 1.5 kV, 200 Ω and 25 μF for 4 ms.
3. Add 500 μL of SOC medium to the cuvette and transfer the *Agrobacterium* cells into tubes.
4. Incubate the *Agrobacterium* at 28 °C for 1–3 h. Spread all the cells on a LB agar plate containing 50 μg/mL of kanamycin and 10 μg/mL of gentamycin, and incubate at 28 °C for 2–3 days.

3.5 Transformation of Arabidopsis Plants

1. Pick up an antibiotic-resistant *Agrobacterium* colony and transfer to 2 mL of LB medium containing 50 μg/mL kanamycin and 10 μg/mL gentamycin and incubate at 28 °C over night.
2. Transfer the above *Agrobacterium* cells to 200 mL of LB medium containing 50 μg/mL kanamycin and 10 μg/mL gentamycin, and grow at 28 °C until the OD_{600} reaches 1.2–1.5.
3. Transfer the *Agrobacterium* cells into a centrifuge tube and centrifuge at 6,000 × *g* for 10 min.

4. Remove the supernatant and resuspend the pellet with infiltration medium to an OD_{600} of 0.8.
5. Invert pots containing *Arabidopsis* plants into the resuspended *Agrobacterium* cells and dip the inflorescences into the medium for 30 s. Use healthy *Arabidopsis* plants for good transformation efficiency. We usually use nine plants in a pot. Transfer the dippeot plant pots into a plastic bag and seal them. Put the plants into the growth chamber and leave overnight. Open the plastic bag and again leave the plants overnight.
6. Take the plant pots out of the plastic bag the following day.
7. Grow plants to harvest the T_1 seeds.

3.6 Isolation of Transgenic Arabidopsis Plants

1. Put 0.25 g of T_1 seeds into a 50 mL tube and add 10 mL of 70 % ethanol (*see* **Note 9**).
2. After shaking the tube briefly, remove the 70 % ethanol.
3. Add 10 mL of bleach solution and leave for 10 min.
4. Decant the solution and wash the seeds with sterile DDW three times.
5. Remove the DDW and add 5 mL of 0.2 % agar to the tube.
6. After inversion of the tube two to three times, spread the seeds on BAM agar plates (100 × 140 mm) containing 20 μg/mL of hygromycin B and 100 μg/mL of cefotaxime sodium salt.
7. Plates are sealed with surgical tape (SUMITOMO 3M limited), put at 4 °C for at least 2 days in the dark to induce germination.
8. Transfer the plates to a growth chamber at 22 °C and grow in the light for 5–10 days.
9. Transformants have true leaves and extended primary roots. They are transferred to soil.
10. Grow the seedlings to harvest the T_2 seeds.

3.7 Check the Phenotype in Re-transformed Arabidopsis Plants

1. Check the phenotype in the T_2 seeds using the initial screening conditions.
2. If the original phenotype is recaptured, analyze the expression level of the fl-cDNA in independent re-transformed lines. Analyze the correlation between the strength of the phenotype and the expression level of the introduced fl-cDNA.

4 Notes

1. Instead of BAM plates, use 1/2× Murashige and Skoog (MS) plates for antibiotic selection. We obtained more resistant plants using these plates.

2. Although a multi-well plate was used to keep each of the seed pools separate and to chemical compounds as shown in Fig. 2, other types of plates can be used. It is best to check in advance how many seedlings can be grown on a plate.
3. We recommend that easily detectable phenotypes are used for mutant selection, such as hypocotyl length, root length, leaf color and lethality.
4. PCR amplification can be carried out using any PCR enzymes. We routinely use EX-Taq (Takara Bio Inc., Shiga Japan), KOD-FX (Toyobo Co., Ltd, Tokyo Japan), and LA-Taq (Takara Bio Inc.) to amplify the fl-cDNAs.
5. Recover PCR products larger than 200 bp. A PCR product of about 200 bp is amplified from an empty vector when the F1 and R1 primer set is used.
6. We have reported that on average 2.6 *Arabidopsis* fl-cDNAs are inserted in an *Arabidopsis* FOX line [6]. As a result, multiple PCR fragments are detected when fl-cDNAs are amplified from the *Arabidopsis* FOX lines. All amplified fragments should be used to generate re-transformed plants to confirm which fl-cDNA causes the phenotype.
7. To characterize the introduced fl-cDNAs, direct sequencing of PCR fragments can be carried out using different internal primers from those used for PCR amplification. It is better to perform sequencing from both ends. We found different fl-cDNAs tandemly inserted into the expression vector in several lines.
8. Carry out ethanol precipitate using the ligation mixture to concentrate the DNA when it is difficult to obtain antibiotic-resistant colonies. Resuspend the pellet with 1 μL of sterile DDW and add 20 μL of *E. coli* competent cells.
9. When we sow 0.25 g of T_1 seeds, we generally obtain about 30–40 transformants.

Acknowledgements

This work was supported by a Special Coordination Fund for Promoting Science and Technology awarded to M.M., K.O. and H.H. This study was also supported by a Grant-in-Aid for Young Scientists (B) from the Ministry of Education, Culture, Sports and Technology of Japan (21780315, 24580098) to M.H. We thank Dr. Hirofumi Kuroda, Ms. Yoko Horii and Dr. Yuko Tsumoto (RIKEN Plant Science Center) for technical support. We appreciate the helpful discussions with Dr. Takanari Ichikawa (Okinawa Institute of Science and Technology Promotion Corporation), Dr. Youichi Kondou (Kanto Gakuin University) and Dr. Arata Yoneda (Nara Institute of Science and Technology).

References

1. Kuromori T, Takahashi S, Kondou Y, Shinozaki K, Matsui M (2009) Phenome analysis in plant species using loss-of-function and gain-of-function mutants. Plant Cell Physiol 50(7): 1215–1231
2. Kondou Y, Higuchi M, Matsui M (2010) High-throughput characterization of plant gene functions by using gain-of-function technology. Annu Rev Plant Biol 61:373–393
3. Nakazawa M, Ichikawa T, Ishikawa A, Kobayashi H, Tsuhara Y, Kawashima M, Suzuki K, Muto S, Matsui M (2003) Activation tagging, a novel tool to dissect the functions of a gene family. Plant J 34(5):741–750
4. An S, Park S, Jeong DH, Lee DY, Kang HG, Yu JH, Hur J, Kim SR, Kim YH, Lee M, Han S, Kim SJ, Yang J, Kim E, Wi SJ, Chung HS, Hong JP, Choe V, Lee HK, Choi JH, Nam J, Park PB, Park KY, Kim WT, Choe S, Lee CB, An G (2003) Generation and analysis of end sequence database for T-DNA tagging lines in rice. Plant Physiol 133(4):2040–2047
5. Tani H, Tani H, Chen X, Nurmberg P, Grant JJ, SantaMaria M, Chini A, Gilroy E, Birch PR, Loake GJ (2004) Activation tagging in plants: a tool for gene discovery. Funct Integr Genomics 4(4):258–266
6. Ichikawa T, Nakazawa M, Kawashima M, Iizumi H, Kuroda H, Kondou Y, Tsuhara Y, Suzuki K, Ishikawa A, Seki M, Fujita M, Motohashi R, Nagata N, Takagi T, Shinozaki K, Matsui M (2006) The FOX hunting system: an alternative gain-of-function gene hunting technique. Plant J 48(6):974–985
7. Seki M, Narusaka M, Kamiya A, Ishida J, Satou M, Sakurai T, Nakajima M, Enju A, Akiyama K, Oono Y, Muramatsu M, Hayashizaki Y, Kawai J, Carninci P, Itoh M, Ishii Y, Arakawa T, Shibata K, Shinagawa A, Shinozaki K (2002) Functional annotation of a full-length Arabidopsis cDNA collection. Science 296(5565):141–145
8. Okazaki K, Kabeya Y, Suzuki K, Mori T, Ichikawa T, Matsui M, Nakanishi H, Miyagishima SY (2009) The PLASTID DIVISION1 and 2 components of the chloroplast division machinery determine the rate of chloroplast division in land plant cell differentiation. Plant Cell 21(6):1769–1780
9. Sato T, Maekawa S, Yasuda S, Sonoda Y, Katoh E, Ichikawa T, Nakazawa M, Seki M, Shinozaki K, Matsui M, Goto DB, Ikeda A, Yamaguchi J (2009) CNI1/ATL31, a RING-type ubiquitin ligase that functions in the carbon/nitrogen response for growth phase transition in Arabidopsis seedlings. Plant J 60(5):852–864
10. Lehar J, Stockwell BR, Giaever G, Nislow C (2008) Combination chemical genetics. Nat Chem Biol 4(11):674–681
11. Luesch H, Wu TY, Ren P, Gray NS, Schultz PG, Supek F (2005) A genome-wide overexpression screen in yeast for small-molecule target identification. Chem Biol 12(1):55–63
12. Yoneda A, Higaki T, Kutsuna N, Kondo Y, Osada H, Hasezawa S, Matsui M (2007) Chemical genetic screening identifies a novel inhibitor of parallel alignment of cortical microtubules and cellulose microfibrils. Plant Cell Physiol 48(10):1393–1403
13. Yoneda A, Ito T, Higaki T, Kutsuna N, Saito T, Ishimizu T, Osada H, Hasezawa S, Matsui M, Demura T (2010) Cobtorin target analysis reveals that pectin functions in the deposition of cellulose microfibrils in parallel with cortical microtubules. Plant J 64(4):657–667
14. Nakazawa M, Matsui M (2003) Selection of hygromycin-resistant Arabidopsis seedlings. Biotechniques 34(1):28–30

Part V

Hormone Transport and Metabolite Profiling

Chapter 20

Quantification of Stable Isotope Label in Metabolites via Mass Spectrometry

Jan Huege, Jan Goetze, Frederik Dethloff, Bjoern Junker, and Joachim Kopka

Abstract

Isotope labelling experiments with stable or radioactive isotopes have long been an integral part of biological and medical research. Labelling experiments led to the discovery of new metabolic pathways and made it possible to calculate the fluxes responsible for a metabolic phenotype, i.e., the qualitative and quantitative composition of metabolites in a biological system. Prerequisite for efficient isotope labelling experiments is a reliable and precise method to analyze the redistribution of isotope label in a metabolic network. Here we describe the use of the CORRECTOR program, which utilizes matrix calculations to correct mass spectral data from stable isotope labelling experiments for the distorting effect of naturally occurring stable isotopes (NOIs). CORRECTOR facilitates and speeds up the routine quantification of experimentally introduced isotope label from multiple mass spectral readouts, which are generated by routine metabolite profiling when combined with stable isotope labelling experiments.

Key words Stable isotope labelling, Stable isotope tracing, Flux, CORRECTOR software, Metabolite profiling

1 Introduction

The investigation of metabolism by mass spectrometry (MS) coupled to a chromatographic separation system, such as gas or liquid chromatography (GC or LC), is an integral part of biological and medical research, including chemical biology approaches where an understanding of metabolic fate can be critical. These investigations can be enhanced by stable isotope labelling experiments which broaden our current knowledge of metabolism and enable evidence-driven mathematical modelling of metabolic pathways. Labelling experiments challenge the investigated biological system with substrates which can be traced via the mass shift of the chosen stable isotope. The metabolic fate of labelled precursors within the metabolic networks can thus be monitored by mass spectrometric detection. For this purpose and

Glenn R. Hicks and Stéphanie Robert (eds.), *Plant Chemical Genomics: Methods and Protocols*, Methods in Molecular Biology, vol. 1056, DOI 10.1007/978-1-62703-592-7_20, © Springer Science+Business Media New York 2014

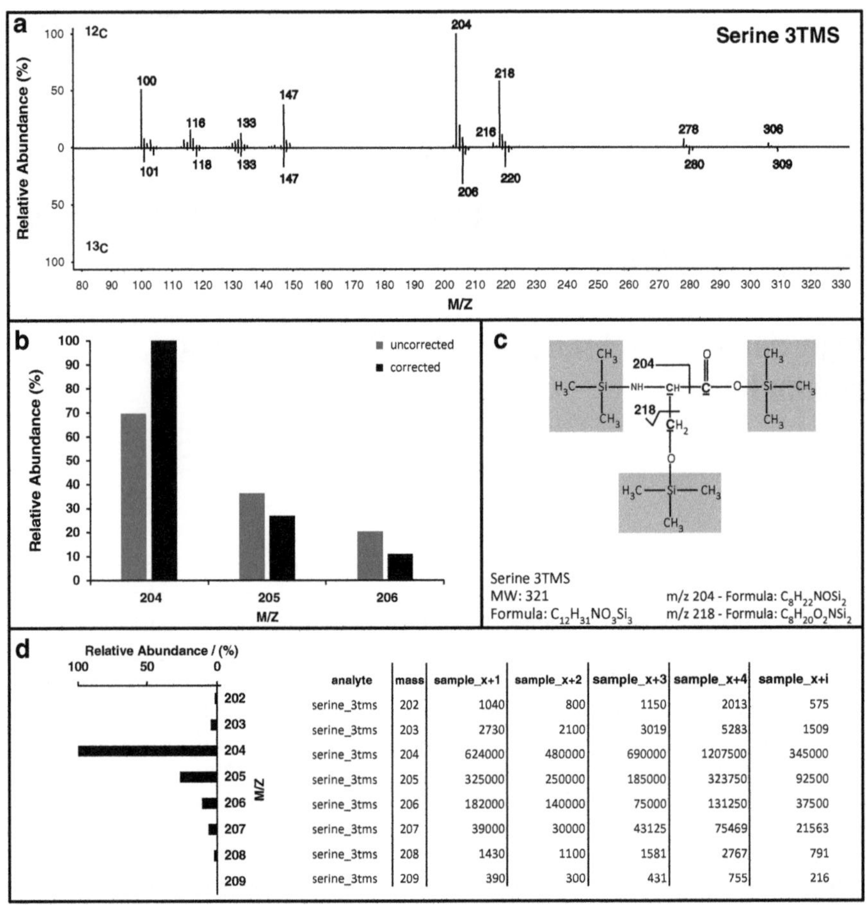

analyte	mass	sample_x+1	sample_x+2	sample_x+3	sample_x+4	sample_x+i
serine_3tms	202	1040	800	1150	2013	575
serine_3tms	203	2730	2100	3019	5283	1509
serine_3tms	204	624000	480000	690000	1207500	345000
serine_3tms	205	325000	250000	185000	323750	92500
serine_3tms	206	182000	140000	75000	131250	37500
serine_3tms	207	39000	30000	43125	75469	21563
serine_3tms	208	1430	1100	1581	2767	791
serine_3tms	209	390	300	431	755	216

Fig. 1 The analyte serine 3TMS illustrates the different aspects of applying GC-MS based metabolite profiling technology for flux or stable isotope tracing studies. Serine 3TMS is the main volatile trimethylsilylated (TMS) derivative of the metabolite serine. (**a**) The head to tail view of a non-labelled ^{12}C-electron impact ionization (EI) time of flight mass spectrum (*top*) and a completely ^{13}C-labelled spectrum (*bottom*) of serine 3TMS reveals the mass fragments and mass shifts that can be used to monitor stable isotope labelling in the chosen compounds. (**b**) The comparison of NOI (naturally occurring isotope) corrected and uncorrected mass isotopomer abundances of the fragment 204 of serine 3TMS indicates the potential bias due to NOIs and thus the requirement for correction. (**c**) The structure and sum formula of serine 3TMS as well as the formula of the serine 3 TMS fragments, *m/z* 204 and 218 [16] demonstrate the potential and limitation of obtaining positional labelling information by GC-EI-MS fragmentation. In the given structure the TMS-groups that cannot be labelled are highlighted by a *grey underlay*. (**d**) Illustrates the format of the input file. For illustration a transposed mass spectrum is aligned to the corresponding abundance values in the input file

depending on the specific technology, mass spectrometers which are directly coupled to chromatography provide metabolite separation and specific mass isotopomer distributions of the complete metabolite structure or of mass fragments which represent defined moieties or substructures of the original metabolite. The mass isotopomers of molecular and fragment ions differ by number and position of the incorporated label into the respective metabolite (Fig. 1).

The preprocessing steps required for multiplexed, parallel labelling studies analyzed by GC-MS technology are (1) alignment of chromatograms, (2) comprehensive identification of mass spectral tags (MST), i.e., the specific molecular and/or mass fragment ions which represent each metabolite, (3) targeted retrieval of mass isotopomer distributions of respective molecular ions and/or mass fragments (*see* **Notes 1–5**), (4) computational correction of the observed mass isotopomer distributions for the bias caused by the occurrence of NOIs, and (5) calculation of the experimentally induced enrichment of the fed stable isotope within each MST.

Steps 1–3 are identical to conventional metabolite profiling experiments and are described elsewhere [1–5]. The CORRECTOR program described here performs and automates steps 4 and 5.

Since a mass spectrometer cannot distinguish between isotopes derived from experimentally labelled substrates and NOIs, the failure to correct for these will lead to significant errors of isotope quantification (*see* Fig. 1b). A refined theory and mathematical tools had been developed previously for this purpose with a primary focus on ^{13}C-labelling [6–10]. The CORRECTOR program is a simple command line application which follows these principles and once configured corrects mass spectral data for the bias caused by NOIs. The CORRECTOR software allows the *in-parallel* analysis of experimentally introduced stable isotopes with a throughput matching typical GC-MS based metabolite profiling experiments. CORRECTOR can be configured beyond ^{13}C-labelling studies to accommodate any other biologically relevant elemental isotope, for example ^{2}H or ^{15}N, and provides different types of ready to use output files which report the corrected mass isotopomer distributions in absolute and scaled values, as well as the respective isotope enrichments. In the following sections, we will describe the efficient use of the Corrector software and the chosen input- and output-data formats which enable combined metabolic flux and pool size profiling (*see* Fig. 2) using GC-MS technology [11].

2 Materials

2.1 Software

Download the CORRECTOR program from http://www-en.mpimp-golm.mpg.de/03-research/researchGroups/01-dept1/Root_Metabolism/smp/CORRECTOR/index.html. In addition to the source code a precompiled version of CORRECTOR for UNIX environments is available from this URL. Data input conventions and returned data output files are tabulator delimited text files. Exemplary data files in respective file formats are part of the download. Standard text editors or spread sheet applications are recommended to view, edit, and interchange data between external programs and CORRECTOR.

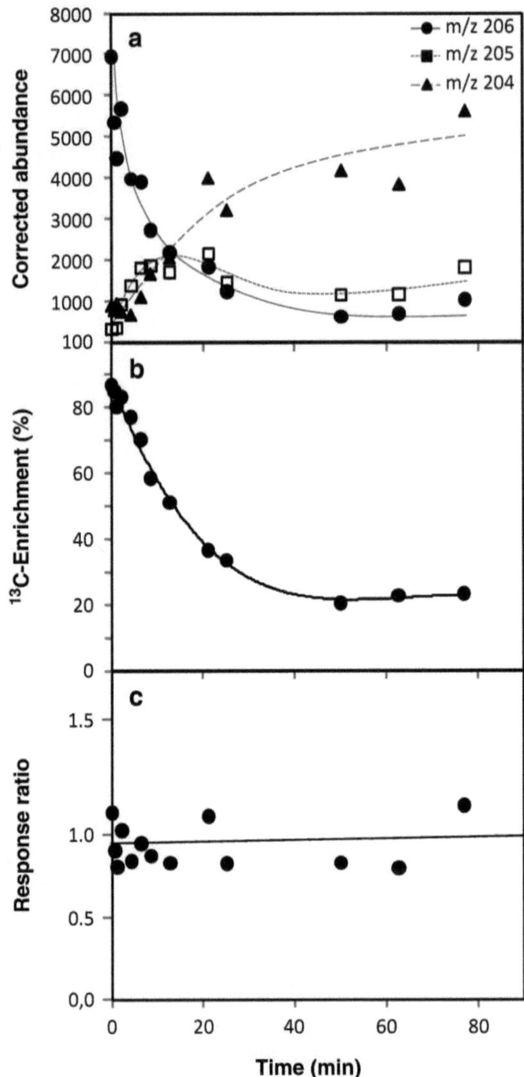

Fig. 2 Exemplary ^{13}C-chase kinetics processed by CORRECTOR of the serine 3TMS fragment $m/z = 204$ (*see* Fig. 1c). Serine was in transition from the fully labelled, $m/z = 206$, towards the non-labelled state, $m/z = 204$, in an experiment similar to [13]. (**a**) CORRECTOR output 1 contains all NOI corrected abundances of the analyzed mass isotopomer distribution(s). Output 2 which contains the corrected and normalized data is not shown. (**b**) Output 3 contains the isotope enrichment calculated from the corrected abundances, here displayed as % ^{13}C-enrichment. (**c**) In some instances information on constant or changed metabolite pool size is required in addition to the isotope labelling information. Pool size information can be obtained from the flux/tracing experiment by calculating the sum of abundances of all observed mass isotopomers. This sum of abundances can be used to calculate the relative change of metabolites or, when calibrated, the absolute pool sizes [11]. Note that in this example the serine pool stayed constant; however, pool sizes of metabolites may change in non-steady state conditions and dynamic labelling experiments

2.2 Required Chemical Information

The sum formula of the analytes or their respective fragments represented by the mass distribution vectors under investigation must be available (*see* Fig. 1c). Furthermore, the isotopic composition of the elements occurring in the analytes is required [12]. The aligned mass spectral data of an isotope labelling experiment, for example generated by the TagFinder software [5], are the standard input (*see* Fig. 1d). When editing these files for use by the CORRECTOR software, namely, either the cfg files included in the download package or the aligned experimental data input, take care that no delimiters other than tabulators are used. Delimiters, such as spaces, semicolons, or others are not supported.

3 Methods

3.1 Configuration of CORRECTOR

1. Import the file isoconfig.cfg from the CORRECTOR download package into a text editor or spread sheet application. If not already present within the provided download file, enter the following information for each additional relevant element with more than one isotope into the file: element symbol, mass number of the lightest isotope of the element, number (A) of all isotopes of the element (m0), the natural abundance (NA) of each isotope in ascending order of the weight of the respective isotope. This information has to be written into one row for each element and each information in this row has to be separated by a tabulator, e.g.: element/m0/A/NA0/NA1/... /NAi. After entering the information save the file in the format of a tabulator delimited text file (see Notes 6, 12 and 13).
2. Import the file corrconfig.cfg from the CORRECTOR download package into a text editor or spread sheet application. The first row of the file represents a comment line and describes the molecular properties of the listed items where each item is characterized by a single row. Starting in the second row the identity and properties required for mass isotopomer correction of all expected analytes and respective fragments can be defined. The properties define the elemental composition of each analyte/fragment and are derived from the respective sum formula. An entry of a specific analyte is defined by the analyte identifier, lowest mass of the analyte/fragment, highest mass of the analyte/fragment, number of additional masses beyond the highest mass that should be included in the correction for NOIs, total sum of all atoms in the analyte/ fragment, atom type, number of atoms of the respective atom type, labelled atom type, and number of labelling positions. If not already present within the provided download file add additional analyte and fragment information by creating new rows. The required row format is as follows: *identifier*

/masslow/masshigh/masssurplus/sum/type$_0$/N_0 ... *type*$_i$/N_i/*type*$_{label}$/N_{label}. After entering the information save the file as tabulator delimited text file (*see* **Notes 7–13**).

3.2 Input Data

The standard input format of CORRECTOR can be created or edited with a text editor or spread sheet application (*see* **Notes 13–17**). An example is given in Fig. 1d. The input convention requires samples to be defined as columns which are described by rows that contain the information of the measured intensities of the aligned nominal masses observed in all samples. The rows describing the mass distribution linked to of each expected analyte/fragment have to be in ascending order according to the nominal masses of the observed mass fragments (Fig. 1d).

3.3 Installation and Software Operation

1. Create a folder "CORRECTOR" (recommended)
2. Unpack the file CORRECTOR_package.zip into this folder
3. Copy the input file containing mass spectral data into the CORRECTOR directory
4. Start a command line tool (e.g., "terminal" using LINUX)
5. Change directory to the CORRECTOR directory
6. Start correction procedure and isotope quantification by running CORRECTOR via:

   ./CORRECTOR_version name_input_file
7. For calculation procedures, output file content and formats (*see* **Notes 18–24** and Fig. 2).

4 Notes

1. For isotope labelling experiments, we recommend to include non-labelled reference samples. These samples should be generated using the same conditions as are applied to the labelled samples and should be processed in parallel to serve as negative controls. The purpose of such negative controls is to ensure that the mass spectral vector of interest is properly corrected for NOIs. The isotope enrichment of the non-labelled samples after correction for NOIs may be higher than expected 0 ± measurement errors [13]. Insufficient correction may indicate that (1) either the natural isotope abundances do not apply to the respective biological samples, (2) the mass detection may suffer from a systematic tuning bias or compound overload artifact, or (3) the chosen fragment for mass isotope distribution analysis may be contaminated by mass signals of unexpected co-eluting compounds or by previously obscured minor mass fragments. The contribution of all of the above

factors should be experimentally minimized before mass isotopomer distribution analysis.

2. Since gas chromatography usually requires the derivatization of metabolites to convert nonvolatile metabolites into volatiles analytes which can pass through the GC-analysis, the elemental composition of the respective analytes derived from the original metabolite structure may differ and depends on the chosen derivatization agent. Reagents should therefore be tested for unexpected deviations of the expected natural isotope composition.

3. All experimental caveats for routine GC-MS metabolite profiling experiments apply. Consider that complex biological matrices and changes in compound abundances can give rise to unexpected co-elution effects. Part of these effects can be corrected by fractionation during sample preparation or by application of enhanced chromatographic resolution technologies such as two dimensional GC × GC-time of flight MS. Also consider that automated mass spectral deconvolution algorithms may distort mass isotopomer distributions in cases where mass isotopomers are partially separated by slightly different chromatographic retention.

4. Accurate mass isotopomer analysis depends on operating MS-detectors within the linear range of the abundance measurement. Low abundance can lead to significant background interference by noise effects, high abundance can cause saturation effects of the most abundant isotopomer(s) within an isotopomer distribution. Both cases will cause inaccurate isotopomer distributions and should be avoided by sample concentrating or dilution. Multiple measurements, for example using different GC-MS split injection settings, of each sample can be recommended in the later case.

5. The detection of multiple fragments of one analyte can provide positional information. Availability of positional information is dependent of the fragmentation path of each molecule and therefore only possible in a limited number of cases. If positional information is required nuclear magnetic resonance measurements are recommended if the target compound is present in/or can be concentrated to sufficiently high amounts.

6. The default *isoconfig.cfg* file from the CORRECTOR download package contains the isotope abundance information reported by [12] for the elements H, C, N, O, Si, P and S. Additions or modifications may be required.

7. An exemplary *corrconfig.cfg* file is provided with the CORRECTOR download package, it represents an adaptation and expansion of the supplementary file 1 published by [13]. This file contains information relevant for polar soluble

metabolites analyzed by routine GC-MS metabolite profiling. Addition of analytes and newly recognized mass fragmentation products may be required.

8. In detail an entry in the *corrconfig.cfg* file of a specific analyte is defined by the analyte identifier that is used to label the respective mass isotopomer distribution within the input file, the lowest and highest possible nominal mass introduced by the chosen mode of stable isotope labelling, namely, "Identifier/masslow/masshigh" given in columns 1–3 (*see* Fig. 1a). For example, the identifier given in corrconfig.cfg "M000029_A132003-101_METB_1298_Amino_acid_Proline_(2TMS)_MP" represents the main silylation product of the metabolite proline. Two rows are defined for this identifier, i.e., the mass fragment 142–146 (masslow − masshigh) and 216–220. The mass ranges can be determined by full isotope labelling experiments (*see* ref. 13) and indicate that both fragments of the analyte proline contain four carbon atoms which can be labelled by ^{13}C-feeding.
9. Because chemical derivatization typically for the GC-MS profiling of polar metabolites, e.g., MSTFA, may introduce trimethylsilyl-moieties, can generate higher mass isotopomers of C (and of H or Si) than the expected full labelling of the metabolite, the "masssurplus" option was introduced. This criterion can be used to enter additional mass isotopomers into the correction algorithm. Such an option can be useful for samples with high isotope enrichments. The default setting of "masssurplus" is zero. The user may vary and optimize different "masssurplus" settings without changing the experimentally fixed "masshigh" value.
10. The properties in *corrconfig.cfg* define the elemental composition of each mass fragment and are derived from the respective sum formula. For example, the *m/z* 142 fragment of the proline (2TMS) analyte has "sum" = 25 atoms and the elemental composition "$type_0/N_0$ … $type_i/N_i$": C 3 H 16 O 0 N 1 S 0 Si 1 and $type_{label}/N_{label}$: C 4, where each element $type_i$ has N_i atoms.
11. Note that in the examples given above carbon is labelled and accordingly the carbon atoms are split into those which are introduced by chemical derivatization, C 3, and those which are derived from the metabolite and thus can be experimentally labelled, C 4. This maximum labelling information needs to be defined at the end of each row. The sum of N_i plus N_{label} atoms should add up to "sum", whereas N_{label} should be equal to the difference, masshigh − masslow. The CORRECTOR program will create an error message if discrepancies are detected for the number of atoms.
12. Identifiers or element type definitions must not contain spaces.

13. The type variables in *corrconfig.cfg* are compared to the type entries in *isoconfig.cfg* and must match. Also the identifiers used to annotate the input data matrix, as was mentioned above, must match exactly to the identifiers defined in the *corrconfig.cfg* file.
14. Within the data matrix of the input file each identifier can only be assigned once to the respective entries within the aligned data matrix. CORRECTOR parses the annotated data matrix for MST identifiers, processes all respective sample entries linked to this identifier and then moves on to the subsequent identifier within the annotated data matrix.
15. If masses within the specified mass range of a vector are not provided by the input file, these missing entries are considered to be zero.
16. The values required for the correction matrix created by CORRECTOR such as the number of total atoms of a given atom type (carbon, oxygen, hydrogen, etc.) are taken from *corrconfig.cfg*, the properties of each atom type in terms of isotopic composition are retrieved from *isoconfig.cfg*.
17. Error messages are produced in cases of nonmatching identifiers present in the *corrconfig.cfg*. Also *duplicated mass values* or *unsorted mass distributions* are reported. The error log will be automatically overwritten by each new application run and should be manually archived during trouble shooting processes.
18. As a result of utilizing the above established matrix calculation concept [6–10], the corrected intensity values of low abundance vector elements or vector elements with an abundance of 0 can drop below zero after correction. Since values below zero are physically not plausible, the algorithm in CORRECTOR sets encountered negative values to zero before using these in a subsequent correction step.
19. CORRECTOR can be executed with the option: -igz [integer], default is 0. Selecting this option will result in ignoring up to [integer] empty or negative higher mass isotopomers for the enrichment calculation (atom percentages) after the first encountered mass isotopomer. This procedure can be useful for experiments with non-homogenous isotopomer distributions, for example, ^{13}C-stable isotope standard addition experiments where an abundant mono-isotopic ^{12}C-mass isotopomer can be accompanied by a series of non-detectable (empty) intermediate mass isotopomers followed by a fully labelled ^{13}C-mass isotopomer. The CORRECTOR default setting, 0, assumes a completed mass distribution vector, if an empty or negative value is encountered. The calculation would stop and thus disregard the present high mass isotopomers of such

non-homogenous labelling experiments. Settings >0 are recommended if non-homogenous labelling is expected.

20. File names can be defined: specifically the definition of the files, output 1 containing the corrected ms data, output 2 containing the normalized corrected ms data, and output 3 containing the isotope enrichments/[atom%] . Default names are: time stamp + out 1–3.
21. The CORRECTOR download provides files in order to test-run new CORRECTOR installations by the command: *./corrector.exe* [Optional: *-igz* [*integer*]] *inputfile* [Optional: *outputfile1 outputfile2 outputfile3*]. Test runs are recommended so as to get familiar with the operating procedures and file formats.
22. After a test run with the provided test files (*input_test.txt*) has been executed, three output files should have been returned by the CORRECTOR program. The *input_test.txt* file contains mass spectral data from a ^{13}C-delabelling experiment [13]. Plants had been grown in a $^{13}CO_2$ atmosphere and were subsequently exposed to ambient atmosphere ($^{12}CO_2$). A kinetic series of samples was taken. The sample chromatogram identifier in the first row was substituted by the time point at which the sample was taken. These data enable a direct plot of the calculated ^{13}C-enrichment over time (Fig. 2).
23. To exemplify the function of the error log, the *input_exemplar_with_errors.txt* file which contains two deliberate mistakes is provided. Firstly, the identifier "Malic_acid_(3TMS)" (rows 789–895) does not match to an identifier in *corrconfig.cfg* file; secondly, in row 898 mass $m/z = 245$ of "M000067_A137001-101_METB_1346.23_Acid_(Dicarboxylic)_Fumaric_acid_(2TMS)" occurs twice. In both cases the correction procedure is not applied due to a flawed input file. By substitution of the false "Malic_acid_(3TMS)" identifier by the correct identifier "M000065_A149001-101_METB_1477.3_Acid_(Dicarboxylic,_Hydroxy-)_Malic_acid_(3TMS)" present in the *corrconfig.cfg* and removal of the duplicated mass information within the fumaric acid mass distribution data, the correction procedure should be performed without further error messages.
24. For some sample types or analytes additional correction procedures might be required, e.g., analytes such as fatty acid methyl ester derived from lipids may need additional correction for proton loss, or the effect of unlabelled original biomass to the mass distribution vector may need consideration [14, 15].

Acknowledgements

This work was supported by the Max Planck society and a grant from the DFG (Deutsche Forschungsgemeinschaft) to JK. The CORRECTOR software tool was supported in part by the European META-PHOR project, FOOD-CT-2006-036220, and the GoFORSYS project (http://www.goforsys.de/) funded by the Bundesministerium für Bildung und Forschung (BMBF). JH and BH were supported by BMBF grants 0315295 and 0315426A.

References

1. Roessner U et al (2000) Simultaneous analysis of metabolites in potato tuber by gas chromatography–mass spectrometry. Plant J 23: 131–142
2. Fiehn O et al (2000) Metabolite profiling for plant functional genomics. Nat Biotechnol 18:1157–1161
3. Lisec J et al (2006) Gas chromatography mass spectrometry-based metabolite profiling in plants. Nat Protoc 1:387–396
4. Erban A et al (2007) Non-supervised construction and application of mass spectral and retention time index libraries from time-of-flight GC-MS metabolite profiles. In: Weckwerth W (ed) Metabolomics: methods and protocols. Humana Press, New York, pp 19–38
5. Luedemann A et al (2008) TagFinder for the quantitative analysis of gas chromatography–mass spectrometry (GC-MS) based metabolite profiling experiments. Bioinformatics 24: 732–737
6. Lee WNP, Byerley LO, Gergner EA (1991) Mass isotopomer analysis: theoretical and practical considerations. J Mass Spectrom 30: 451–458
7. Fernandez CA et al (1996) Correction of ^{13}C mass isotopomer distributions for natural stable isotope abundance. J Mass Spectrom 31:255–262
8. Wittmann C, Heinzle E (1999) Mass spectrometry for metabolic flux analysis. Biotechnol Bioeng 62:739–750
9. van Winden WA et al (2002) Correcting mass isotopomer distributions for naturally occurring isotopes. Biotechnol Bioeng 80:477–479
10. Wahl AS, Dauner M, Wiechert W (2003) New tools for mass isotopomer data evaluation in ^{13}C flux analysis: mass isotope correction, data consistency checking, and precursor relationships. Biotechnol Bioeng 85:259–268
11. Huege J et al (2011) Modulation of the major paths of carbon in photorespiratory mutants of synechocystis. PLoS One 6(1):e16278. doi:10.1371/journal.pone.0016278
12. Rosman KJR, Taylor PDP (1998) Isotopic compositions of the elements 1997. Pure Appl Chem 70:217–235
13. Huege J et al (2007) GC-EI-TOF-MS analysis of in vivo carbon-partitioning into soluble metabolite pools of higher plants by monitoring isotope dilution after ($^{13}CO2$)-labelling. Phytochemistry 68:2258–2272
14. Allen DK, Shachar-Hill Y, Ohlrogge JB (2007) Compartment-specific labelling information in metabolic flux analysis of plants. Phytochemistry 68:2197–2210
15. Allen DK, Ratcliffe RG (2009) Quantification of isotope label. In: Schwender J (ed) Plant metabolic networks. Springer, New York, pp 105–149
16. Leimer KR, Rice RH, Gehrke CW (1977) Complete mass spectra of the pertrimethylsilylated amino acids. J Chromatogr 141:355–375

Chapter 21

^{1}H NMR-Based Metabolomics Methods for Chemical Genomics Experiments

Daniel J. Orr, Gregory A. Barding Jr., Christiana E. Tolley, Glenn R. Hicks, Natasha V. Raikhel, and Cynthia K. Larive

Abstract

Metabolomics and chemical genomics studies can each provide unique insights into plant biology. Although a variety of analytical techniques can be used for the interrogation of plant systems, nuclear magnetic resonance (NMR) provides unbiased characterization of abundant metabolites. An example methodology is provided for probing the metabolism of *Arabidopsis thaliana* in a chemical genomics experiment including methods for tissue treatment, tissue collection, metabolite extraction, and methods to minimize variance in biological and technical sample replicates. Additionally, considerations and methods for data analysis, including multivariate statistics, univariate statistics, and data interpretation are included. The process is illustrated by examining the metabolic effects of chemical treatment of Arabidopsis with Sortin 1, also known as vacuolar protein *sort*ing *in*hibitor 1. Sortin 1 was applied to Arabidopsis seedlings to examine metabolic effects in a chemical genomics experiment and to demonstrate the utility of metabolomics in conjunction with other "omics" techniques.

Key words Metabolomics, NMR spectroscopy, Sample preparation, Chemical genomics, Metabolic profiling, Tissue extraction, Normalization, Sortin1

1 Introduction

1.1 Chemical Genomics and Metabolomics

The objectives of metabolomics studies are the global measurement of small molecule metabolites in a biological system and detection of changes in the levels of these metabolites in response to an abiotic or biotic perturbation [1, 2]. Metabolomics has proven to be an important tool for understanding the responses of plants and other organisms to genetic and environmental perturbations, especially when considered in conjunction with other "omics" data. Thus far, metabolomics measurements have not been extensively applied to chemical genomics experiments; however, these techniques are increasingly used to understand complex problems in plant biology. From an experimental perspective the field of toxicology, which focuses on the adverse effects of

Glenn R. Hicks and Stéphanie Robert (eds.), *Plant Chemical Genomics: Methods and Protocols*, Methods in Molecular Biology, vol. 1056, DOI 10.1007/978-1-62703-592-7_21,

chemicals on biological systems, is similar in many respects to chemical genomics. The strength of metabolomics for toxicology studies, as well as its greatest potential in chemical genomics, is the ability to reveal biochemical details that lead to mechanistic insights following chemical treatment [3, 4]. Plant biology would greatly benefit if the potential of chemical genomics to provide tunable and reversible responses is realized [5]. One difficulty with this type of experiment is that the plant response to a minimal chemical dose may not produce a visible phenotype. The chemical treatment may, however, induce a metabolic phenotype; thus metabolomics measurements have a role as biological readouts in chemical genomic studies.

1.2 Metabolic Profiling by Nuclear Magnetic Resonance

Selection of the analytical instrument best suited to an individual metabolomics experiment is an important consideration. Nuclear magnetic resonance (NMR) and mass spectrometry (MS) are the most frequently used techniques and both have distinct advantages in metabolomics studies. The greater sensitivity of MS makes it well suited for this application, especially for measurements of secondary metabolites or discovery of novel biomarkers; however, a detailed discussion of MS-based metabolomics is outside the scope of this article and the interested reader is referred to several recent reviews on this topic [2, 6–9]. The advantages of NMR for metabolomics studies are that it provides an inherently quantitative analysis, is relatively unbiased to differences in sample composition or matrix, and has better inter-laboratory or inter-instrument reproducibility. These attributes make NMR essential to chemical screening applications in toxicology and also well suited for chemical genomics experiments [10]. Molecules can be quantified by NMR using any resolved resonance over a dynamic range of $\sim 10^5$. Its tolerance for complex matrices allows NMR experiments to be conducted without involved sample preparation or coupling to chromatography as is typically required for MS analysis.

As shown in Fig. 1, ^{1}H-NMR allows observation of many metabolite classes including carbohydrates, organic and amino acids. The ability of NMR to quantitatively detect spectral features that are directly related to metabolite structure is a key advantage in metabolite identification. Chemical shift and coupling information from one-dimensional spectra provide structural data suitable for assignment of metabolites for which standards are available. This information can be augmented by two-dimensional NMR experiments making the structure elucidation of uncommon and novel metabolites feasible and providing confidence in the assignment of metabolite resonances [11].

1.3 Chemical Genomics Using Sortin1

Genes involved in endomembrane protein trafficking are essential to development in Arabidopsis [12–14]. As a result, gene inactivation mutants could provide only a basic understanding of protein

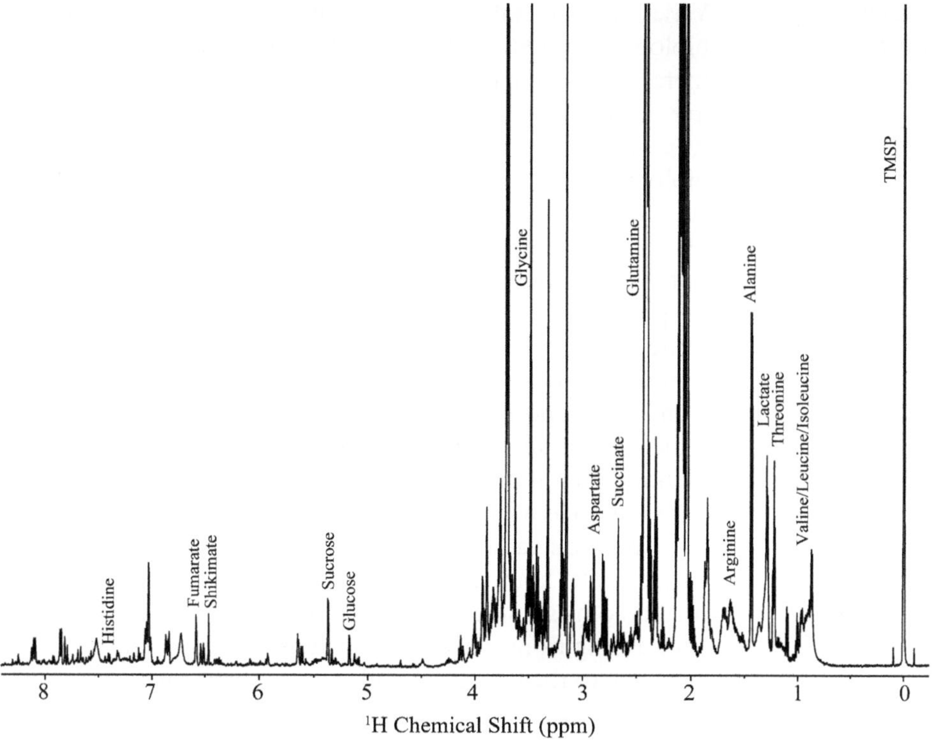

Fig. 1 Representative ^{1}H-NMR spectrum of a tissue extract from 7-day-old *Arabidopsis thaliana* seedlings of the Columbia ecotype. This spectrum was measured for an untreated control sample not treated with Sortin1. *Labels* indicate selected metabolite resonances

trafficking and a chemical genomic approach was sought. Fourteen compounds from the ChemBridge DIVERSetE compound library were isolated in *Saccharomyces cerevisiae* by their ability to cause vacuolar secretion of carboxypeptidase Y [5]. Three of these compounds are biologically active in Arabidopsis; the most potent compound is referred to as Sortin1 (vacuolar protein *sort*ing *in*hibitor 1) [5]. Sortin1 has a limited effect on flavonoid biosynthesis and inhibits the transport of flavonoid derivatives in Sortin1-hypersensitive mutants [15].

Sortin1 was used to perturb vacuolar protein sorting of Arabidopsis seedlings and the effect on endogenous metabolites was observed using ^{1}H NMR. This experiment provides a specific example of an NMR-based metabolomics protocol for chemical treatment of Arabidopsis seedlings, sample collection, tissue extraction, NMR data acquisition, and data analysis methods. Selection of an appropriate sample collection method, and extraction buffer system and protocol typically requires sample specific optimization. However, our methods for observing the effects of Sortin1 can serve as an example for chemical genomics experiments in Arabidopsis and, with appropriate modification, other plants.

1.4 Tissue Preparation and Extraction

Variation in metabolomics samples can originate from both the biological system and the analytical methods used to acquire metabolomics data. Biological variance can be minimized by ensuring consistent growth conditions, nutrient availability, temperature, humidity, and light exposure. One way to reduce the biological variance would be to grow multiple biological replicates consisting of several plants each. For example, adult Arabidopsis plants can be grown in a 9 cm × 13 cm pot with eight plants per pot. To minimize biological variance, tissue from an entire pot can be pooled and each pot treated as an individual sample. When using Arabidopsis seedlings, 50 seedlings can be grown on each agar plate as described in Subheading 2.2, **item 4**, and all seedlings from a plate pooled as a single biological replicate. Multiple biological replicates should be grown at the same time and under identical conditions, providing a means to measure biological variance as well as a larger tissue pool for analysis.

Analytical variance can arise from differences in tissue harvesting, sample homogenization, extraction, and analysis. In most metabolomics studies the biological variance far outweighs the variance arising from steps in the analysis protocol. The analytical variance due to the extraction and analysis steps can be evaluated using multiple samples taken from a single homogeneous tissue pool. Tissue should be flash-frozen in liquid nitrogen as quickly as possible during harvest. Complete and uniform tissue homogenization can be performed using automated homogenizers [16] or by mortar and pestle [17]; however, care should be taken to ensure similar particle sizes for all samples being compared. Following homogenization, samples should be freeze-dried. Because the wet weight of harvested tissue can vary, dry tissue mass is more reliable for calculating metabolite concentrations. Variance induced by weighing should be minimized. For accurate mass measurements use a balance capable of measuring or less 0.1 mg and calibrate the balance prior to use. Tissue should be weighed into the Eppendorf tube used for extraction (Subheading 3.1 **step 4**); the mass of the tube should not be tared but rather weighed empty and then subtracted from the total mass of the tissue + tube to yield the sample mass.

Extraction solvent choice is another important aspect of plant metabolomics experiments. Although a variety of extraction solvents have been examined for different organisms, they may not all work equally well on the targeted plant samples [17]. A suitable extraction protocol should provide efficient and reproducible extraction of the metabolites of interest, minimize co-extraction of interfering matrix components such as lipids and proteins, and yield high-quality NMR spectra. Solvents that produce high extraction efficiency are not always ideal, for example, Sekiyama and coworkers found that hexafluoroacetone/4-(2-hydroxyethyl)-1-piperazineethanesulfonic acid-d_{18} contributed to line broadening, causing loss of signals otherwise detectable in common deuterated solvents [18].

1.5 Data Normalization

Despite the care a researcher might take to minimize analytical variance, inconsistencies in the sample pool should still be expected. One source of variance in NMR-based metabolomics measurements is pH-dependent changes in chemical shift, which are commonly encountered in biological samples even when a buffered extraction solvent is used [19]. Errors introduced during weighing or volumetric transfer, and variation in extraction efficiency and data acquisition parameters can all affect measured metabolite concentrations. Sum normalization is a simple and effective means to normalize for differences in dilution, pipetting errors, and measurement parameters like instrument gain. A common practice in NMR-based metabolomics is binning, a process that divides the spectrum into small regions (or bins) which are then integrated. Bins are selected that are sufficiently small that they contain a limited number of metabolites but are large enough to account for small changes in chemical shift due to variation in pH. Typical bin widths are 0.02–0.04 ppm [20]. Constant sum normalization compares the value measured for an individual peak or bin relative to sum of all integrals from that spectrum, excluding the resonances of the solvent, buffer, or any contaminants [21]. Sum normalization does not adjust the magnitude of variance in each bin so when sum normalized data is used for multivariate analysis techniques based on covariance, such as principal component analysis (PCA), bias toward the most abundant components in the sample is expected. The development of new methods of normalization and spectral alignment is a dynamic area of research and sum normalization may soon be replaced by more effective methods [22–24].

1.6 Statistical Analysis

PCA is a multivariate statistical method that groups sets of data based on their similarities (scores, Fig. 2a), and identifies points within the data set that are responsible for the most variance (loadings, Fig. 2b). The input values for the PCA analysis shown in Fig. 2 were the sum normalized integrals of spectral bins taken from ^{1}H NMR data measured for extracts of Sortin1 treated and control Arabidopsis seedlings. The scores plot in Fig. 2a shows a dose-dependent grouping of samples from Sortin1 treated plants distinct from the controls. Labels in the loadings plot (Fig. 2b) indicate the NMR chemical shift in ppm of the bins that have the greatest degree of variance in the orthogonal principal components 1 or 2. The longer the line connecting the label to the center of the cluster in the loadings plot, the greater the contribution of that bin to separation observed in the scores plot. For example, metabolites with high loading values in this experiment include sucrose (3.514 ppm) and glutamine (2.412 and 2.106 ppm) as shown in Fig. 2b. An advantage of PCA is that it is a model-free approach providing a description of the variance in a data set without requiring a priori knowledge. PCA also has significant limitations. Over-interpretation of the separation of sample groups in PCA scores

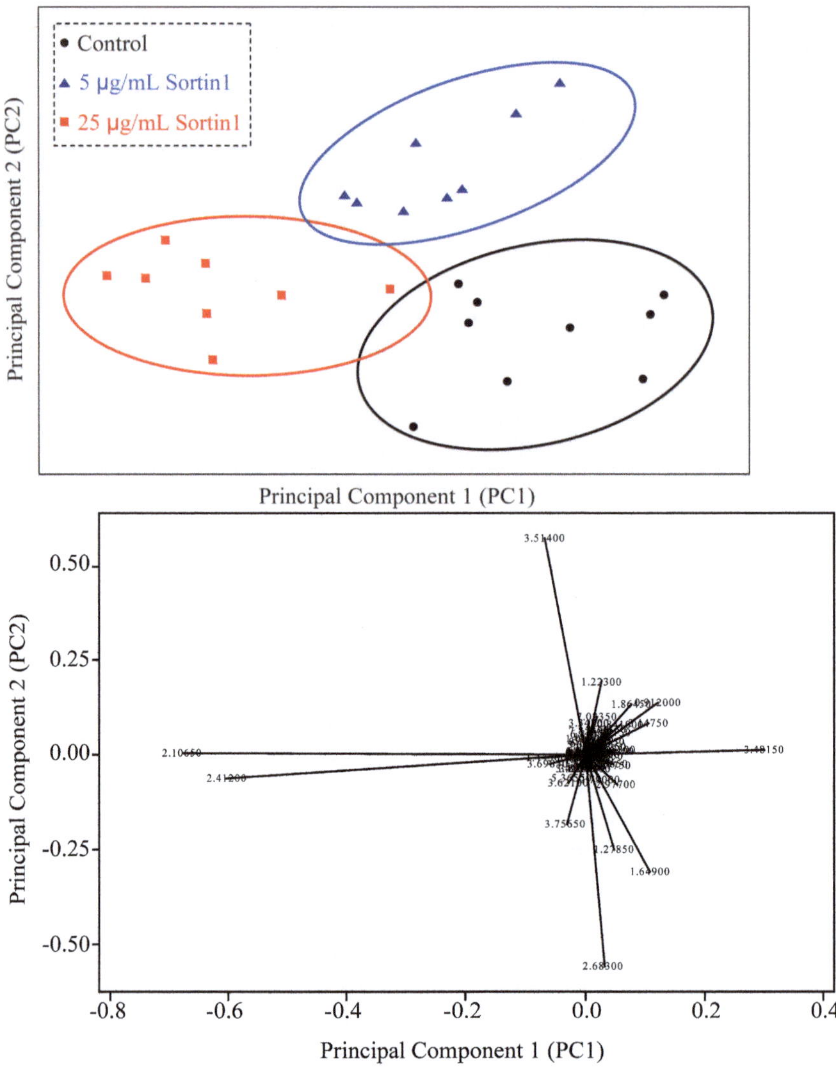

Fig. 2 (**a**) PCA scores plot of ^{1}H NMR spectra measured for extracts of 7-day-old *A. thaliana* seedlings treated with 0 μg/mL (*filled circle*), 5 μg/mL (*filled triangle*), or 25 μg/mL (*filled square*) of Sortin1. Spectra were divided 0.02 ppm bins from 0.5 to 9.0 ppm excluding the resonance of HOD. (**b**) PCA loadings plot for principal components 1 and 2

plots as an indication of statistical significance is one of several common mistakes in interpretation of PCA results [25].

PCA and other multivariate statistical analysis techniques are useful for segregation of sample groups; however, the real potential of metabolomics is through the observation of changes in the levels of specific metabolites that can be correlated with biochemical pathways and provide a mechanistic understanding of the chemical treatment. This is best achieved via univariate analysis. Although PCA loadings plots highlight those resonances responsible for the separation of sample classes, metabolite fingerprinting, or identification

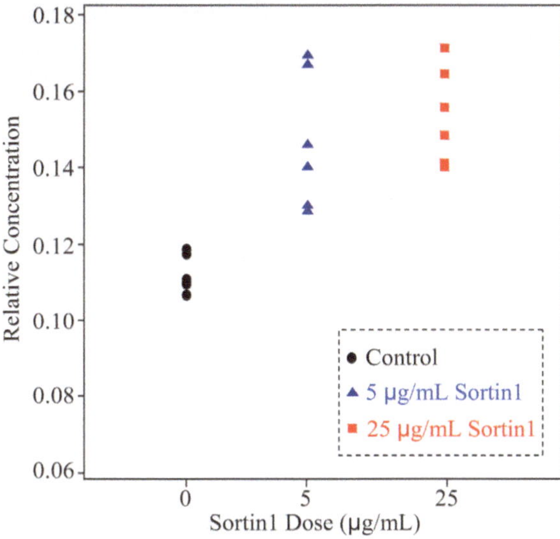

Fig. 3 Relative concentration of glutamine in 7-day-old *A. thaliana* seedlings treated with 0 μg/mL (*filled circle*), 5 μg/mL (*filled triangle*), or 25 μg/mL (*filled square*) of Sortin1

and quantification of as many metabolites as possible, may be more informative [26]. In the chemical genomics data shown in Fig. 2, the prominence of glutamine in the loadings plot (Fig. 2b) suggests that there are significant differences in its concentration in plants treated with Sortin1 compared to the controls. Because NMR integrals are directly proportional to the concentration of the nuclei giving rise to each peak, it is simple to integrate a well-resolved resonance and compare it to an internal or external standard to obtain a relative concentration. In complex samples such as plant tissue extracts, resonance overlap often limits spectral resolution, complicating the analysis. Peak fitting algorithms, in which mathematical parameters are used for peak picking, and integration improve the accuracy of integration and extend analysis to partially overlapped resonances that cannot be accurately integrated manually [27]. The glutamine resonances at 2.14 ppm in the Sortin1 chemical genomics data set were subjected to peak fitting to determine accurate integrals which were normalized relative to the 3-trimethylsilylpropionate-d_4 concentration, and plotted as shown in Fig. 3. The relative concentration of glutamine increases in the hypocotyls of seedlings treated with Sortin1 compared to the control.

2 Materials

2.1 Chemicals

Deuterium oxide (D_2O), deuterated chloroform ($CDCl_3$), sodium-3-trimethylsilylpropionate-d_4 (TMSP), acetic acid-d_4, and ammonium deuteroxide-d_5 were purchased from Cambridge Isotope

Laboratories, Inc. (Andover, Massachusetts, USA). Pure water (18 MΩ) was obtained by filtration with a Millipore filtration system (Millipore, Billerica, MA, USA). Methanol (MeOH) was HPLC grade (≥99 % purity) purchased from Fisher Scientific (Pittsburgh, PA, USA). Ammonium hydroxide was obtained from EMD Chemicals (Merck, Darmstadt, GER). Tween 40 (~90 % palmitic acid) and Ethanol (200 proof) were purchased from Sigma-Aldrich (Milwaukee, WI, USA).

2.2 Plant Samples

1. Arabidopsis seeds are sterilized by rinsing with 70 % ethanol and shaking vigorously for 30 s. The ethanol is removed with a sterile pipette and a 10 % solution of polysorbate detergent (Tween 40) is added. The seeds are soaked in the detergent for 10 min. Detergent is removed after soaking by rinsing the seeds four times with sterile water.
2. Following sterilization, seeds are suspended in a 0.15 % agar solution and stored in the dark at 4 °C for 48 h to break seed dormancy and help ensure an even germination time.
3. A common method for achieving reproducible dosage of a chemical for plant studies is to mix the compound into the growth medium. The chemical of interest should be added to the media after it has been autoclaved and allowed to cool to approximately 50 °C to avoid thermal degradation. Placing sterilized media in a hot water bath is good way to keep the media from solidifying long enough to allow careful dosing and thorough mixing prior to pouring plates. In this experiment, the final concentration of Sortin1 was 25 μg/mL for the high-dose treatment, 5 μg/mL for the low-dose treatment, and 0 μg/mL for the controls. Six replicate plates of each dose were prepared (*see* **Note 1**).
4. Seedlings are spread on plates of growth medium (prepared containing the appropriate concentration of chemical as described in Subheading 2.2 **item 3**) for germination. A total of 50 seeds are spread on each plate.
5. Seedlings are grown for 7 days under an 8 h light/16 h dark cycle.
6. After 7 days, seedlings are removed from the agar and washed twice with deionized water. For the data shown, the hypocotyls were separated from roots and then snap-frozen with liquid nitrogen. Collection of specific tissues can make changes in metabolite concentrations easier to observe because different tissues may vary in their response to the perturbation. In this experiment the endogenous metabolites in the roots and the hypocotyls were not uniformly altered by Sortin1 treatment. Pooling of tissue from hypocotyls with that of the roots would thus dilute the observed changes in hypocotyl metabolism.

3 Methods

3.1 Tissue Sample Preparation

1. Homogenize tissue into a fine powder using a mortar and pestle (*see* **Note 2**)
2. Split into two aliquots (if there is sufficient sample) so that only half is dried for NMR analysis. Storage of wet tissue at −80 °C allows for additional measurements using enzyme assays and provides a back-up in case dried tissue is contaminated or lost (*see* **Note 3**).
3. Freeze-dry tissue aliquots prior to analysis to help ensure tissue stability during storage and reduce the intensity of the water resonance in the sample extracts.
4. Record the mass of a 2.0 mL Eppendorf tube, weigh approximately 30 mg of dried tissue into the tube and record the exact weight of the tissue + tube. Technical replicates may not be necessary as biological replicates are often suitable for evaluating the overall reproducibility.
5. A 1.5 mL aliquot of 80/20 MeOH/H_2O is added to each Eppendorf tube. Samples are placed in a 70 °C water bath for enzyme denaturation [28] and stirred with a spin vein for 15 min for metabolite extraction.
6. Samples are centrifuged at 12,000 × *g* for 4 min and 1.0 mL of the supernatant transferred to a new 1.5 mL Eppendorf tube which is subsequently centrifuged under vacuum until dry (Savant SpeedVac SC110, Fisher Scientific). Samples are stored at −20 °C until analyzed.
7. The dried samples are reconstituted in 700–800 μL of D_2O buffered at a pH meter reading of 7.15 (pD of 7.55) [29] using an ammonium-d_4 acetate-d_3 buffer which contains 1 mM trimethylsilyl propionate-d_4 (TMSP) or sodium-2,2-dimethyl-2-silapentane-5-sulfonate-d_6 (DSS). For best results, samples should be reconstituted and analyzed on the same day. Although we often choose neutral pH conditions for analysis, it may be desirable to use a buffer at other pH values to resolve particular resonances with pH-dependent chemical shifts. If possible, all buffer components should be deuterated to allow the use of the highest possible receiver gain setting and to minimize spectral overlap with targeted metabolites. TMSP is a common ^{1}H chemical shift reference but it may not be suited to all samples because the chemical shift of TMSP is sensitive to pH near the pK_a of its carboxylate group, and because it adsorbs to the glass of NMR tubes over time [30]. DSS is also a useful internal standard that is especially well-suited to measurements at low pH because its sulfonate group is not subject to protonation at the pH values used in typical metabolomics studies (*see* **Note 3**).

8. An aliquot of 100 μL $CDCl_3$ is added to extract hydrophobic matrix compounds such as lipids and waxes and samples are placed on an orbital shaker for 5 min. Extracts are centrifuged at 5,000 × *g* for 4 min to break the emulsion and separate the solvents into two phases. A 600–750 μL aliquot of the D_2O layer is removed and placed into a high-quality (e.g., Wilmad 535 PP-7) 5 mm NMR tube for analysis by high-field NMR.

3.2 NMR Methods

1. For the data shown, ^{1}H-NMR spectra were acquired using a 14.1T Bruker Avance spectrometer operating at 600 MHz and equipped with a 5 mm inverse multinuclear broadband probe. The temperature of the samples was maintained at 298.2 K. Because resolution of overlapped resonances is often a limiting factor in NMR analysis of biological samples, experiments should be conducted using the highest frequency magnet available to the investigator (*see* **Note 4**).
2. Manual shimming is performed for each sample. If the spectrometer is equipped with automated shimming routines, this may provide an efficient route to the acquisition of high-quality spectra. Spectra measured with very different levels of magnetic field homogeneity can be grouped by the PCA according to how well the spectra were shimmed. Therefore, it is important to set a line width threshold that should be achieved for each sample that can be consistently attained and provides sufficient resolution. For the plant extracts analyzed using our instrument, a width at half-height less than 0.8 Hz (as measured for DSS or TMSP) can be readily achieved.
3. The intensity of the HOD resonance of the solvent is reduced using the Bruker defined pulse program WET (water suppression through enhanced T_1 effects) [31]. Alternative methods such as selective saturation or WATERGATE [32] may also provide suitable suppression of the HOD resonance. WET suppression is often our method of choice because it is more selective than presaturation; the spectra typically have easily corrected baselines, and experiments are not subject to artifacts from the spin echo sometimes observed using WATERGATE [31]. To achieve the best possible signal-to-noise ratio (S/N) and minimize baseline distortion near the solvent resonance, the 90° pulse length and power level of the shaped WET pulses should be optimized for each sample.
4. For the data reported herein, free induction decays (FIDs) are acquired over 3.27 s into 32,786 points. A 9.0 μs pulse is used to excite a spectral width of 7,184 Hz and a relaxation delay of 1.5 s is used. When optimizing these parameters for new experiments, accurate integration requires that at least four acquired data points span the resonance above the half-height. Acquisition of spectra for absolute quantitation requires repetition times exceeding $5 \times T_1$, or the measurement of the T_1

relaxation times of each resonance so that integrals can be adjusted to compensate for incomplete relaxation. Shorter repetition times can be used without compensation for incomplete T_1 relaxation provided that the integrals are used for relative quantitation, which is sufficient for most metabolomics experiments.

5. In the ^{1}H NMR spectra acquired for the Sortin1 data, 1,024 scans were coadded following 16 dummy scans. To determine the number of scans, a balance between S/N and the length of the experiment must be reached. For best results, an $S/N \geq 200$ is needed for accurate NMR quantitation, but the long experiments required to achieve this level for low abundance metabolites may not be feasible when a large number of samples must be analyzed.

3.3 NMR Data Processing and Normalization

1. ^{1}H NMR spectra are processed using MestReNova version 7.0 (MestReNova 7.0, Santiago de Compostela, Spain). All spectra are stacked in a single file and processed simultaneously to ensure consistent treatment.
2. An exponential function equivalent to 0.3 Hz line broadening is applied prior to Fourier transformation. Line broadening improves S/N at the expense of resolution and the ideal amount and type of apodization can depend on the sample and the experimental parameters used. FIDs are zero-filled (to 65,536 data points in these experiments) to increase the digital resolution of the measured spectra. Alternatively, linear prediction can be used to calculate data points that can be added to the FID.
3. The chemical shift reference (TMSP or DSS) is set to 0.0 ppm in each spectrum to ensure spectral alignment across the data set.
4. Phasing is automatically applied and then manually adjusted as required. In this study, spectra were manually baseline corrected using the multipoint baseline correction function of MestReNova. To ensure that baseline correction is reproducible for the entire data set, baseline correction is performed on a single spectrum and applied to all spectra in the data set. Each spectrum is then individually inspected as some minor baseline adjustment may be required, especially in the region near the residual solvent resonance.
5. Peak assignments are achieved by comparison to an in-house library of spectra measured for common metabolites acquired under identical buffer conditions as the plant samples, querying public metabolite databases including the Madison Metabolomics Consortium Database [33] and the Human Metabolome Database [34], and inspection of calculated spectra generated using ACDlabs software version 12 (Advanced Chemistry Development, Inc. Toronto, ON, Canada) for

possible metabolite structures. Assignments are confirmed using 2D NMR experiments including TOCSY [35] and HMBC [36] and by spiking standards into the sample matrix.

3.4 Principal Component Analysis

1. Following data acquisition NMR spectra are binned into 0.02 ppm regions from 0.50 to 9.00 ppm, integrated using MestReNova and exported to Excel (Microsoft Corp., Redmond, WA, USA) or another spreadsheet program.
2. Binned regions corresponding to the residual HOD signal (4.54–5.02 ppm), CHD_3COOD (1.88–2.06 ppm), CH_3OH (3.34–3.30 ppm), and the internal chemical shift standard are removed and the sum of the remaining spectral regions calculated
3. Constant sum normalization is applied by dividing each integral region by the sum of all regions for each spectrum.
4. The normalized data are exported to Minitab version 16 statistical software for multivariate analysis (Minitab Inc., State College, PA, USA).
5. Principal components are calculated by covariance for 10 principal components. In this study the scores plot for principal components 1 and 2 is used to assess the overall grouping of samples. For the Sortin1 data shown in Fig. 2, PC 1 represents 83 % of the explained variance and PC 2 explains 8 % of the variance. The explained variance plot is used to evaluate the number of principal components that are relevant for a given data set. Loadings plots are used to identify integral regions which make significant contribution to the explained variance.

3.5 Univariate Analysis

1. Following multivariate analysis, the NMR resonances of individual metabolites are identified for quantification.
2. Metabolite resonances are integrated using the global spectral deconvolution (GSD) feature from MestReNova. GSD is a peak analysis algorithm that automatically estimates noise and baseline, removes baseline distortions and artifacts, performs peak recognition, and reports integrals for identified peaks while accounting for estimated baseline [37].
3. The resulting peak table is exported to Excel where the integral values of the identified metabolite resonances are selected for further analysis.
4. The integrals measured for each resonance are sum normalized (using the same methods and values as in Subheading 3.5) and the average and standard deviation calculated for biological replicates of the same treatment. Significance testing can be performed to identify dose-dependent changes induced relative to the control.

4 Notes

1. Some organic compounds have poor solubility in agar solutions. In our experience addition of 0.1 % dimethyl sulfoxide (DMSO) to the growth medium can improve solubility and does not have appreciable effect on observed phenotypes.
2. Consistent and thorough homogenization is important and a sample ground to a fine powder results in better extraction efficiency. Researchers new to metabolomics analysis but with a background in other biological analysis methods will find that homogenization for metabolite analyses should produce particle sizes similar to that used for RNA extraction.
3. The addition of sodium azide to the extraction buffer as a bacteriostat is commonly reported. We find this is unnecessary when extracts are dried, stored frozen and reconstituted on the same day as analysis.
4. While NMR is much more tolerant of salt content than mass spectrometry, probe tuning is salt-dependent and the NMR probe should be tuned and matched for each sample. The presence of paramagnetic cations in some plant sample extracts may produce significant line broadening. The addition of deuterated ethylenediaminetetraacetic acid (EDTA-d_{16}) to the reconstitution buffer can improve spectral resolution.

Acknowledgements

DJO, GAB, and CEM gratefully acknowledge support by the National Science Foundation Integrative Graduate Education Research and Training Program grant DGE-0504249. CKL and DJO acknowledge additional support from POM Wonderful, LLC.

References

1. Nicholson JK, Lindon JC, Holmes E (1999) 'Metabonomics': understanding the metabolic responses of living systems to pathophysiological stimuli via multivariate statistical analysis of biological NMR spectroscopic data. Xenobiotica 29:1181–1189
2. Fiehn O (2002) Metabolomics—the link between genotypes and phenotypes. Plant Mol Biol 48: 155–171
3. Robertson DG (2005) Metabonomics in toxicology: a review. Toxicol Sci 85:809–822
4. Robertson DG, Watkins PB, Reily MD (2011) Metabolomics in toxicology: preclinical and clinical applications. Toxicol Sci 120: S146–S170
5. Zouhar J, Hicks GR, Raikhel NV (2004) Sorting inhibitors (Sortins): chemical compounds to study vacuolar sorting in Arabidopsis. Proc Natl Acad Sci U S A 101:9497–9501
6. Fiehn O (2008) Extending the breadth of metabolite profiling by gas chromatography coupled to mass spectrometry. Trends Analyt Chem 27:261–269
7. Koek MM, Jellema RH, van der Greef J, Tas AC, Hankemeier T (2011) Quantitative metabolomics based on gas chromatography

mass spectrometry: status and perspectives. Metabolomics 7:307–328

8. Lei Z, Huhman DV, Sumner LW (2011) Mass spectrometry strategies in metabolomics. J Biol Chem 286:25435–25442
9. Scalbert A, Brennan L, Fiehn O, Hankemeier T, Kristal BS, van Ommen B, Pujos-Guillot E, Verheij E, Wishart D, Wopereis S (2009) Mass-spectrometry-based metabolomics: limitations and recommendations for future progress with particular focus on nutrition research. Metabolomics 5:435–458
10. Keun HC, Ebbels TMD, Antti H, Bollard ME, Beckonert O, Schlotterbeck G, Senn H, Niederhauser U, Holmes E, Lindon JC, Nicholson JK (2002) Analytical reproducibility in H-1 NMR-based metabonomic urinalysis. Chem Res Toxicol 15:1380–1386
11. Sumner LW, Amberg A, Barrett D, Beale MH, Beger R, Daykin CA, Fan TWM, Fiehn O, Goodacre R, Griffin JL, Hankemeier T, Hardy N, Harnly J, Higashi R, Kopka J, Lane AN, Lindon JC, Marriott P, Nicholls AW, Reily MD, Thaden JJ, Viant MR (2007) Proposed minimum reporting standards for chemical analysis. Metabolomics 3:211–221
12. Sanderfoot AA, Pilgrim M, Adam L, Raikhel NV (2001) Disruption of individual members of Arabidopsis syntaxin gene families indicates each has essential functions. Plant Cell 13:659–666
13. Lukowitz W, Mayer U, Jurgens G (1996) Cytokinesis in the Arabidopsis embryo involves the syntaxin-related KNOLLE gene product. Cell 84:61–71
14. Rojo E, Gillmor CS, Kovaleva V, Somerville CR, Raikhel NV (2001) VACUOLELESS1 is an essential gene required for vacuole formation and morphogenesis in Arabidopsis. Dev Cell 1:303–310
15. Rosado A, Hicks GR, Norambuena L, Rogachev I, Meir S, Pourcel L, Zouhar J, Brown MQ, Boirsdore MP, Puckrin RS, Cutler SR, Rojo E, Aharoni A, Raikhel NV (2011) Sortin1-hypersensitive mutants link vacuolar-trafficking defects and flavonoid metabolism in Arabidopsis vegetative tissues. Chem Biol 18:187–197
16. Lee DY, Fiehn O (2008) High quality metabolomic data for *Chlamydomonas reinhardtii*. Plant Methods 4:13
17. Kaiser KA, Barding GA, Larive CK (2009) A comparison of metabolite extraction strategies for H-1-NMR-based metabolic profiling using mature leaf tissue from the model plant *Arabidopsis thaliana*. Magn Reson Chem 47: S147–S156
18. Sekiyama Y, Chikayama E, Kikuchi J (2011) Evaluation of a semipolar solvent system as a step toward heteronuclear multidimensional NMR-based metabolomics for C-13-labelled bacteria, plants, and animals. Anal Chem 83: 719–726
19. Davis RA, Charlton AJ, Godward J, Jones SA, Harrison M, Wilson JC (2007) Adaptive binning: an improved binning method for metabolomics data using the undecimated wavelet transform. Chemometr Intell Lab Syst 85: 144–154
20. Holmes E, Foxall PJD, Nicholson JK, Neild GH, Brown SM, Beddell CR, Sweatman BC, Rahr E, Lindon JC, Spraul M, Neidig P (1994) Automatic data reduction and pattern recognition methods for analysis of H-1 nuclear magnetic resonance spectra of human urine from normal and pathological states. Anal Biochem 220:284–296
21. Craig A, Cloareo O, Holmes E, Nicholson JK, Lindon JC (2006) Scaling and normalization effects in NMR spectroscopic metabonomic data sets. Anal Chem 78:2262–2267
22. De Meyer T, Sinnaeve D, Van Gasse B, Tsiporkova E, Rietzschel ER, De Buyzere ML, Gillebert TC, Bekaert S, Martins JC, Van Criekinge W (2008) NMR-based characterization of metabolic alterations in hypertension using an adaptive, intelligent binning algorithm. Anal Chem 80:3783–3790
23. Dieterle F, Ross A, Schlotterbeck G, Senn H (2006) Probabilistic quotient normalization as robust method to account for dilution of complex biological mixtures. Application in H-1 NMR metabonomics. Anal Chem 78: 4281–4290
24. Anderson PE, Mahle DA, Doom TE, Reo NV, DelRaso NJ, Raymer ML (2011) Dynamic adaptive binning: an improved quantification technique for NMR spectroscopic data. Metabolomics 7:179–190
25. Broadhurst DI, Kell DB (2006) Statistical strategies for avoiding false discoveries in metabolomics and related experiments. Metabolomics 2:171–196
26. Kaiser KA, Merrywell CE, Fang F, Larive CK (2008) In: Holzgrabe U, Wawer I, Diehl B (eds) NMR spectroscopy in pharmaceutical analysis. Elsevier, Oxford, pp 233–267
27. Barding GA Jr, Salditos R, Larive CK (2012) Quantitative NMR for bioanalysis and metabolomics. Anal. Bioanal. Chem. 404:1165-1179
28. Fiehn O, Kopka J, Trethewey R, Willmitzer L (2000) Identification of uncommon plant metabolites based on calculation of elemental

compositions using gas chromatography and quadruple mass spectrometry. Anal Chem 72:3573–3580

29. Glasoe PK, Long FA (1960) Use of glass electrodes to measure acidities in deuterium oxide. J Phys Chem 64:188–190
30. Larive CK, Jayawickrama D, Orfi L (1997) Quantitative analysis of peptides with NMR spectroscopy. Appl Spectrosc 51:1531–1536
31. Smallcombe SH, Patt SL, Keifer PA (1995) WET solvent suppression and its applications to LC NMR and high-resolution NMR spectroscopy. J Magn Reson A 117:295–303
32. Liu ML, Mao XA, Ye CH, Huang H, Nicholson JK, Lindon JC (1998) Improved WATERGATE pulse sequences for solvent suppression in NMR spectroscopy. J Magn Reson 132: 125–129
33. Cui Q, Lewis IA, Hegeman AD, Anderson ME, Li J, Schulte CF, Westler WM, Eghbalnia HR, Sussman MR, Markley JL (2008) Metabolite identification via the Madison Metabolomics Consortium Database. Nat Biotechnol 26:162–164
34. Wishart DS, Knox C, Guo AC, Eisner R, Young N, Gautam B, Hau DD, Psychogios N, Dong E, Bouatra S, Mandal R, Sinelnikov I, Xia J, Jia L, Cruz JA, Lim E, Sobsey CA, Shrivastava S, Huang P, Liu P, Fang L, Peng J, Fradette R, Cheng D, Tzur D, Clements M, Lewis A, De Souza A, Zuniga A, Dawe M, Xiong Y, Clive D, Greiner R, Nazyrova A, Shaykhutdinov R, Li L, Vogel HJ, Forsythe I (2009) HMDB: a knowledgebase for the human metabolome. Nucleic Acids Res 37:D603–D610
35. Braunschweiler L, Ernst RR (1983) Coherence transfer by isotropic mixing: application to proton correlation spectroscopy. J Magn Reson 53:521–528
36. Bodenhausen G, Ruben DJ (1980) Natural abundance N-15 NMR by enhanced heteronuclear spectroscopy. Chem Phys Lett 69: 185–189
37. Cobas C, Seoane F, Dominguez S, Sykora S, Davies A (2011) A new approach to improving automated analysis of proton NMR spectra through Global Spectral Deconvolution 8GSD. Spectrosc Eur 23:26–30

Chapter 22

Determination of Auxin Transport Parameters on the Cellular Level

Jan Petrášek, Martina Laňková, and Eva Zažímalová

Abstract

The accumulation of radioactively labelled compounds in cells is frequently used for the determination of activities of various transport systems located at the plasma membrane, including the system for carrier-mediated transport of plant hormone auxin. The measurements of auxin transport could be performed on the tissue level as well, but for more precise quantitative analysis of activity of individual auxin carriers the model of plant cell cultures represents an invaluable tool. Here, we describe the method for the determination of the activities of auxin influx and efflux carriers in plant cells grown in a suspension using radiolabelled synthetic auxins 2,4-dichlorophenoxyacetic acid (2,4-D) and naphthalene-1-acetic acid (NAA). By making use of specific inhibitors of active auxin influx and efflux, as well as cell lines overexpressing or silencing particular auxin carriers, this method allows the determination of kinetic parameters of auxin flow across the plasma membrane and the activity of those carriers.

Key words Auxin transport assay, Plant cell cultures, Auxin influx, Auxin efflux, Radiolabelled auxin

1 Introduction

Distribution of auxin (IAA; indole-3-acetic acid) in the plant body is important for a plethora of morphoregulatory events during plant development [1]. Besides its passive diffusion across the plasma membrane, the cell-to-cell transport of IAA is assisted by several groups of auxin influx and efflux carriers [2]. In principle, two indirect methods could be used for tracking of auxin flow *in planta*: the determination of applied radiolabelled IAA distribution [3–5] and the noninvasive auxin flux measurements with IAA-selective microelectrode [6]. These two indirect methods are suitable for auxin transport assays at the organ/tissue level, but they cannot be used for the determination of activities of individual transporters at a single cell level. For this purpose, the accumulation of radiolabelled auxin by plant cells cultured in liquid medium represents the most suitable tool, and it helped to define

Glenn R. Hicks and Stéphanie Robert (eds.), *Plant Chemical Genomics: Methods and Protocols*, Methods in Molecular Biology, vol. 1056, DOI 10.1007/978-1-62703-592-7_22, © Springer Science+Business Media New York 2014

the chemiosmotic hypothesis of IAA transport across the plasma membrane [7]. According to this model, molecules of IAA that dissociate in the more alkaline intracellular environment cannot be transported out of the cell by passive diffusion and so they are trapped inside the cell. Taken together, the transport of the dissociated form of IAA is assisted by both auxin influx and efflux carriers.

Here we describe a simple method for the indirect determination of activities of auxin influx and efflux carriers using suspension-cultured *Nicotiana tabacum* L., cv. Bright Yellow (BY-2) [8] and *Arabidopsis thaliana* L., cv. Landsberg erecta [9] cells. Such a method was originally described by Delbarre et al. for tobacco cv. Xanthi XHFD8 cells [10] and modified by Petrášek et al. for tobacco cv. Virginia Bright Italia (VBI-0) [11] and BY-2 [12, 13] cells as well as for Arabidopsis cells [13]. Instead of determining the accumulation of the native auxin IAA, the method uses the radiolabelled synthetic auxins 2,4-D and NAA. According to Delbarre et al. [10], the accumulation of 2,4-D reflects predominantly the activity of auxin influx carriers (and is a poor substrate for auxin efflux carries), while the accumulation of NAA (taken up into cells mainly by passive diffusion) is determined by the activity of auxin efflux carriers. In addition, the specific inhibitor of active auxin influx, 3-chloro-4-hydroxyphenylacetic (CHPAA) [14, 15], and the specific inhibitor of active auxin efflux, 1-naphthylphthalamic acid (NPA) [16], can be applied immediately after the addition of radiolabelled auxin to follow the activity of auxin influx and efflux carriers more precisely. The protocol presented here describes cell culture cultivation and equilibration as well as an auxin accumulation assay performed in cell lines of tobacco and Arabidopsis. Indeed, the potential of this method is fully revealed when using transformed cell lines either overexpressing or silencing gene(s) for particular auxin carrier(s). In combination with more or less specific inhibitor(s) of the activity or localization of auxin carriers, properties of these carriers can be monitored.

2 Materials

Prepare all water solutions using standard distilled (deionized) water, there is no need of ultra pure water. Use analytical grade chemicals and solvents. Sterilize culture media and stock solutions as well as equipment for handling with cells (*see* **Note 1**).

2.1 Plant Material

1. 7-day-old cell suspension of N. tabacum L., cv. Bright Yellow (BY-2) cell line (*see* **Note 2**).
2. 7-day-old cell suspension of A. thaliana L., cv. Landsberg erecta cell line (*see* **Note 2**).

2.2 Media, Buffers and Incubation Solutions

1. Modified Murashige–Skoog [17] medium for suspension-cultured cells: 3 % (w/v) sucrose, 4.3 g/L MS basal salt mixture (Sigma Chemical Company, St. Louis, MO, USA, cat. n. M5524), 100 mg/L myo-inositol (Duchefa Biochemie B.V., The Netherlands, cat. n. I0609), 1 mg/L thiamine (Sigma Chemical Company, St. Louis, MO, USA, cat. n. T3902), 200 mg/L KH_2PO_4, and 0.2 mg/L (0.9 μM) 2,4-D (Sigma Chemical Company, St. Louis, MO, USA, cat. n. D8407), pH 5.8. For 1 L, weigh 4.3 g MS salt mixture and transfer it into glass or plastic beaker with around 800 mL of distilled water, add 30 g sucrose, 100 mg myo-inositol, 200 mg KH_2PO_4, 1 mL stock solution of thiamine, and 2 mL stock solution of 2,4-D (*see* **Note 3**). Stir thoroughly with magnetic stirrer at room temperature, adjust pH with 3 M KOH to 5.8 and bring volume to 1 L. Autoclave at 121 °C for 20 min under 0.1 MPa (*see* **Note 4**).
2. Uptake buffer: 20 mM 2-(N-Morpholino)-ethanesulfonic acid (MES; Duchefa Biochemie B.V., The Netherlands, cat. n. M1503.0025), 10 mM sucrose, 0.5 mM $CaSO_4 \cdot 2H_2O$, pH 5.7. For 1 L, weigh 3.9 g MES and transfer it into glass or plastic beaker with around 800 mL of distilled water, add 3.4 g sucrose and 86 mg $CaSO_4 \cdot 2H_2O$. Stir thoroughly with magnetic stirrer at room temperature for at least 30 min, adjust pH with 3 M KOH to 5.7 and bring volume to 1 L. Autoclave at 121 °C for 20 min under 0.1 MPa (*see* **Note 4**).
3. Liquid scintillation cocktail for aqueous samples, e.g., Ecolite(+)™ (MP Biomedicals, cat. n. 882475).
4. Pure ethanol (96 %, v/v) and pure ethanol for UV spectroscopy (96 %, v/v).
5. Dimethyl sulfoxide (DMSO) (Sigma Chemical Company, St. Louis, MO, USA, cat. n. D4540)

2.3 The Equipment for Cell Culture Cultivation and Auxin Accumulation Assay

1. Autoclave (e.g., Systec DX-2, Systec GmbH, Labor-Systemtechnik, Wettenberg, Germany).
2. Laminar flow cabinet for handling with cells during the inoculation and equilibration. Fume hood for manipulations with radioactive material (beta radiation).
3. Sterilized standard laboratory equipment: metal tweezers and spatulas, glass Erlenmeyer flasks (250, 100 and 50 mL) covered with aluminum foil for the incubation of cell suspensions, cylinders (50 and 100 mL), pipette tips (for 1–5 mL) with cut top.
4. Stepper pipette for repetitive pipetting of ethanol onto individual cell cakes, manual pipette for 1–5 mL for handling with cell suspensions, electronic motorized pipettes for handling with radiolabelled solutions and inhibitors (ranges 0.2–10 μL and 10–250 μL) and taking cell samples during the auxin accumulation assay (range 50–1,000 μL) (e.g., Labopette, Hirschmann

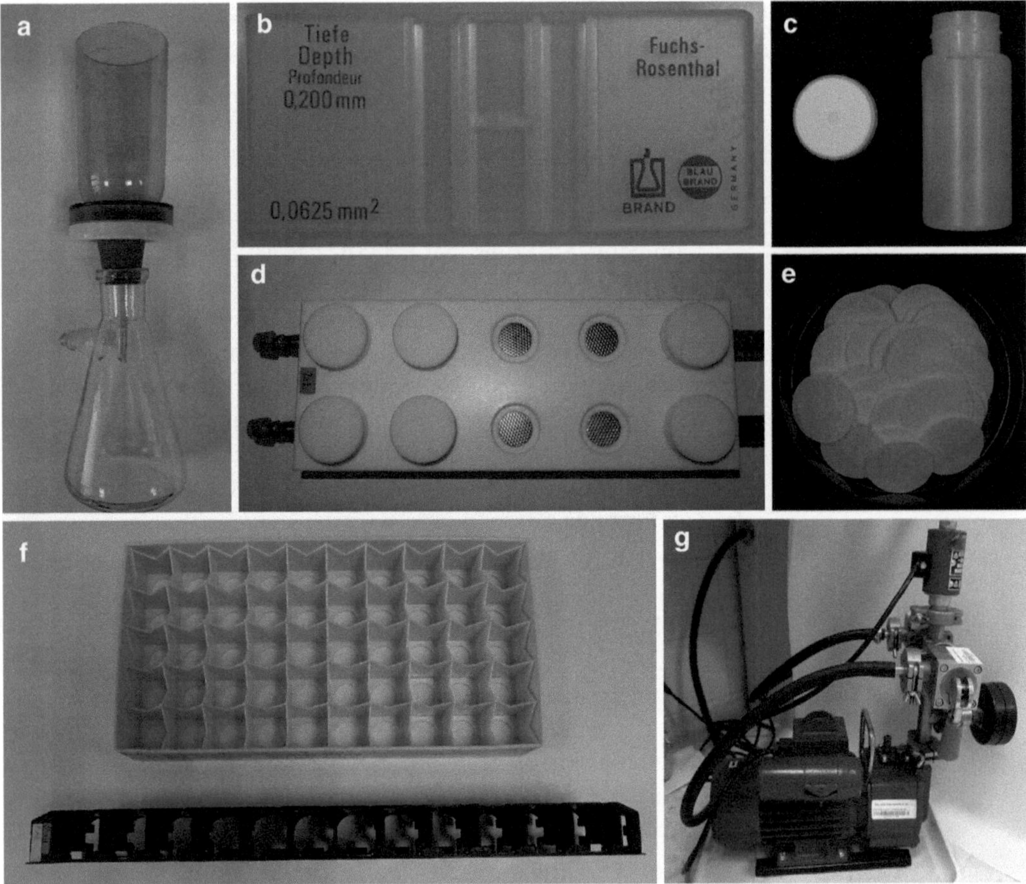

Fig. 1 Equipment used for the auxin transport assays in plant cell suspensions. (**a**) Sterile filter holder with nylon net filter placed at the suction Büchner suction flask for filtering cells during the equilibration procedure. (**b**) Fuchs-Rosenthal counting chamber for the determination of density of cell population used for the accumulation assay. (**c**) Disposable scintillation vial with screw cap for collecting individual cell cakes at the cellulose filter papers. (**d**) Vacuum filtration manifold with stoppers, four opened positions can accommodate filter paper for four technical repetitions. (**e**) Cellulose filter paper discs. (**f**) Racks for the arrangement (*top*) and counting (*bottom*) of scintillation vials. (**g**) Vacuum pump holding valve directing "vacuum" to the manifold

Laborgeräte, Germany) (*see* **Note 5**). Tips for stepper pipette and electronic motorized pipettes (all three ranges).

5. Orbital incubator (e.g., IKA KS501, IKA Labortechnik, Staufen, Germany) placed at 27 °C (tobacco) or 24 °C (Arabidopsis) in darkness in the culture room. Alternatively, air-conditioned incubator (e.g., Sanyo Gallenkamp, Schoeller Instruments Inc.) could be used.
6. Fuchs-Rosenthal counting chamber (size of the chamber 4×4×0.2 mm; e.g., Brand, Germany) (*see* **Note 6** and Fig. 1b).
7. Routine upright or inverted light microscope, ideally both.
8. Sterile glass or plastic filter holder with nylon net filter with pore size 20 μm and diameter 45 mm (Millipore 20 μm NY20,

cat. n. NY2004700). The filter holder placed at the Büchner (suction) flask (Fig. 1a) is used for filtering cells during the cell equilibration procedure (*see* **Note 7**).

9. Vacuum filtration manifold with stoppers (Fig. 1d) accommodating cellulose filter paper discs (Fig. 1e), 80 g/m^2, diameter 25 mm (*see* **Note 8**).
10. Vacuum pump (Fig. 1g) (*see* **Note 9**).
11. Disposable scintillation vials (e.g., Ratiolab, Germany, cat. n. 5810100) with screw caps (Fig. 1c) for collecting individual cellulose filter paper discs with filtered cell cakes (*see* **Note 10**).
12. Racks for the arrangement and transportation of scintillation vials (Fig. 1f top) (e.g., Tricarb Servo-tray Packard, cat. no. 6008129), racks for measuring individual vials in the scintillation counter (Fig. 1f bottom) (*see* **Note 11**).
13. Dispenser for scintillation cocktail (e.g., Dispensette III, Dispensette Organic, Packard BioScience Company).
14. Liquid scintillation analyzer (e.g., Packard Tri-Carb 2900TR, Packard Instrument Co., Meriden, CT, USA).
15. Laboratory timer (four-channel) for measuring time during the accumulation assay.

2.4 Radiochemicals and Inhibitors

1. Tritium-labelled ^{3}H[NAA] (cat. n. ART 0610), specific radioactivity 20 Ci/mmol (American Radiolabeled Chemicals, St. Louis, MO, USA) (*see* **Note 12**).
2. Tritium-labelled ^{3}H[2,4-D] (cat. n. ART 0559), specific radioactivity 20 Ci/mmol (American Radiolabeled Chemicals, St. Louis, MO, USA) (see Note 12).
3. NPA (OlChemIm Ltd, Olomouc, Czech Republic, cat. n. 0183041-3), make 10 mM stock solution (10 mM) in ethanol or DMSO, the choice depends on the particular experiment.
4. CHPAA (Sigma Chemical Company, St. Louis, MO, USA, cat. n. 224529), make 10 mM stock solution (10 mM) in ethanol or DMSO, the choice depends on the particular experiment.

3 Methods

Carry out all procedures at room temperature (20–25 °C), unless otherwise stated.

3.1 Cell Culture Preparation

1. Inoculate 2–4 mL of 7-day-old BY-2 cells or Arabidopsis cells into 100 mL of modified MS cultivation medium (*see* **Note 13**). Use mechanical pipette and tips with cut top to transfer cells under sterile conditions into the fresh medium. The laminar flow cabinet should be sterilized with UV for 20 min in advance.

2. Cultivate cells in darkness under continual shaking at 27 °C (BY-2) or 24 °C (Arabidopsis) on an orbital incubator with 150 rpm for BY-2 and 130 rpm for Arabidopsis cells (orbital diameter 30 mm) (*see* **Note 14**). The length of cultivation depends on the purpose of the experiment.

3.2 Cell Culture Equilibration

1. At the desired time point (usually 24 or 48 h after inoculation) filter cells using filter holder placed at the Büchner flask (Fig. 1a). Do not use under pressure; just leave cells to settle at the bottom as the medium is passing through nylon net filter with 20 μm pore size. Transfer resulting cell cake with sterile spatula into the uptake buffer. The volume of uptake buffer for this washing step is usually half of the volume of medium used during the cultivation (i.e., from two flasks with 100 mL cell culture make one flask with cells in 100 mL of uptake buffer). Using a mechanical pipette take 1 mL of suspension in the uptake buffer and transfer it into a test tube for the determination of cell density (*see* **Note 15**).
2. Incubate cells in the uptake buffer under continuous orbital shaking for 45 min at room temperature.
3. Determine the cell density using the Fuchs-Rosenthal counting chamber (Fig. 1b) (*see* **Note 16**).
4. Filter cells again in the same way as in the first washing step and resuspend them in the fresh uptake buffer. Increase or decrease the volume of the uptake buffer depending on the density of cell population determined in the previous step (*see* **Note 17**).
5. Incubate cells in the uptake buffer under continuous orbital shaking for an additional 1.5 h at room temperature. Thereafter, cell suspension is ready for the auxin accumulation assay, under the conditions of a MES-buffered extracellular environment (pH 5.7).

3.3 Auxin Accumulation Assay

1. Divide 100 mL of equilibrated cell suspension of desired cell density into two 50 mL aliquots in 250 mL Erlenmeyer flasks using a graduated cylinder. Shake with the flask to prevent cells from sedimenting. Mark clearly each flask with an ethanol-resistant felt tip lab marker and keep them constantly shaking on an orbital shaker. The first flask will be used for the [^{3}H]2,4-D accumulation assay, the second for the [^{3}H]NAA accumulation assay (*see* **Note 18**). The amount of material in each flask is enough to obtain two auxin accumulation curves (with and without inhibitor) with 5 sampling points within a 25–30 min accumulation period. Prepare also four empty 100 ml Erlenmeyer flasks and mark them (e.g., [^{3}H]2,4-D, [^{3}H]2,4-D + CHPAA, [^{3}H]NAA, [^{3}H]NAA + NPA).

2. Open four positions in the vacuum filtration manifold (Fig. 1d) (placed ideally in fume hood) and leave the others closed with stoppers. Fill racks for the arrangement and transportation of scintillation vials (Fig. 1f top) with disposable scintillation vials (Fig. 1c) (*see* **Note 11**). Place four cellulose filter paper discs (Fig. 1e) onto the filtration manifold at opened positions and moisturize them using uptake buffer in a wash bottle.
3. Turn on the vacuum pump (Fig. 1g) and check the suction; moisturized filters should be adhering to the metallic bottom of the filtration manifold position.
4. Using electronic motorized pipette (0.2–10 μL), add the appropriate amount of ^{3}H[2,4-D] into the first flask with 50 ml of equilibrated culture, usually it is around 0.8–1 μL/10 mL to give a final concentration of 2 nM (*see* **Notes 12** and **19**). Start timer together with the addition of radioactivity. Immediately (as soon as possible) take four 0.5 mL aliquots consecutively with motorized pipette (500–1,000 μL) and filter them one per one through a cellulose filter using filtration manifold opened positions. Enter the time of filtration of each individual repetition into a log. Since one is busy with pipetting the optimal way is with the assistance of a colleague who writes down the time of filtration for each sample (or, alternatively and for very experienced researchers, one can use previously set time points for filtering samples). Transfer the cellulose filters with cell cakes into scintillation vials. Wash empty positions in the manifold briefly with uptake buffer (using wash bottle), apply another four cellulose filters and moisturize them. The cell suspension with radiolabel must be shaken constantly.
5. Make two 20 ml aliquots from the ^{3}H[2,4-D] sample. Add 20 μl of the 10 mM ethanolic stock of CHPAA (the final concentration is 10 μM) into the first flask and the corresponding amount of ethanol into the other one (20 μl). Write down the time of addition of an inhibitor.
6. Repeat the previous two steps with ^{3}H[NAA] and NPA. Write down carefully the time of addition of radiolabel, filtering of samples and addition of NPA (*see* **Note 20**).
7. Take four additional time points for each sample (i.e., [^{3}H]2,4-D, [^{3}H]2,4-D + CHPAA, [^{3}H]NAA, [^{3}H]NAA + NPA) during the 25–30 min incubation under continuous shaking at 25 °C. At the end of the run there are four racks with vials with filters with cell cakes. Sampling immediately after the addition of the radiolabel is common for two [^{3}H]2,4-D samples (with and without CHPAA) and [^{3}H]NAA samples (with and without NPA).

8. Add 500 μL of pure ethanol into all vials using a stepper pipette for repetitive pipetting, to extract the radioactivity from cell cakes, and incubate for 30 min.
9. Add 4 ml of liquid scintillation cocktail for aqueous samples with a dispenser into all vials and close them with screw caps carefully.
10. Mark screw caps with numbers and write down carefully how numbers correspond to individual samples. Here for [^{3}H]2,4-D you obtain samples n. 1–20, for [^{3}H]2,4-D + CHPAA samples n. 21–36, for [^{3}H]NAA samples n. 37–56 and for [^{3}H] NAA + NPA samples n. 57–72.
11. Move all vials into racks for measuring the radioactivity in the scintillation analyzer (Fig. 1f bottom). Shake racks with vials 20 min by rigorous shaking (300–400 rpm).
12. Measure the radioactivity in the scintillation counter; the setup for measuring tritium should be used (2 min each vial) with automatic correction for quenching and recalculation from cpm to dpm. Include blank samples (vials with scintillation cocktail without radioactivity) before and after measured samples. Obtained dpm values represent raw data reflecting the amount of radioactivity accumulated in cells.
13. Collect all data in the spreadsheet and correct for cell surface radioactivity by subtracting dpm values obtained for aliquots of cells collected immediately after the addition of [^{3}H]2,4-D or [^{3}H]NAA. Express obtained data in pmol of [^{3}H]2,4-D or [^{3}H] NAA per million of cells and make plots (Fig. 2) (*see* **Note 21**).

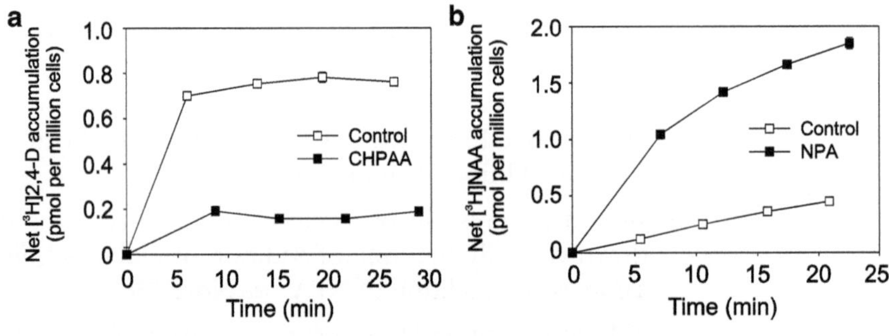

Fig. 2 Auxin accumulation kinetics in tobacco BY-2 cells obtained after the recalculation of the radioactivity to pmol of radiolabelled auxin per million cells. Error bars represent standard errors ($n = 4$) reflecting the error of technical repetitions. When not visible they are smaller than the symbol. (**a**) [^{3}H]2,4-D accumulation in control cells and cells treated with 10 μM CHPAA. The inhibitor strongly decreases the accumulation of [^{3}H]2,4-D. The difference between the two curves reflects the activity of auxin influx carriers. It could be quantified at different time points or compared among various samples. (**b**) [^{3}H]NAA accumulation in control cells and cells treated with 10 μM NPA. The inhibitor strongly increases the accumulation of [^{3}H]NAA (decreases its active efflux). The difference between the two curves reflects the activity of auxin efflux carriers. It could be quantified at different time points or compared among various samples

4 Notes

1. Thorough sterilization of material and solutions helps greatly to standardize the whole technique, because plant cell cultures are sensitive to the contamination from the environment (bacteria, fungi). Even invisible contamination could interfere with the growth of cell cultures and with the activities of auxin carriers in particular.
2. For success of the measurement of auxin accumulation, it is absolutely necessary to work with high quality cell suspensions. Both tobacco BY-2 and Arabidopsis cell suspensions are normally yellowish, dense cultures at the end of subculture interval that normally takes 7 days. For the inoculation of cells for auxin accumulation measurements, avoid using suspensions that do not grow optimally. This should be controlled without opening the culture flask using an inverted microscope during the first 3 days of cultivation. From day 4, when the cell culture becomes too dense for direct observation through the flask, it is better to take a sample under sterile conditions.
3. MS salts are quite hygroscopic; the bottle should not be left opened unless necessary. For the storage of MS salts at 4 °C seal the cap carefully with Parafilm. Prepare the stock solution of thiamine by dissolving 100 mg of thiamine in 100 ml of distilled water. This stock could be stored in 2 ml aliquots in Eppendorf tubes in −20 °C. Prepare the stock of 2,4-D by dissolving 10 mg of 2,4-D in 1 mL pure ethanol. After addition of distilled water to the final volume 100 mL (0.45 μM 2,4-D) heat the solution under continuous stirring for at least 4 h to dissolve 2,4-D completely.
4. Pay special attention when autoclaving media and buffers; always use the program containing slow release of the pressure at the end of the run. Autoclaves not equipped with this option are not suitable for this purpose. Avoid repetitive autoclaving of the same buffer or medium.
5. It is advisable to use electronic motorized pipettes for all manipulations with radiolabelled auxins, inhibitors applied before or during the accumulation assay, and-most importantly-for taking samples of cell suspension during the accumulation run. Making use of electronic motorized pipettes greatly reduces differences between sampling repetitions.
6. The counting chamber is originally intended for counting much smaller cells than plant cells (e.g., red blood cells). However, it could be used also for plant cells by counting all cells in the chamber and careful calibration using dilution series.

7. Nylon net filters could be used repeatedly. Always take care to have excess of clean and sterile filters available.
8. The vacuum filtration manifold is primarily used for rapid filtration of cell suspensions and collecting cell cakes at the filter paper. For this purpose, four positions are opened and could accommodate filter papers for collecting four repetitions. Other positions are closed with stoppers; obviously, they can be used as well-in case of the need for a greater number of repetitions. On the other hand, the more positions that are open, the less suction is applied at individual positions. Take care to standardize the suction by closing always the same number of positions. For some applications it might be important to determine the radioactivity in the uptake buffer. For this purpose you can use another vacuum filtration manifold (e.g., Millipore® model 1225).
9. The level of oil in the vacuum pump should be checked before the experiment. Moreover, oil should be replaced after some time (according to manufacturer's manual) to prevent unwanted changes in the efficiency of the suction. The collecting bottle (e.g., Wolf's bottle) should be inserted between the filtration manifold and the vacuum pump to trap the liquid that is filtered out of cell suspension. The content of this bottle should be discarded regularly.
10. Making use of disposable scintillation vials greatly saves time for washing of glass vials. However, always consider the method of disposal of radioactive waste. Since the disposal of whole flasks with scintillation liquid inside is costly, pouring the scintillation liquid out of vials after the experiment is finished and disposing of vials and liquid separately is much cheaper. For handling of radioactive waste always carefully follow local regulations.
11. Always pay very close attention to the unambiguous labelling of individual racks with stickers to clearly label individual runs. The point is that by doing several accumulated runs (e.g., control and three treated samples from whose cell cakes are harvested consecutively), it is very easy to mix up the samples.
12. Ethanol stock solutions of radiolabelled auxins should be stored in a deep freezer at −80 °C. This helps to slow down decomposition of the molecules significantly. From the original stock, ethanol dilutions (1:2) are prepared that are used for the addition into the equilibrated cell culture at the beginning of accumulation run. These stock solutions are stored also in −80 °C. Be very careful after taking stocks from deep-freeze. It is essential that before opening, the stock vial's temperature is the same as room temperature! For thawing, always keep vials in darkness. Avoid unnecessary thawing.
13. The inoculation density is critical for the successful growth of suspension-cultured cells. For routine propagation of BY-2 and

Arabidopsis cells with a 7-day-long subculture interval, stationary cells are diluted (at the day 7) 1:50 (2 mL/100 mL of medium in 250 mL Erlenmeyer flask) or 1:30 (1 mL/30 mL of medium in 100 mL Erlenmeyer flask). Never fill flasks with more medium than specified above. Depending on the purpose of auxin transport assays it might be necessary to make inoculate at higher density (up to 4 mL/100 mL in 250 mL flask or 2 mL/30 ml flask) and prepare more flasks. The reason is that one can obtain reproducible results of auxin accumulation assays only with a defined cell density range-this is critical especially during the first 2 days of the subculture interval, when the density of culture is not very high (*see* **Note 16**).

14. The temperature during cultivation of cells should not exceed 27–28 °C for BY-2 cells and 25 °C for Arabidopsis cells. Cells are also very sensitive to prolonged standing without shaking. Try to minimize this time as much as possible.
15. Take the small sample (1 mL) of cell suspension very carefully. Hold the flask with one hand and swirl it continuously to keep cells suspended while taking the sample with the other hand. Use a pipet tip with cut end to widen it. By doing this, the cell population is homogeneous and the sample contains the same density of cells as in the flask.
16. Determination of density of the cell population is crucial for the reproducibility of the accumulation assays. Suitable cell density for BY-2 cells ranges between 5 and 8×10^5 cells/mL, for Arabidopsis it is slightly more, normally around $14–16 \times 10^5$ cells/mL. Count cells always in at least ten aliquots of each sample. Arabidopsis cells are smaller than tobacco cells, and they are usually much more difficult to count. Consider this when planning the experiment. Expressing the resulting dpm values per million cells is a very convenient way of the comparison between treatments. To obtain reliable results and relevant values for comparisons it is also absolutely crucial to adjust the density of the samples exactly and to roughly the same density. This is the reason why counting of cells should be done during the equilibration procedure so that the correction is done before the accumulation assay by diluting or concentrating equilibrated cultures.
17. Determination of cell densities could be done also during the second washing period (1.5 h).
18. From this point it is not necessary to work under sterile conditions. Wear laboratory gloves when working with the radioactivity. There are always special regulations for handling radioactivity. For tritium used in this protocol it is convenient do at a dedicated working place on a bench or ideally in the fume hood (with beta-radiation certificate). Monitoring the

surface of the bench and other work places should be done regularly to control accidental contamination. Tritium-labelled auxins and their degradation products can be removed from the surface of the bench with water or water and ethanol, all flasks used during the assay are pre-washed under streaming water, then left in the container with water and kitchen detergent for few hours, rinsed and then washed normally.

19. Because of radioactive decay of tritium, the radioactivity should be measured when doing the pre-dilution (1:2) of original stock solutions, corrected for the half-life of tritium, and then up-to-date molar concentrations of radiolabelled auxin solution should be specified.
20. The optimal concentrations of inhibitors of auxin influx (CHPAA) and efflux (NPA) were determined previously [15, 18].
21. There are many ways of presentation of results of this assay. In principle, resulting values could be expressed in dpm of a particular auxin per sample, per cell number, per unit of cell surface, or per cell volume (for the latter two, cell dimensions should be determined using image analysis). For biological purpose, however, it is more informative to express auxin amounts in pmol, which can be calculated from the dpm and specific (molar) radioactivity of the particular labelled auxin preparation. Expressing results as pmols of a particular auxin per cell number reflects the ratio between the number of auxin molecules and number of cells (Fig. 2).

As auxins can be metabolized even during the accumulation assay [10], the correction for the metabolic conversions of labelled auxin should be done. Metabolism can be checked, for example, by HPLC profiling of samples taken during the accumulation run. Check the metabolism also for long-term stock solutions of radiolabelled auxins at least every year.

Acknowledgement

This work was supported by the Czech Science Foundation, projects GAP305/11/2476 (JP) and GAP305/11/0797 (EZ).

References

1. Vanneste S, Friml J (2009) Auxin: a trigger for change in plant development. Cell 136:1005–1016
2. Petrášek J, Friml J (2009) Auxin transport routes in plant development. Development 136:2675–2688
3. Lewis DR, Muday GK (2009) Measurement of auxin transport in *Arabidopsis thaliana*. Nat Protoc 4:437–451
4. Goldsmith MHM (1977) Polar transport of auxin. Annu Rev Plant Physiol Plant Mol Biol 28:439–478

5. Peer WA, Murphy AS (2007) Flavonoids and auxin transport: modulators or regulators? Trends Plant Sci 12:556–563
6. Mancuso S, Marras AM, Magnus V, Baluska F (2005) Noninvasive and continuous recordings of auxin fluxes in intact root apex with a carbon nanotube-modified and self-referencing microelectrode. Anal Biochem 341:344–351
7. Rubery PH, Sheldrake AR (1974) Carrier-mediated auxin transport. Planta 118: 101–121
8. Nagata T, Nemoto Y, Hasezawa S (1992) Tobacco BY-2 cell-line as the Hela-cell in the cell biology of higher-plants. Int Rev Cytol 132:1–30
9. May M, Leaver C (1993) Oxidative stimulation of glutathione synthesis in *Arabidopsis thaliana* suspension cultures. Plant Physiol 103:621–627
10. Delbarre A, Muller P, Imhoff V, Guern J (1996) Comparison of mechanisms controlling uptake and accumulation of 2,4-dichlorophenoxy acetic acid, naphthalene-1-acetic acid, and indole-3-acetic acid in suspension-cultured tobacco cells. Planta 198:532–541
11. Petrášek J, Elčkner M, Morris DA, Zažímalová E (2002) Auxin efflux carrier activity and auxin accumulation regulate cell division and polarity in tobacco cells. Planta 216:302–308
12. Petrášek J, Zažímalová E (2006) The BY-2 cell line as a tool to study auxin transport. In: Biotechnology in Agriculture and Forestry. Tobacco BY-2 Cells: From Cellular Dynamics to Omics, Nagata, T., Matsuoka, K., Inzé, D. (eds.), Springer-Verlag, Berlin Heidelberg, 58:107–115
13. Petrášek J, Mravec J, Bouchard R, Blakeslee JJ, Abas M, Seifertová D et al (2006) PIN proteins perform a rate-limiting function in cellular auxin efflux. Science 312:914–918
14. Parry G, Delbarre A, Marchant A, Swarup R, Napier R, Perrot-Rechenmann C et al (2001) Novel auxin transport inhibitors phenocopy the auxin influx carrier mutation aux1. Plant J 25:399–406
15. Laňková M, Smith R, Pešek B, Kubeš M, Zažímalová E, Petrášek J et al (2010) Auxin influx inhibitors 1-NOA, 2-NOA, and CHPAA interfere with membrane dynamics in tobacco cells. J Exp Bot 61:3589–3598
16. Rubery PH (1990) Phytotropins-receptors and endogenous ligands. Symp Soc Exp Biol 44:119–146
17. Murashige T, Skoog F (1962) A revised medium for rapid growth and bio assays with tobacco tissue cultures. Physiol Plant 15: 473–497
18. Petrášek J, Černá A, Schwarzerová K, Elčkner M, Morris DA, Zažímalová E (2003) Do phytotropins inhibit auxin efflux by impairing vesicle traffic? Plant Physiol 131:254–263

Chapter 23

Analyzing the In Vivo Status of Exogenously Applied Auxins: A HPLC-Based Method to Characterize the Intracellularly Localized Auxin Transporters

Sibu Simon, Petr Skůpa, Petre I. Dobrev, Jan Petrášek, Eva Zažímalová, and Jiří Friml

Abstract

Exogenous application of biologically important molecules for plant growth promotion and/or regulation is very common both in plant research and horticulture. Plant hormones such as auxins and cytokinins are classes of compounds which are often applied exogenously. Nevertheless, plants possess a well-established machinery to regulate the active pool of exogenously applied compounds by converting them to metabolites and conjugates. Consequently, it is often very useful to know the in vivo status of applied compounds to connect them with some of the regulatory events in plant developmental processes. The in vivo status of applied compounds can be measured by incubating plants with radiolabeled compounds, followed by extraction, purification, and HPLC metabolic profiling of plant extracts. Recently we have used this method to characterize the intracellularly localized PIN protein, PIN5. Here we explain the method in detail, with a focus on general application.

Key words Auxin, Auxin metabolism, BY-2 tobacco cells, PIN proteins, HPLC, Plant cell cultures

1 Introduction

The plant hormone auxin is an important regulator of plant growth and development. To initiate auxin-mediated responses, a local concentration maximum of auxin is formed at its point of action by a polarized cellular localization of PIN-FORMED (PIN) protein efflux carriers [1]. In *Arabidopsis thaliana*, the PIN group of auxin efflux carriers has eight members, of which PIN5, 6, and 8 have a short hydrophilic loop compared to that of other members of the family [2]. Our recent study reported that the short PINs are localized in endomembranes and that the transport behavior of PIN5 in particular is not intended for maintenance of this concentration gradient, but for auxin homeostasis [3]. This differential functional importance of short PINs, in particular PIN5, is illustrated by tracking the metabolic transformation of exogenously added tritium-labeled

Glenn R. Hicks and Stéphanie Robert (eds.), *Plant Chemical Genomics: Methods and Protocols*, Methods in Molecular Biology, vol. 1056, DOI 10.1007/978-1-62703-592-7_23,

indole-3-acetic acid ([^{3}H]IAA) in a PIN5-overexpressing cell suspension culture of *Nicotiana tabacum* L. cv. Bright Yellow-2 (BY-2) [4]. Utilization of BY-2 suspension grown cells facilitates uniform analysis of target processes at the level of a single cell, avoiding possible interference of different tissue variations when using whole plants or seedlings. BY-2 suspension grown cells are successfully used for studying different plant-specific processes, including characterization of various biologically important compounds for their transport potential across the plasma membrane (PM) [5].

High-performance liquid chromatography (HPLC) is a powerful tool to separate, purify, identify and quantify a target compound from a mixture of compounds. Plants contain numerous low-molecular-weight compounds, including plant hormones, but their quantitative representation is yet to be explored, bringing a high demand on the trace analysis of plant hormones. HPLC is successfully utilized in plant hormone analysis [6, 7] to track endogenous auxins and metabolites generated from exogenously applied compounds.

The protocol presented here is mainly focused on the HPLC-based analysis of exogenously applied radiolabeled IAA in BY-2 to track the metabolites generated in this system. We demonstrate the applicability of this method to characterize the PIN5 auxin efflux carrier by comparing the metabolic profiles of PIN5 overexpressed and control lines. However, a similar approach can be adapted to study transport or metabolic processing of any exogenously applied compound.

2 Materials

2.1 Plant Material

7-day-old *N. tabacum* L. Bright Yellow-2 (BY-2) cell lines transformed with the dexamethasone-inducible PIN5 gene (*see* **Note 1**).

2.2 Growth Media, Buffers, and Other Solutions

1. Modified Murashige–Skoog (MS) medium for suspension-cultured cells: 3 % (w/v) sucrose, 4.3 g/L MS basal salt mixture (Sigma Chemical Company, St. Louis, MO, USA, cat. n. M5524), 100 mg/L myo-inositol (Duchefa Biochemie B.V., The Netherlands, cat. n. I0609), 1 mg/L thiamine (Sigma Chemical Company, St. Louis, MO, USA, cat. n. T3902), 200 mg/L KH_2PO_4, and 0.2 mg/L (0.9 μM) 2,4-dichlorophenoxyacetic acid (2,4-D) (Sigma Chemical Company, St. Louis, MO, USA, cat. n. D8407), pH 5.8. For 1 L, weigh 4.3 g MS salt mixture and transfer it into glass or plastic beaker with around 800 mL of distilled water, add 30 g sucrose, 100 mg myo-inositol, 200 mg KH_2PO_4, 1 mL stock solution of thiamine, and 2 mL stock solution of 2,4-D. Stir thoroughly with magnetic stirrer at room temperature, adjust pH with 3 M KOH to 5.8 and bring volume to 1 L. Autoclave at 121 °C for 20 min under 0.1 Mpa.

2. Uptake buffer: 20 mM 2-(*N*-morpholino)-ethanesulfonic acid (MES; Duchefa Biochemie B.V., The Netherlands, cat. n. M1503.0025), 10 mM sucrose, 0.5 mM $CaSO_4 \cdot 2H_2O$, pH 5.7. For 1 L, weigh 3.9 g MES and transfer it into glass or plastic beaker with around 800 mL of distilled water: add 3.4 g sucrose and 86 mg $CaSO_4 \cdot 2H_2O$. Stir thoroughly with magnetic stirrer at room temperature for at least 30 min, adjust pH with 3 M KOH to 5.7 and bring volume to 1 L. Autoclave at 121 °C for 20 min under 0.1 MPa.
3. Modified Bieleski's solution: a mixture of methanol, distil. H_2O, and formic acid in 15:4:1 proportion (v/v/v) (*see* **Note 2**).
4. 1 M formic acid (HCOOH) (Sigma chemical company, St. Louis, USA).
5. Methanol (Sigma chemical company, St. Louis, USA).
6. 15 % acetonitrile (ACN) in water, v/v, Merck, Darmstadt, Germany.
7. Dimethyl sulfoxide (DMSO) (Sigma Chemical Company, St. Louis, USA).
8. 1 mM dexamethasone (dex) stock solution prepared in DMSO.
9. Liquid scintillation cocktail (Flo-Scint III, Perkin Elmer Life and Analytical Sciences, Shelton, CT, USA).
10. Liquid nitrogen.

2.3 Equipment and Other Tools Used in This Protocol

1. Sterilized standard laboratory equipment for handling BY-2 suspension cultures, such as metal tweezers, measuring cylinders (100 mL), Erlenmeyer flasks (250 mL), aluminum foil, pipette (1 mL), pipette tips (1 mL) with cut tips to facilitate easy suspension pipetting (*see* **Note 3**), Eppendorf tubes (2 mL), milling balls (9 mm diameter), 15 mL falcon tubes, and rubber bulb.
2. Autoclave (e.g., Systec DX-2, Systec GmbH, Labor-Systemtechnik, Wettenberg, Germany).
3. Laminar flow cabinet for sterile handling of the cell suspension during inoculation and equilibration.
4. Orbital incubator (IKA KS501, IKA Labortechnik, Staufen, Germany) maintained at 27 °C in the dark.
5. Sterile glass filter holder with cellulose filter (55 mm diameter, 65 g/m^2, Sartorius Stedim Biotech GmbH, Germany) and nylon net filter with pore size 20 μm. The filter holder is connected to the Büchner suction flask with a spring clamp (Millipore-XX1504700) (*see* **Note 4**).
6. Vacuum pump (VRO 1.5–1.2, MEZ, Mohelnice, Czech Republic) to collect the BY-2 cells using the glass filter.
7. Tissue homogenizer (Retsch MM301 Ball mill, Retsch GmbH, Germany).

8. Centrifuge (any thermo regulated centrifuge with rotors for 2 mL tubes that can spin at speeds sufficient to generate 20,000 × *g*).
9. Centrifugal vacuum evaporator (Alpha-RVC, Martin Christ Gefriertrocknungsanlagen GmbH, Osterode am Harz, Germany).
10. SPE Oasis MCX column, 6 cc/150 mg (Waters Corporation, Milford, MA).
11. Supelco Visiprep SPE vacuum manifold (Sigma-Aldrich, Germany) (Fig. 1a).
12. HPLC vials (Fig. 1b).
13. Radioactive detector (Ramona 2000 flow-through radioactivity detector, Raytest GmbH, Straubenhardt, Germany) (Fig. 1c).
14. HPLC (Series 200, Perkin Elmer, Norwalk, CT, USA) (Fig. 1c).
15. Laboratory timer for measuring the treatment time.
16. Laboratory digital scales.

2.4 Radiochemicals

Tritium-labeled indole-3-acetic acid ([^{3}H] IAA), specific radioactivity 20 Ci/mmol (American Radiolabeled Chemicals, St. Louis, MO, USA) (*see* **Note 5**).

2.5 Chromatography Software Program

Winnie32 (version 2.10, Raytest Isotopenmessgeräte GmbH, Germany).

3 Methods

3.1 Cell Culture Preparation

Inoculate 2 mL of 7-day-old PIN5-overexpressing BY-2 cells into fresh 100 mL BY-2 culture medium. Add a volume of Dex to a concentration of 1 μM into two flasks and the same volume of DMSO in control flasks (*see* **Note 6**). All manipulations up to the [^{3}H]IAA incubation (*see* Sect. 3.3) should be performed in sterile conditions in a laminar flow cabinet. Place flasks on the orbital incubator, continuously shaking with 150 rpm at 27 °C for 24 h (orbital diameter 30 mm) (*see* **Note 7**).

3.2 Cell Culture Equilibration

Twenty-four hours after inoculation/*PIN5* gene induction or at desired time point (*see* **Note 8**), filter the cells with the nylon net filter arranged in the filter holder in the Büchner flask. Apply mild vacuum pressure with the rubber bulb connected to the Büchner flask for easy removal of the medium (*see* **Note 9**). If there are multiple flasks for one variant, filter them together and make a single batch of filtered cells. Using a sterile spatula, transfer the filtered cells to the uptake buffer. The volume of uptake buffer is approximately half the volume of BY-2 culture medium used for induction. Resuspend the cells completely in the uptake buffer by

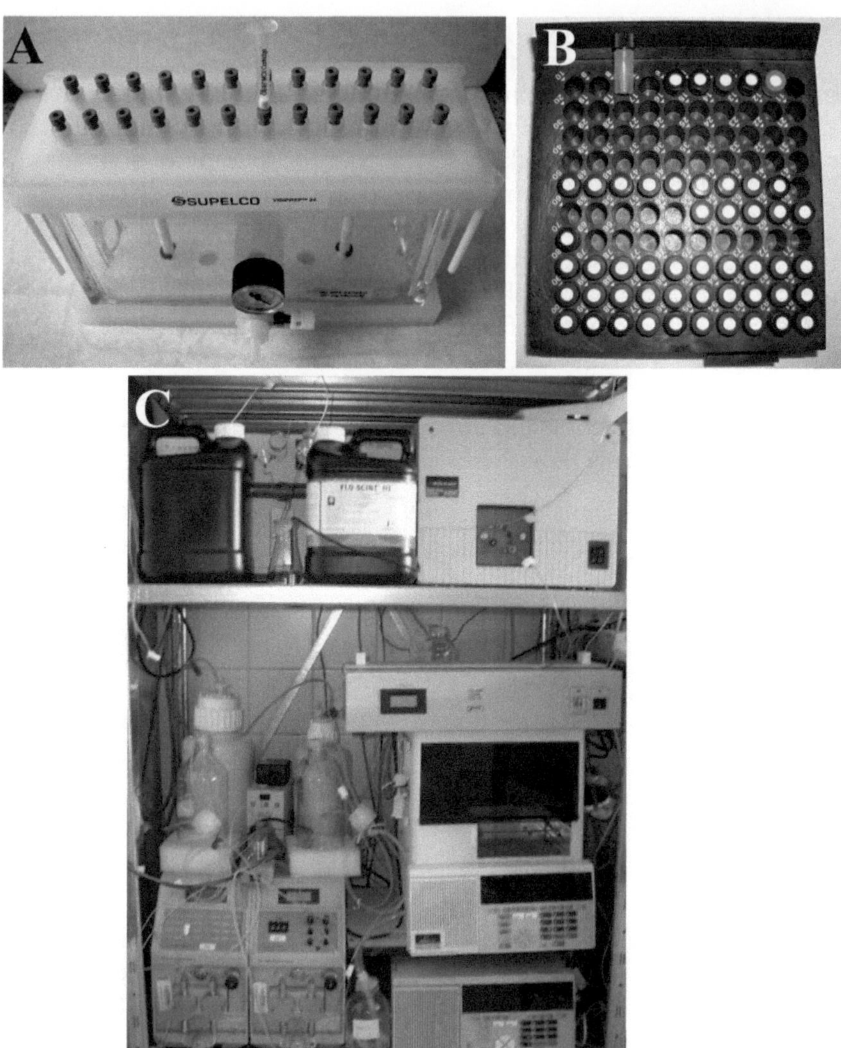

Fig. 1 Equipment used for harvesting and measuring metabolites. (**A**) SPE Oasis MCX column is connected to Supelco Visiprep SPE vacuum manifold. The vacuum promoted flow of the sample from the column can be controlled by turning the gray regulator. (**B**) HPLC vials to collect final elute of metabolites and HPLC rack. (**c**) HPLC detector unit. The *upper part* is the radioactive detector (Ramona 2000 flow-through radioactivity detector, Raytest GmbH, Straubenhardt, Germany), the *lower part* is the HPLC (Series 200, Perkin Elmer, Norwalk, CT, USA)

hand stirring; transfer 1 mL of the suspension to a test tube to determine the cell density. Leave the rest of the suspension under continuous orbital shaking for 45 min at room temperature.

After determination of the cell density (*see* **Note 16** Petrášek et al., Chap. 22) wash the cells again as explained above and resuspend them in a certain volume of fresh uptake buffer required to achieve an equivalent cell density for all variants, based on the density calculated by cell count (*see* **Note 10**). Leave the cells in the orbital shaker for another 1.5 h at room temperature.

3.3 Incubation of Cells with Radiolabeled IAA

Divide the equilibrated cells in different aliquots and add labeled IAA to give final concentration of 20 nM. Incubate the cells for 2.5 h at room temperature (*see* **Note 11**). After the incubation, filter the cells with a cellulose filter in the filter holder into the Büchner flask. Use the vacuum pump to drain out the entire medium from the cells. Weigh 200 mg of collected cells on a piece of aluminum foil using a digital balance. Flash-freeze the weighed cells in liquid nitrogen. At this point the cells can be stored at −80 °C until further analysis. Alternatively, the cells might be quickly transferred by spatula to a durable Eppendorf tube to be frozen.

3.4 Extraction and Purification of Auxin Metabolites

Extraction and purification of auxin metabolites was based on previously published method [8].

1. Transfer the cells stored in aluminum foil at −80 °C to a durable 2 mL Eppendorf tube and add one milling ball to each tube. Add 500 μl of modified Bieleski's solution and homogenize the cells using a tissue homogenizer (30 Hz, 4 min). Leave the samples overnight at −20 °C for extraction (*see* **Note 12**).
2. After overnight extraction, centrifuge the samples at 20,000 × *g* for 15 min at 4 °C. Transfer the supernatant to a fresh 2 mL Eppendorf tube. Leave this sample in a centrifugal vacuum evaporator at 40 °C and 15 mbar, until the water fraction obtained is 0.1–0.2 mL (*see* **Note 13**).
3. Dilute the obtained water fraction in 0.5 mL 1 M formic acid.
4. Install the MCX column on the Supelco Visiprep SPE vacuum manifold and connect it to a vacuum pump. Keep 15 mL collection tubes inside the Supelco Visiprep SPE vacuum manifold to collect all waste solutions (*see* **Note 14**).
5. Apply 5 mL distilled water to the column and let it drain by the vacuum force created by the pump. Apply 5 mL of 1 M formic acid to the column for activation. Apply the metabolites diluted in 1 M formic acid to the activated column. Wash the Eppendorf tube with 0.5 mL formic acid and apply it to the column. Wash the MCX column with 5 mL of formic acid. Elute the metabolites into a new 15 mL tube from the column with 5 mL methanol.
6. Evaporate the samples in SpeedVac to dryness.
7. Add 50 μl of acetonitrile (ACN) to each sample. Vortex mildly for complete dissolution of the metabolites. Centrifuge at 20,000 × *g* for 15 min at 4 °C.
8. Transfer the supernatant to HPLC vials and load them in the HPLC rack for analysis.

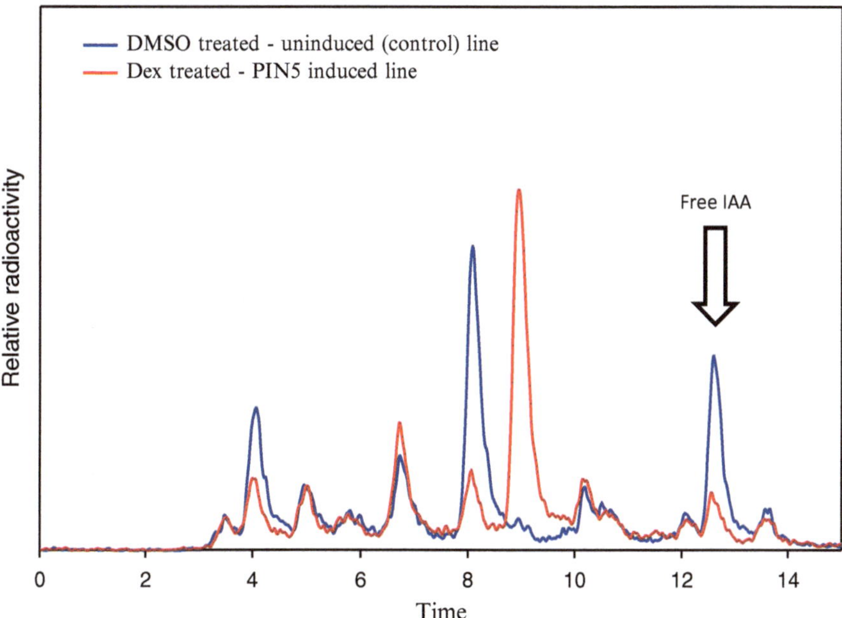

Fig. 2 HPLC chromatogram showing altered IAA metabolic profile of *PIN5* overexpressed (Dex) and control (DMSO) BY-2 cell lines. Free IAA content in PIN5 over expressed variant reduced significantly. The metabolic fate of free IAA was changed due to its compartmentalization by the activity of PIN5 to the endoplasmic reticulum and conversion with different enzymatic system

3.5 HPLC Analysis

The radioactive metabolites of [^{3}H]IAA were separated on HPLC using the following conditions: HPLC column Luna C18(2), 150×4.6 mm, 3 μm; mobile phase A: 40 mM CH_3COONH_4, pH 4.0; mobile phase B: CH_3CN–CH_3OH, 1/1, v/v. Flow rate was 0.6 mL/min with linear gradient 30–50 % B for 10 min, 50–100 % B for 1 min, 100 % B for 2 min, 100–30 % B for 1 min. The column elute was monitored by a Ramona 2000 flow-through radioactivity detector (Raytest) after online mixing with 3 volumes (1.8 mL/min) of liquid scintillation cocktail (Flo-Scint III, Packard BioScience Co.). The rough chromatogram data from the HPLC reading was recalculated to peak area percent values using Winnie32 software. The radioactive metabolites of [^{3}H]IAA were identified by comparison of their retention times with authentic standards.

The graphical data from the HPLC run clearly shows differential metabolic profiles of PIN5 in overexpressing (induced) and control (uninduced) lines, demonstrating the shift in metabolic profile of auxin (Fig. 2). It could be concluded that metabolic fate of free IAA is changed due to its compartmentalization by PIN5 to the compartment with different enzymatic system.

4 Notes

1. We used a Dex-inducible PIN5 construct to transform BY-2 cells, but it is possible to use any other gene construct to characterize its role in regulating metabolic transformation of any exogenously applied compound. Nonetheless, the choice of inducible gene system is advisable, because it allows us to eliminate the error introduced to the system, if the cells are allowed to manifest any changes ensuing from the function of the protein in question (*see* **Note 7**).
2. Bieleski's solution can be prepared in large volumes like 1 l and should be stored at −20 °C.
3. This small detail should not be omitted, not only for the sake of unhindered pipetting of homogenous and undiluted cell suspension, but also more importantly, to avoid direct harm to the pipetted cells caused by compression in the tip.
4. This filter set can be used for both BY-2 cell equilibration and radioactive compound treated BY-cell filtration as well. But it is more acceptable to use a separate filter set for handling radioactive compound treated cells. After finishing the work with the filter sets used for handling tritium treated cells, dismantle them and wash all parts in running water for at least 2 h and then wash them in a labware washer (*see* **Notes 18** and **19** in Petrášek et al., Chapter 22, regarding proper handling of tritiated compounds and for advisable working condition while working with radioactivity).
5. Ethanol stock solution of radiolabelled IAA should be stored at −80 °C in dark glass vials with tight fitting cap. They should be kept at room temperature for at least 30 min prior to use. Because IAA is light sensitive, it should be kept in dark, not only the [^{3}H]IAA stock solution should be in dark glass vials, but both the stock solution and cells incubated with [^{3}H]IAA also should be preferably shaded from light.
6. The final concentration of Dex or any other inducer should be pre-tested and optimized for the particular cell line. Nevertheless, concentrations higher than 5 μM of Dex should be avoided
7. All BY-2 culture manipulations should be performed in strict sterile conditions with autoclaved pipette tips. Depending on the experiment, calculate the number of inoculating flasks required to get enough treated cells for analysis. In general, 150–200 mg cells can be expected from inoculation of a 2 mL 7-day-old culture in 100 mL new medium, grown for 25 h.
8. Our experimental setup used cells 24 h after induction/inoculation. Nevertheless, it is reasonable to note that the choice of

this timing was related to properties of our model system, particularly to the expression and early functional targeting candidate protein of the protein in question. That is, we used the timing because our cells had been expressing PIN5 at a transcriptional level as early as 12 h after induction, as proven by RT-PCR. Twenty-four hours after induction, relatively homogenous targeting to the endomembrane system was demonstrated by immunolocalization. Importantly, 24 h after induction the phenotype of induced cells still did not display any remarkable changes, so the cellular dimensions and phenotypes of induced cells and non-induced cells were comparable.

9. If the suspension volume is low, the use of a rubber bulb is not necessary for filtration. But a large amount of cell sediment on the filter can reduce the filtration flow, which can be overcome by mild vacuum created by a rubber bulb.
10. Cell density depends on the expectancy of the outcome of the experiment. If comparison of the different metabolites in different variants is desired, equivalent cell densities in all variants are not so important. However, for quantitative analysis of different metabolites in different variants, similar cell densities for all variants should be obtained. If cells incubated in desired compounds at precise concentrations are with different cell densities, the cell population will have different uptake kinetics leading to differential kinetics of metabolism.
11. The incubation time also depends on the specific needs of the experiment. When long term metabolites are investigated, the treatment time is prolonged; when immediate effects are analyzed, the treatment time is restricted to a few minutes. The IAA metabolic machinery in BY-2 is very fast, so 2–3 h incubation is sufficient to see significant effects.
12. It is possible to keep them in a normal −20 °C freezer after closing the Eppendorf tubes properly.
13. Since this step takes 7–8 h, it is better to plan this step overnight to save time.
14. Be careful to check if the Supelco Visiprep SPE vacuum manifold is installed properly, especially that the waste draining pipes from the top part are inserted into 15 mL tubes to avoid sample loss and contamination of the set up.

Acknowledgements

This work was supported by grants from the Czech Ministry of Education (MSM0021622415), the Grant Agency of the Academy of Sciences of the Czech Republic (IAA601630703), and the Odysseus program of Research Foundation-Flanders (JF), Grant

Agency of the Czech Republic, projects n. P305/11/0797 (JP), and P305/11/0797 (EZ). We thank Annick Bleys for critical reading of this chapter.

References

1. Benková E, Michniewicz M, Sauer M, Teichmann T, Seifertová D, Jürgens G, Friml J (2003) Local, efflux-dependent auxin gradients as a common module for plant organ formation. Cell 115:591–602
2. Křeček P, Skůpa P, Libus J, Naramoto S, Tejos R, Friml J, Zažímalová E (2009) The PIN-FORMED (PIN) protein family of auxin transporters. Genome Biol 10:249
3. Mravec J, Skůpa P, Bailly A, Hoyerová K, Křeček P, Bielach A, Petrášek J, Zhang J, Gaykova V, Stierhof Y-D, Dobrev PI, Schwarzerová K, Rolčík J, Seifertová D, Luschnig C, Benková E, Zažímalová E, Geisler M, Friml J (2009) Subcellular homeostasis of phytohormone auxin is mediated by the ER-localized PIN5 transporter. Nature 459:1136–1140
4. Nagata T, Nemoto Y, Hasezawa S (1992) Tobacco BY-2 cell line as the "HeLa" cell in the cell biology of higher plants. Int Rev Cytol 132:1–30
5. Petrášek J, Zažímalová E (2006) The BY-2 cell line as a tool to study auxin transport. In: Nagata T, Matsuoka K, Inzé D (eds) Tobacco BY-2 cells: from cellular dynamics to omics. Springer, Berlin, pp 107–117
6. Dobrev PI, Havlíček L, Vágner M, Malbeck J, Kamínek M (2005) Purification and determination of plant hormones auxin and abscisic acid using solid phase extraction and two-dimensional high performance liquid chromatography. J Chromatogr A 1075:159–166
7. Pěnčík A, Rolčík J, Novák O, Magnus V, Barták P, Buchtík R, Salopek-Sondi B, Strnad M (2009) Isolation of novel indole-3-acetic acid conjugates by immunoaffinity extraction. Talanta 80:651–655
8. Dobrev PI, Kamínek M (2002) Fast and efficient separation of cytokinins from auxin and abscisic acid and their purification using mixed-mode solid-phase extraction. J Chromatogr A 950:21–29

Index

A

Abscisic acid (ABA) 20, 36, 37, 64
ACC. *See* 1-Aminocyclopropane-1-carboxylic acid (ACC)
ACC synthase 65
ACS inhibitor quinazolinones (acsinones) 65
Agonist 36, 38–41
1-Aminocyclopropane-1-carboxylic acid (ACC) 65, 66, 71, 73, 74, 76
Aminoethoxyvinylglycine (AVG) 65, 71, 73–75
Antagonist 36, 38–41, 57, 64
Anther 112
Arabidopsis Biological Resource Center (ABRC) 55
Arabidopsis thaliana 3–9, 13, 19–30, 46, 51, 52, 64, 65, 80, 82–84, 88, 96, 104, 115, 117, 135, 160, 192, 202, 227, 242, 255
Argonaute (AGO) 96
ATP-BINDING CASSETTE TYPE B19 (ABCB19) (formerly PGP19) 58
Autofluorescence 108, 163
Autofluorescent compounds 5
Automation workstation 7
Auxin
- accumulation assay 242–248, 251
- efflux 58, 59, 242, 248, 255, 256
- efflux inhibitor 1-naphthylphthalamic acid (NPA) 242, 252
- influx 59, 134, 241, 242, 248, 252
- influx inhibitor 3-chloro-4-hydroxyphenylacetic (CHPAA) 242, 252
- metabolism 263
- metabolites 260
- radiolabelled 241, 242, 248–250, 252
- transport 52, 56, 58, 59, 117, 121, 241–252, 255–264
- transport assay 241, 244, 251
- transport inhibitor 58, 59, 121

AUXIN-BINDING PROTEIN1 (ABP1) 58
AUXIN-RESISTANT1 (AUX1) 59
AVG. *See* Aminoethoxyvinylglycine (AVG)

B

Benzo-1,2,3-thiadiazole-7-carbothioic acid *S*-methyl ester (BTH) 46
β-Glucuronidase (GUS) 20
Bioactive
- clusters 159–167
- compound 21, 64, 146

Bioactivity database
- AffinDB 146
- BindingDB 146, 153
- ChEBI 146
- DrugBank 146, 153
- KEGG LIGAND 146
- PDBind 146
- PubChem 146–154
- STITCH 146
- SuperTarget 146

Bioassay 20, 27, 29, 146, 149, 195, 197
Bioconversion 197–198
Bioinformatic 146
Bleaching 14, 16, 140
*B*rassino*p*ride (BRP) 64
Brefeldin A (BFA) 58, 59, 132–134, 136, 137, 140, 160, 163
2,5-Bromo-4-chloro-3-indoyl-β-D-glucuronide (X-gluc) 47
BUM 58, 59

C

Carbohydrate 103, 105, 107, 226
Cell culture
- *Arabidopsis thaliana* L., cv. Landsberg erecta 242
- *Nicotiana tabacum* L., cv. Bright Yellow (BY-2) 242

Cellulose 105, 202, 244, 245, 247, 257, 260
Cell wall 103–109, 196
Chemical
- 5271050 .... 117, 122
- 5403629 117
- 6220480 117
- Cobtorin 202
- endosidin 111, 160
- endosidin 1 (ES1) 56, 160
- endosidin 3 (ES3) 57
- galvestine-1 80
- gravacin 117, 122
- pyrabactin 64

Glenn R. Hicks and Stéphanie Robert (eds.), *Plant Chemical Genomics: Methods and Protocols*, Methods in Molecular Biology, vol. 1056, DOI 10.1007/978-1-62703-592-7, © Springer Science+Business Media New York 2014

Chemical (*cont.*)
sortin ... 126
sortin 1 ... 134, 136, 226–227, 229–232, 235, 236
vacuolar protein *sorting in* hibitor 1 (sortin1) ... 227
Chemical biology ... 3, 132, 133, 135, 145, 213
Chemical Computing Group (CCG) MOE 2010.10 ... 171
Chemical derivatization ... 220
Chemical disruptors ... 111
Chemical genetics ... 20–22, 63–76, 80, 191–198, 202
Chemical libraries
ChemBridge ... 37, 65, 117, 126, 227
DiverSet ... 37, 65, 117, 126, 227
fluorescent ... 112, 117, 135
Chemical library design ... 20, 65, 117, 118
Chemical space ... 145, 170, 171, 182, 195
Cheminformatic ... 145–155
Chemotype ... 116, 118, 120, 121
Chloroplast ... 79–92
^{13}C-labelling ... 214, 215, 221
Clustering
binning ... 154
hierarchical ... 154
MDS ... 154
Compartments
endomembrane ... 54, 117, 132, 140, 159
Golgi ... 121, 125, 159
prevacuolar compartment (PVC) ... 125, 134
tonoplast ... 125, 137, 159
vacuole ... 125, 134, 137
vesicle ... 159–161
Compound
availability ... 4, 193, 196, 197
bioavailability ... 196, 197
biotransformation ... 194
dose response ... 116, 120
frequent hitters ... 195, 196
purity ... 183, 193
solubility ... 194, 237
Concanamycin A ... 134, 136, 137
Confocal microscope ... 12, 14, 54, 108, 114, 121, 162
Cycloheximide ... 20, 22, 28, 132, 134, 136, 140
Cytochalasin D ... 134

D

*D*anger-*a*ssociated *m*olecular *p*atterns (DAMPs) ... 45
Data
analysis ... 15, 24–26, 28, 74, 75, 151, 227
format ... 148, 215
normalization ... 229
Database
ChemBank ... 146, 153
ChemDB ... 146
ChemMine ... 146, 147, 152, 154
NCI ... 146
PubChem ... 146, 147, 152–154
SciFinder ... 123
selection ... 153
ZINC ... 146
Dendrogram ... 154, 165, 166
Derivatization ... 197, 219, 220
Descriptors ... 147, 149–153, 155, 171, 173, 174, 176, 177, 184, 186
Dicer-like (DCL) protein ... 95
3, 5-dicholoroanthranilic acid (DCA) ... 46
Dimethyl sulfoxide (DMSO) ... 5, 7, 9, 21, 26, 28, 37, 39–41, 47, 48, 53, 65, 70–74, 81–90, 92, 97–100, 104, 106–109, 113, 117–121, 126, 127, 136, 140, 162, 163, 194, 204, 237, 243, 245, 257, 258, 261
Diversity ... 26, 45, 46, 59, 108, 128, 146, 149, 150, 170, 171, 194
DMSO. *See* Dimethyl sulfoxide (DMSO)
D-Optimal Design ... 177, 178, 186, 187
Dose-response ... 116, 120, 169
Dot blot ... 127, 128
Drug filter ... 65, 67

E

Effectors ... 45, 191, 193
Endocytosis ... 58, 132, 134, 135, 137, 139, 140, 160
Endomembrane ... 54, 59, 80, 111–114, 116, 117, 121–122, 131–140, 159–167, 226, 255, 263
Endomembrane cycling ... 111–114
Ethylene ... 45, 63–76

F

Fingerprint search ... 152
FM4-64 ... 54, 135, 136, 139, 140
Full length cDNA ... 72, 201–209
Full-length cDNA OvereXpressor gene (FOX) hunting system ... 201–209

G

Gain-of-function mutants ... 52, 57, 201, 202
Galactolipid ... 79–92
Gas chromatography (GC) ... 73, 74, 219
Gene silencing ... 95, 96
GFP. *See* Green fluorescent protein (GFP)
Gravitropic ... 20, 58, 116, 118–122, 196
Gravitropism ... 58, 120
Green fluorescent protein (GFP) ... 5, 58, 96, 99–100, 121, 132, 137, 139, 140, 161, 164, 167

H

Hamming distance 154, 165
Hemicellulose 105
Herbicide 89, 92, 169, 191, 192, 198
High performance liquid chromatography (HPLC) 193, 194, 213, 232, 252, 255–263
High throughput chemical screen 36, 195
High throughput screening 3, 4, 20, 45–49, 83–84, 86–88, 91, 103–109, 146, 169
High throughput screening (enzymatic),
Hit
  selection 193, 194
  triage 191–198
HPLC. *See* High performance liquid chromatography (HPLC)

I

IAA. *See* Indole-3-acetic acid (IAA)
Image
  analysis 12, 139, 160, 252
  in vivo 107–109, 138
Immunolocalization 140, 263
Indole-3-acetic acid (IAA) 121, 241, 242, 256, 258, 260–263
Inhibitory concentration 50 (IC50) 21, 81, 86, 87, 90
INRA-Versailles Genome Resource Centre 55–56
Isopropyl-β-D-thiogalactopyranoside (IPTG) 83
Isotope tracing 214

J

Jasmonate 20, 22, 25, 26
Jasmonic acid 37, 45, 80

K

Karrikins 56

L

Latrunculin B 134, 136
Lethality
  embryonic 131
  gametophytic 131
Likeness filter 149–150
Lipinski Rule of Five 149, 155
LIPOXYGENASE 2 (LOX2) 20–22, 24–29
Low throughput screening 11
Luciferase 5, 7, 19–29, 65, 70, 71, 75, 96, 99
Luciferin 23, 25, 26, 28, 29, 66, 71, 72, 97, 99
Luminescence 5, 7–9, 21, 23, 25, 27–29, 71, 72

M

Mass isotopomers 214–217, 219–221
Mass spectrometry (MS) 214–217, 219–221
Metabolic profiling 226
Metabolite 213–222, 225–233, 235–237, 256, 259–261, 263
Metabolite database
  Human Metabolome Database 235
  Madison Metabolomics Consortium Database 235
Metabolomics 225–237
Methyl Jasmonate 20, 22, 25
microRNAs (miRNAs) 95, 96
Monogalactosyldiacylglycerol (MGDG) 80, 81, 83, 85, 86, 88, 90, 91
Multi-drug approach 131–140
Multi-tiered screen 115
Murashige and Skoog (MS) medium 13, 23, 37, 52, 84, 88, 117, 135, 204, 208
Mutant
  ethylene overproducer (eto1-4) 65
  knock out 51
  null 51, 63
  pyrabactin resistant (pyr) 64
  vacuolar protein sorting (vps) 125

N

Naphthalene acetic acid (NAA) 134, 136, 140, 242, 245–248
Nicotiana tabacum (tobacco) 112, 242, 256
Nicotine adenine dinucleotide (NAD)-dependent histone deacetylase (SIR2) 64
1-N-naphthylphtalamic acid (NPA) 59, 134, 136, 242, 245–248, 252
Nottingham Arabidopsis Stock Centre (NASC) 26, 55
NPA. *See* 1-N-naphthylphtalamic acid (NPA)
Nuclear magnetic resonance (NMR)
  data processing 235–236
  spectroscopy 243

O

Oxyluciferin 29

P

Pathogen 45–47
Pathogen/Microbe-Associated Molecular Patterns (PAMPs/MAMPs) 45
Pathway
  alkaline phosphatase (ALP) pathway 125
  carboxypeptidase Y (CPY) 125
  endocytic 132, 140

Pathway (*cont.*)
 metabolic ... 213
 secretory ... 134, 140, 161
 signaling ... 22, 145, 202
PCA. *See* Principal component analysis (PCA)
Peak assignments ... 235
Pectin ... 105, 202
Phosphatidylethanolamine ... 85, 91
Phospholipase C (PLC) ... 83, 85, 88, 91, 92
Phytagel ... 52, 59, 60
Phytohormone ... 35, 52, 56–58, 63–76
PIN5 ... 255, 256, 258, 261–263
PIN-FORMED (PIN) ... 58, 160, 255
Plant
 cell culture ... 249
 defense inducer ... 45–49
 immunity ... 45
PLC. *See* Phospholipase C (PLC)
Pollen ... 56, 57, 111–114, 116, 160
Principal component analysis (PCA) ... 150, 154, 171, 173–177, 184, 185, 229, 230, 234, 236
Protoplast ... 3
PYR/PYL/Regulatory Component of ABA Receptor (RCAR) ... 64

Q

Quantitative structure-activity relationship (QSAR) ... 146, 147, 151, 153, 170, 183

R

RBRC. *See* RIKEN Bioresource Center (RBRC)
Reactive oxygen ... 45
Redundancy ... 64, 96, 115, 131, 160
Reverse genetics ... 51–61
RIKEN Bioresource Center (RBRC) ... 56, 202, 203, 205
RNA-induced silencing complex (RISC) ... 96
Robot (liquid handling) ... 4, 5, 9, 37, 38, 42, 118

S

Salicylic acid ... 45
SAR. *See* Structure activity relationship (SAR)
Screening
 data ... 146, 151
 phenotype based ... 63, 64, 68
 target based ... 63
Secretory ... 132–134, 140, 161
Seed germination ... 7, 14, 16, 20, 24, 56, 64, 89, 96
Seed sterilization ... 4, 6, 52, 53, 98
SENDAI Arabidopsis Seed Stock Center (SASSC) ... 56
Short-interfering RNAs (siRNAs) ... 95–100
Similarity search
 Tanimoto coefficient ... 152–154
 Tversky index ... 152
Simplified Molecular Input Line Entry Specification (SMILES) ... 147, 148
siRNAs. *See* Short-interfering RNAs (siRNAs)
Small molecule ... 3, 5, 7, 19–29, 45–49, 51–61, 63–65, 68–70, 75, 83, 114, 116, 117, 145–151, 159–161, 165, 169–187, 191, 192, 225
Small RNA. *See* Short-interfering RNAs (siRNAs)
Software
 ChemMine Tools ... 147, 150, 152–155
 CORRECTOR ... 215–223
 Image J ... 13, 16, 53, 55, 68, 70, 72, 118, 121, 139
 installation ... 218
 OpenEye FILTER 2.0.2 ... 171
 PubChem ... 123, 146, 147, 150–154
 R ... 147, 150–152, 155, 167
 TagFinder ... 217
 Umetrics MODDE 9.0 ... 171
 Umetrics SIMCA-P+ 12.0.1 ... 171
Specific fluorescence ... 15
Stable isotope labeling ... 213–222
Statistical molecular design (SMD) ... 169–187
Sterilization ... 4, 6, 13, 15, 52, 53, 67, 90, 98, 104, 107, 161, 162, 232, 249
Strigolactones ... 56
Structure activity relationship (SAR) ... 122, 123, 146, 170, 187, 194, 195
Sub-library ... 120
SYNTAXIN OF PLANTS 61 (SYP61) ... 134, 160
Synthetic elicitor ... 46
Synthon ... 172, 184

T

Target identification ... 122, 193, 195, 197
T-DNA ... 201, 203
The *Arabidopsis* Information Resource (TAIR) ... 55, 56
Time-profiling ... 11–16
Tissue preparation ... 228
Tobacco Bright Yellow (BY-2) ... 3
Transcriptome ... 74–75
Transformation ... 37–39, 202–204, 207, 208, 235, 255, 262
Triclosan (5-chloro-2-(2,4-dichlorophenoxy) phenol) ... 92
Triple response ... 65, 68–70, 72, 75
Tyrphostin A23 (TyrA) ... 133, 134, 136, 140

U

UDP-galactose 80, 82, 83, 85–87, 90–92
UDP-galactose:1,2-sn-diacylglycerol galactosyltransferase 1 (MGD1) 80
Univariate analysis 230, 236

W

Wortmannin 133, 134, 136, 137

Y

Yeast 35–42, 64, 83, 115, 116, 125–128, 202, 204

MIX
Papier aus verantwortungsvollen Quellen
Paper from responsible sources
FSC® C105338

If you have any concerns about our products,
you can contact us on
ProductSafety@springernature.com

In case Publisher is established outside the EU,
the EU authorized representative is:
Springer Nature Customer Service Center GmbH
Europaplatz 3, 69115 Heidelberg, Germany

Printed by Libri Plureos GmbH
in Hamburg, Germany